QUINOA

Botany, Production and Uses

Dedicated to the loving memory of my parents, who departed for the heavenly abode on 16 January 2001

Atul Bhargava

QUINOA

Botany, Production and Uses

Atul Bhargava

and

Shilpi Srivastava

www.cabi.org

CABI is a trading name of CAB International

CABI	CABI
Nosworthy Way	38 Chauncey Street
Wallingford	Suite 1002
Oxfordshire OX10 8DE	Boston, MA 02111
UK	USA
Tel: +44 (0)1491 832111	Tel: +1 800 552 3083 (toll free)
Fax: +44 (0)1491 833508	Tel: +1 (0)617 395 4051
E-mail: info@cabi.org	E-mail: cabi-nao@cabi.org
Website: www.cabi.org	

A catalogue record for this book is available from the British Library, London, UK.

Library of Congress Cataloging-in-Publication Data

Bhargava, Atul, 1975-
  Quinoa : botany, production and uses / Atul Bhargava, Shilpi Srivastava.
    p. cm.
  Includes bibliographical references and index.
  ISBN 978-1-78064-226-0 (alk. paper)
1. Quinoa. I. Srivastava, Shilpi. II. Title.

SB177.Q55B43 2013
664′.7--dc23

                                    2013009503

ISBN-13: 978 1 78064 226 0

Commissioning editor: Sreepat Jain
Editorial assistant: Alexandra Lainsbury
Production editor: Simon Hill

Typeset by SPi, Pondicherry, India
Printed and bound in the UK by CPI Group (UK) Ltd, Croydon, CR0 4YY.

Contents

PART IV – QUALITATIVE ASPECTS, ECONOMICS AND MARKETING

Contributors

Didier Bazile, UPR47, GREEN, CIRAD (Centre de Coopération Internationale en Recherche Agronomique pour le Développement), TA C-47/F, Campus International de Baillarguet, 34398 Montpellier Cedex 5 – France. didier.bazile@cirad.fr

Atul Bhargava, Amity Institute of Biotechnology, Amity University Uttar Pradesh (Lucknow Campus), Mango Orchard Campus, Gomti Nagar Extension, Lucknow-227105, India. atul_238@rediffmail.com

Francisco Fuentes, Facultad de Recursos Naturales Renovables, Universidad Arturo Prat. Avenida Arturo Prat 2120, Iquique, Chile. francfue@gmail.com

Enrique A. Martínez, Centro de Estudios Avanzados en Zonas Áridas (CEAZA), La Serena, Chile. enrique.a.martinez@ceaza.cl

Ángel Mujica, Universidad Nacional del Altiplano, Escuela de Postgrado, Avenida del Ejército 329, Puno, Peru. amhmujica@yahoo.com

Pablo Olguín, Instituto de Geografía, Laboratorio de Biodiversidad, Instituto de Geografía, Pontificia Universidad Católica de Valparaíso, Valparaíso, Chile. pablo.olguinm@gmail.com

Shilpi Srivastava, Amity Institute of Biotechnology, Amity University Uttar Pradesh (Lucknow Campus), Mango Orchard Campus, Gomti Nagar Extension, Lucknow-227105, India. shilpi_bt@rediffmail.com

Andrés Zurita-Silva, Centro de Estudios Avanzados en Zonas Áridas (CEAZA). Av. Raúl Bitrán s/n. Casilla 554, La Serena, Chile.

Preface

The term 'green revolution' was coined in the 1960s after improved varieties of wheat dramatically increased yields in test plots in northwest Mexico. The green revolution, attributed to Norman Borlaug, an American scientist, ushered in an era of agricultural surpluses in many areas of the world. The subsistence agriculture practised in many regions was replaced by intensive agriculture, which required farm mechanization and increased inputs in the form of labour, high-yielding varieties, chemical fertilizers and pesticides. Many countries that had imported food grains before the green revolution became exporters of these commodities. The green revolution forever changed the way agriculture was conducted worldwide, benefiting the people of many countries in need of increased food production.

However, the failure of the green revolution in several areas and the negative impacts associated with it have forced mankind to look for alternatives. The accelerated inputs have resulted in intolerable pressure on fragile agroecosystems. The situation is compounded by a shrinking portfolio of species and an emphasis on a handful of major crops that have narrowed the number of species on which global food security depends. The consequences of crop failures from unforeseen stresses, pests and diseases are potentially catastrophic. Therefore, there is an urgent need to broaden the species portfolio for agriculture and food security. The need of today is a gradual shift from input-intensive to environmentally sound sustainable agriculture. This requires a shift in focus towards increasing production by using agriculturally marginal lands. Underutilized crops can be the answer to this burning problem.

Underutilized crops are those that were formerly widely grown and consumed but that have fallen, or are falling, into disuse. Many underutilized species are extremely important for food production in various parts of the world. They are well adapted to marginal lands, have a role in traditional medicine and constitute an important part of the local diet, providing valuable nutritional elements often lacking in staple crops. Underutilized crops also represent an important source of revenue for local economies and are part of the rich cultural and traditional heritage of communities around the world. These crops are also important sources of resistance genes for biotic and abiotic stress breeding that can also be used for the genetic improvement of commonly used crops. Quinoa is one such underutilized crop that has recently gained attention for its ability to cope with different environmental stresses and its potential for helping to solve the widespread problem of malnourishment.

Quinoa (*Chenopodium quinoa*), an important crop of the Andean region of South America, is being grown in a wide range of environments in South America, North America, Europe and Asia. Quinoa is an important food source for human consumption, is well adapted to extreme environmental conditions and has potential industrial applications. The plant is nutritionally important because its seeds have a high protein content and an abundance of essential amino acids, vitamins, carbohydrates, minerals and natural antioxidants. This crop has recently attracted worldwide attention due to its ability to produce high-protein grains under ecologically extreme conditions, making it an important crop for the diversification of future agricultural systems in different parts of the world. The increasing demand for quinoa and insufficient supply from quinoa-producing countries of South America necessitated the introduction of the crop to newer areas. Experiments in Europe, Asia and Africa have confirmed that the crop can be successfully cultivated there. At present, quinoa is grown commercially throughout the western regions of South America, including Bolivia, Chile, Ecuador and Peru for domestic markets and emerging export markets in Japan, Australia, Europe and North America. With the emerging quinoa market, the consumer trend towards quinoa is expected to keep increasing, with international support from both political and industry organizations in Europe and Asia. Quinoa has the potential to shed its underutilized status and become an important industrial and food crop of the 21st century. Quinoa has been selected by the FAO as one of the crops destined to offer food security in the 21st century.

This book is the first comprehensive work on quinoa aimed at researchers throughout the world who are working on this crop. The purpose of the book is to provide an introduction to the concept and applications of quinoa. The book is divided into four sections spread over 14 chapters. Part I introduces the readers to the crop, its history and distribution. Part II discusses the taxonomical position, based on morphological, cytological and molecular data. Part III is dedicated to agrotechnology and includes chapters on botany, crop production and management, pathology and breeding approaches. Part IV sheds light on the nutritional significance of the crop, and also contains detailed discussion of the antinutritional components.

The book has detailed chapters that have been written in an easy-to-read, succinct format and that incorporate information from in-depth study of classical work as well as recent research. The book presents the novelty and complexity of the topics in a lucid, methodical way, taking care to keep the level of discussion simple and free from technical jargon. The chapters are straightforward, and contain figures to provide clarity. The authors have tried to present complex and updated information in a manner that familiarizes the reader with the important concepts and tools of recent researches on the crop.

This detailed reference work traces the crop from its origin to the present day, and incorporates up-to-date information about this underutilized crop. The book will help to stimulate interest in the crop among academics and researchers, both novices and experts, and among those who are increasingly focused on exploiting this underutilized crop.

Acknowledgements

First and foremost, the authors wish to convey their sincere thanks to all the researchers working on quinoa throughout the world. It is their valuable work that has enabled us to bring out a treatise on this fantastic crop.

We are lost for words to describe the contribution of Dr Francisco Fuentes in the preparation of this book. He has wholeheartedly supported us on several aspects along with Drs Enrique Martinez and Didier Bazile. Dr Fuentes has provided us with several fabulous images that are now a part of this book. We also wish to sincerely thank Drs Ángel Mujica, Andrés Zurita-Silva and Pablo Olguín M., along with Francisco, Enrique and Didier, for sharing their valuable work with us.

It is our great pride and pleasure to express our deep sense of gratitude and heartiest veneration to Prof. (Dr) Deepak Ohri, Professor and Deputy Dean, Department of Research, Amity University Uttar Pradesh Lucknow Campus (formerly Scientist and Deputy Director, Division of Genetics and Plant Breeding, National Botanical Research Institute, Lucknow) for his inspiring guidance, untiring supervision and mentorship during the early days of our research career on quinoa. In fact, it was he who taught Atul Bhargava about quinoa and inculcated in him a deep interest in the crop.

We feel privileged in expressing our deep gratitude to Dr Sudhir Shukla, Senior Scientist, Division of Genetics and Plant Breeding, National Botanical Research Institute, Lucknow, whose valuable, rich and constant supervision, enlightened guidance, moral encouragement and guardianship have made us what we are today. Dr Shukla's joy and enthusiasm for research has been contagious and motivational, and has propelled us to complete this task against all sorts of odds.

On a personal note, we are grateful to our brother (Akhilesh Bhargava), sister-in-law (Meenakshi Bhargava) and niece (Anushka Sharma) for their extreme patience and perseverance while this book was being written. Our family has supported us in our overburdened schedule and has been the source of our happiness and contentment.

We also thank our friends and colleagues at Amity namely, Mr Ajay Kumar Singh, Dr Prachi Srivastava and Dr Rachna Chaturvedi for their stimulating company and support at every level. Mr Ajay Kumar Singh has been with us through thick and thin and has wholeheartedly helped us in completing this endeavour. Sincere thanks are also due to Dr Kajal Srivastava, Lecturer, Amity School of Languages, Amity University Lucknow Campus for checking the manuscript thoroughly for English language and suggesting suitable improvements. The authors wish to thank Mr Santosh Kumar Pal and Ms Rita Pal (BCS In-Silico Biology, Lucknow) for working hard to improve the clarity of figures and redrawing many of them.

We gratefully acknowledge the indispensable help of Dr Sreepat Jain, Commissioning Editor (CABI, UK) and Alexandra Lainsbury, Editorial Assistant (CABI, UK) in the preparation of the manuscript and for showing the way whenever the authors needed their guidance. Both have cheerfully inspired us to make changes in text and art even very late in the production stage to bring out the best to the readers.

Atul Bhargava
Shilpi Srivastava

1 Introduction

There are an estimated 7000 plant species that have been used as crop plants at some point in human history (FAO, 1998). However, today only 150 plant species are cultivated; just 12 of these provide approximately 75% of the world's food and four produce over 50% of the world's food (Bermejo and León, 1994). Prescott-Allen and Prescott-Allen (1990) state that common figures range from seven plant species providing 75% of human nutrition to 30 plant species providing 95% of human nutrition. These commonly utilized crops are intensively cultivated and require farm mechanization and increased inputs in the form of labour, high-yielding varieties, chemical fertilizers and pesticides (Bhargava *et al.*, 2008, 2012). These accelerated inputs have resulted in intolerable pressure on fragile agroecosystems. Modern agriculture has increased homogeneity and mono-crop cultivation, resulting in loss of agrobiodiversity and frequent crop losses due to infestation by pathogens. The need of present times is a gradual shift from input-intensive to environmentally sound sustainable agriculture. Modelling of traditional farming systems to modern needs with increased organic linkages might be a good option for sustainability of the agricultural production system and maintenance of agroecological stability (Bhargava *et al.*, 2008). This would also require a shift in focus towards increasing production by using agriculturally marginal lands for crops that are less exploited but that have immense potential for diverse uses (Partap *et al.*, 1998).

The emphasis on a handful of major crops has narrowed the number of species on which global food security depends. The consequences of crop failures from unforeseen stresses, pests and diseases can be catastrophic (Prescott-Allen and Prescott-Allen, 1990). The past three decades have seen a wide range of research interests on underutilized crops and a number of significant programmes have been undertaken in both developing and developed countries to promote underutilized species for agricultural systems, as alternative crops or as sources of new products.

1.1 Underutilized Crops

Underutilized or neglected crop species are often indigenous ancient crop species that are still used at some level within the local, national or even international communities, but have the potential to contribute further to the mix of food sources (Mayes *et al.*, 2011). These species appear to have considerable potential for use yet their potential is barely exploited, if not totally neglected, in agricultural production. Many underutilized crops were once more widely grown but are today falling into disuse for a variety of agronomic, genetic, economic and cultural reasons (Hammer *et al.*, 2001). Farmers and consumers are using these crops less because they are not competitive with other crop species in the same agricultural environment. Orphan, abandoned, new, underutilized, neglected, lost, under-used, local, minor, traditional, forgotten, alternative, niche, promising, underdeveloped: these and other terms are often used as synonyms for underutilized species (Padulosi and Hoeschle-Zeledon, 2004). Underutilized crops are often known as 'new crops', not because they are 'new' but because they have been taken up by agricultural researchers and commercial companies for a new market. The main features of the underutilized crops are that they are:

- important in local consumption and production systems;
- highly adapted to agroecological niches and marginal areas;
- represented by ecotypes or landraces;
- cultivated and utilized drawing on indigenous knowledge;
- characterized by fragile or non-existent seed supply systems;
- hardly represented in *ex situ* gene banks and
- ignored by policy makers and excluded from research and development agendas (Padulosi and Hoeschle-Zeledon, 2004).

Moreover, the limited information available on many important and frequently basic aspects of neglected and underutilized crops hinders their development and sustainable conservation (Hammer *et al.*, 2001).

Many wild and underutilized plants have potential for more widespread use and could contribute to food security, agricultural diversification and income generation (Vietmeyer, 1986; Anthony *et al.*, 1995). Neglected and underutilized crops represent an important source of revenue for local economies and are part of the rich cultural and traditional heritage of communities around the world (IAEA, 2004). In addition to this, these crops are important sources of resistance genes for biotic and abiotic stress breeding that can also be used for the genetic improvement of crops. Compared with the major crops, they require relatively low inputs and, therefore, contribute to sustainable agricultural production. Underutilized crops have great potential to alleviate hunger directly, through increasing food production in challenging environments where major crops are severely limited, through nutritional enhancement to diets focused on staples and through providing the poor with purchasing power, helping them buy the food that is available (Mayes *et al.*, 2011).

1.2 *Chenopodium* as an Underutilized Plant

Among a number of underutilized species, members of the genus *Chenopodium* (family Amaranthaceae) are most promising since they have the ability to thrive and flourish under stressful conditions (Bhargava *et al.*, 2003, 2006a; Jacobsen *et al.*, 2003a) as well as on soils with minimum agricultural inputs. Many complex adaptive modifications related to breeding system, seed dispersal and their germination account for the success of the members of this genus in colonizing disturbed habitats (Williams and Harper, 1965; Dostalek, 1987). With a shift in focus towards production on agriculturally marginal lands, *Chenopodium* has a significant role to play both as a nutritious food crop and as a cash crop. The genus *Chenopodium*, commonly known as 'goosefoot', comprises about 250 species (Giusti, 1970) that include herbaceous, suffrutescent and arborescent perennials, although most species are colonizing annuals (Wilson, 1990). Some well-known species include *C. quinoa*, *C. pallidicaule*, *C. berlandieri* ssp. *nuttalliae*, *C. ambrosioides*, *C. murale* and *C. amaranticolor*. *Chenopodium* spp. have been cultivated for centuries as a leafy vegetable and subsidiary grain crop in different parts of the world (Risi and Galwey, 1984). Although only three species (*C. quinoa*, *C. pallidicaule* and *C. berlandieri* subsp. *nuttalliae*) are reported to be cultivated (Heiser and Nelson, 1974; Wilson, 1980; Bhargava *et al.*, 2006a, 2007), the leaves and tender stems of numerous other species are consumed as food and fodder (Tanaka, 1976; Kunkel, 1984; Partap, 1990; Moerman, 1998; Partap *et al.*, 1998). The foliage of *Chenopodium* is an inexpensive and rich source of protein, carotenoids and vitamin C (Koziol, 1992; Prakash *et al.*, 1993; Bhargava *et al.*, 2006a). The protein has a balanced amino acid spectrum with high lysine (5.1–6.4%) and methionine (0.4–1.0%) contents (Prakash and Pal, 1998; Bhargava *et al.*, 2006a).

1.3 Quinoa

Of all the new-world crops, *Chenopodium quinoa* Willd., commonly known as 'quinoa', is one of the most underutilized, given its superb seed protein composition and yield potential. It is principally a grain crop, harvested and consumed in a manner similar to that for cereal grains, although its leaves are also used as a potherb (Maughan *et al.*, 2007). Quinoa is not a true cereal grain, but rather is a pseudocereal, which is dicotyledonous. In contrast, cereals are monocotyledonous (Valencia-Chamorro, 2003). Quinoa has risen from a neglected subsistence crop of indigenous farmers to become a major export of the Andean nations of Bolivia and Peru within the past 20 years (Jellen *et al.*, 2011). The emergence of quinoa to prominence in organic food markets of the developed world has led to scientists giving increasing attention to the crop's unique nutritional benefits, and potentially novel abiotic stress-tolerance mechanisms.

Quinoa is a native of the Andean region and has been cultivated in the region for around 7000 years (Garcia, 2003). Quinoa was known by a number of names in local languages. The people of the Chibcha (Bogota) culture called quinoa 'suba' or 'supha', while the Tiahuancotas (Bolivia) called it 'jupha' and the inhabitants of the Atacama desert knew it by the name 'dahue' (Pulgar-Vidal, 1954). León (1964) is of the view that the names 'quinoa' and 'quinua' were used in Bolivia, Peru, Ecuador, Argentina and Chile. The crop has been an important food grain source in the Andean region since 3000 BC (Tapia, 1982) and occupied a place of prominence in the Inca Empire only next only to maize (Cusack, 1984). However, after the conquest of the region by the Spaniards in 1532, other crops, such as potato and barley, relegated quinoa to the background (Bhargava *et al.*, 2006a). However, the sporadic failure of green revolution in the Andes and enormous destruction of other crops by droughts, once again brought native crops like quinoa to the forefront as it showed much less fall in the yields in severe conditions (Cusack, 1984). In the mid-1970s, the exceptional nutritional characteristics of quinoa were discovered and its popularity began to increase (Maughan *et al.*, 2007). Andean countries established small but effective breeding programmes and several new varieties were released. Efforts were made to collect diverse landraces to prevent genetic erosion, resulting in national quinoa germplasm banks in many Andean countries, with the largest banks being in Bolivia and Peru (Maughan *et al.*, 2007).

Quinoa is grown in a wide range of environments in the South American region (especially in and around the Andes), at latitudes from 20°N in Colombia to 40°S in Chile, and from sea level to an altitude of 3800 m (Risi and Galwey, 1989). Recently it has been introduced in Europe, North America, Asia and Africa. Many European countries are members in the project entitled 'Quinoa – A multipurpose crop for EC's agricultural diversification', which was approved in 1993 (Bhargava *et al.*, 2006a). The American and European tests of quinoa have yielded good results and demonstrate the potential of quinoa as a grain and fodder crop (Mujica *et al.*, 2001; Casini, 2002; Jacobsen, 2003; Bhargava *et al.*, 2006a).

1.3.1 Nutritional importance of quinoa

The nutritional excellence of quinoa has been known since ancient times in the Inca Empire. The importance that quinoa could play in nutrition has been emphasized not only in developing countries but also in the developed world. Quinoa seeds have a higher nutritive value than most cereal grains and contain high-quality protein and large amounts of carbohydrates, fat, vitamins and minerals. Perisperm, embryo and endosperm are the three areas where reserve food is stored in quinoa seed (Prego *et al.*, 1998).

The mean protein content reported for quinoa grain is 12–23% (González *et al.*, 1989; Koziol, 1992; Ruales and Nair, 1994a, 1994b; Ando *et al.*, 2002; Karyotis *et al.*, 2003; Abugoch, 2009), which is higher than that of barley, rice or maize, and is comparable to that of wheat (USDA, 2005; Abugoch, 2009). Moreover, the essential amino acid balance is excellent because of a wide range of amino acids, with higher lysine (5.1–6.4%) and methionine (0.4–1%) contents (Prakash and Pal, 1998; Bhargava *et al.*, 2003, 2006a; Abugoch, 2009).

Quinoa protein can supply around 180% of the histidine, 274% of the isoleucine, 338% of the lysine, 212% of the methionine + cysteine, 320% of the phenylalanine + tyrosine, 331% of the threonine, 228% of the tryptophan and 323% of the valine recommended by FAO/WHO/UNU in protein sources for adult nutrition (Vega-Gálvez *et al.*, 2010). Starch is the most important carbohydrate in quinoa grains, making up approximately 58.1–64.2% of the dry matter (Repo-Carrasco *et al.*, 2003). Quinoa starch consists of two polysaccharides: amylose and amylopectin. The amylase content of quinoa starch varies between 3% and 20%, while the amylose fraction of quinoa starch is quite low (Abugoch, 2009). The starch of quinoa is highly branched, with a minimum degree of polymerization of 4600 glucan units, a maximum of 161,000 and a weighted average of 70,000 (Praznik *et al.*, 1999). Granules of quinoa starch have a polygonal form, with a diameter of 2 μm, being smaller than starch of the common grains (Vega-Gálvez *et al.*, 2010). The total dietary fibre of quinoa is near that of cereals (7–9.7% by difference, db), and the soluble fibre content is reported between 1.3% and 6.1% (db) (Ranhotra *et al.*, 1993; USDA, 2005).

The ash content of quinoa (3.4%) is higher than that of rice (0.5%), wheat (1.8%) and other traditional cereals (Cardozo and Tapia, 1979). Quinoa grains contain large amounts of minerals like Ca, Fe, Zn, Cu and Mn (Repo-Carrasco *et al.*, 2003). Calcium (874 mg/kg) and iron (81 mg/kg) in the seeds are significantly higher than most commonly used cereals (Ruales and Nair, 1992). Minerals like P, K and Mg are located in the embryo, while Ca and P in the pericarp are associated with pectic compounds of the cell wall (Konishi *et al.*, 2004). The abundant mineral content makes the grains valuable for children and adults who can benefit from calcium for bones and from iron for blood functions (Konishi *et al.*, 2004).

The oil content in quinoa ranges from 1.8 to 9.5%, with an average of 5.0–7.2% (DeBruin, 1964; Koziol, 1990) that is higher than that of maize (3–4%). Quinoa oil is rich in essential fatty acids such as linoleate and linolenate (Koziol, 1990) and has a high concentration of natural antioxidants like α-tocopherol and γ-tocopherol (Repo-Carrasco *et al.*, 2003). The antioxidant activity of quinoa could be of particular interest to medical researchers and needs more attention (Bhargava *et al.*, 2006a).

Few reports are available on the vitamin content of quinoa grain. Ruales and Nair (1992) reported appreciable amounts of thiamin (0.4 mg/100 g), folic acid (78.1 mg/100 g) and vitamin C (16.4 mg/100 g). Koziol (1992) gave riboflavin and carotene content as 0.39 mg/100 g and 0.39 mg/100 g respectively, and concluded that quinoa contains substantially more riboflavin (B_2), α-tocopherol (vitamin E) and carotene than wheat, rice and barley. In a 100 g edible portion, quinoa supplies 0.20 mg vitamin B_6, 0.61 mg pantothenic acid, 23.5 μg folic acid and 7.1 μg biotin (Koziol, 1992). Recent reports have also confirmed that quinoa is rich in vitamins A, B_2 and E (Repo-Carrasco *et al.*, 2003).

However, several antinutritional substances such as saponins, phytic acid, tannins and protease inhibitors have been found in quinoa seed, which can have a negative effect on the performance and survival of monogastric animals when it is used as the primary dietary energy source (Vega-Gálvez *et al.*, 2010).

The leaves of quinoa contain ample amount of ash (3.3%), fibre (1.9%), vitamin E (2.9 mg α-TE/100 g) and Na (289 mg/100 g) (Koziol, 1992). Prakash *et al.* (1993) reported that leaves have about 82–190 mg/kg of carotenoids, 1.2–2.3 g/kg of vitamin C and 27–30 g/kg of proteins. A recent study on the leaf quality parameters in quinoa has shown that the leaves contain ample amount of carotenoids (230.23–669.57 mg/kg), which was higher than that reported for spinach, amaranth and *C. album* (Gupta and Wagle, 1988; Prakash and Pal, 1991; Shukla *et al.*, 2003; Bhargava *et al.*, 2006b, 2007).

1.3.2   Stress tolerance

Quinoa exhibits high levels of resistance to several of the predominant adverse factors such as soil salinity, drought (Jensen *et al.*, 2000; González *et al.*, 2009, 2011; Jacobsen *et al.*, 2009; Fuentes and Bhargava, 2011), frost (Jacobsen *et al.*, 2005, 2007), diseases and pests (Jacobsen *et al.*, 2003a; Bhargava *et al.*, 2003). Due to its durability under adverse climate conditions, quinoa may be one of the options for food production under various adverse abiotic constraints (FAO, 1998).

Quinoa is a halophytic species that is regarded as having an unusually high tolerance to salinity. Some varieties of the crop show remarkable resistance to salt during germination. Many varieties of this crop can grow in salt concentrations as high as those found in seawater (40 mS/cm) (Jacobsen *et al.*, 2001, 2003a; Wilson *et al.*, 2002; Jacobsen, 2007; Delatorre-Herrera and Pinto, 2009; Adolf *et al.*, 2012). These characteristics make it an attractive crop for regions where salinity has been recognized as a major agricultural problem (Prado *et al.*, 2000). Quinoa has several mechanisms that aid in successful acclimatization of the plant to saline environments. In the cotyledonous stage, high adaptability to soil salinity is probably due to improved metabolic control based on ion absorption, osmolyte accumulation and osmotic adjustment (Ruffino *et al.*, 2010). Quinoa also accumulates salt ions in its tissues and thereby adjusts leaf water potential, enabling the plant to maintain cell turgor and to limit transpiration under saline conditions (Jacobsen *et al.*, 2001; Hariadi *et al.*, 2011). In addition, quinoa is able to maintain K^+/Na^+ and Ca^{2+}/Na^+ selectivity under saline conditions (Rosa *et al.*, 2009).

The drought resistance of quinoa is attributed to morphological characters such as a deep, extensively ramified root system, reduction of leaf area through leaf dropping, small and thick walled cells adapted to large losses of water, and the presence of vesicles containing calcium oxalate that are hygroscopic in nature and reduce transpiration (Canahua, 1977; Jensen *et al.*, 2000; Jacobsen *et al.*, 2003a). Physiological characteristics indicating drought resistance include low osmotic potential, low turgid weight/dry weight ratio, low elasticity and an ability to maintain positive turgor even at low leaf water potentials (Andersen *et al.*, 1996). It has been observed that the stomatal conductance of quinoa remains relatively stable with low but ongoing gas exchange under very dry conditions and low leaf water potentials (Vacher, 1998). Quinoa maintains high leaf water use efficiency to compensate for the decrease in stomatal conductance and thus optimizes carbon gain with a minimization of water losses. Jensen *et al.* (2000) studied the effects of soil drying on leaf water relations and gas exchange in quinoa. The study showed that high net photosynthesis and specific leaf area (SLA) values during early vegetative growth resulted in early vigour of the plant, supporting early water uptake and tolerance to a following drought. The leaf water relations were characterized by low osmotic potentials and low turgid weight/dry weight ratios during later growth stages sustaining a potential gradient for water uptake and turgor maintenance under high evaporation demands. Garcia *et al.* (2003) calculated the seasonal yield response factor (Ky) for quinoa and observed that it was lower than that of groundnut and cotton. This low Ky value for quinoa indicated that a minor drought stress does not result in a large yield decrease.

The frost resistance of quinoa has been recognized for many years (Rea *et al.*, 1979). The species exhibits 100% germination even at 2°C and no serious effect on the plant at temperatures close to −3°C (Bois *et al.*, 2006). The main mechanism for the frost resistance of quinoa seems to be that it tolerates ice formation in the cell walls and the subsequent dehydration of the cells, without suffering irreversible damage (Jacobsen *et al.*, 2003a). The presence of soluble sugars, such as fructans, sucrose and dehydrins, may be good indicators of frost tolerance in quinoa (Jacobsen *et al.*, 2003a, 2005). Results have shown that quinoa seeds germinate rapidly even at low temperatures, with the base temperature for germination lower than 0°C for 9 cultivars out of 10 (Bois *et al.*, 2006).

Hail and snow are sporadic and localized in the Andean region and sometimes causes irreversible damage, especially when the crop is near to maturity (Jacobsen *et al.*, 2003a). Cultivars of quinoa exist with good tolerance to hail, mainly due to a minor leaf angle and greater thickness and resistance of leaves and stem. Flooding occasionally occurs in rainy years on flat areas and produces root rot, greatly reducing yield (Jacobsen *et al.*, 2003a). Wind affects crop productivity by causing plants to fall, especially in the arid region of the altiplano and in some inter-Andean valleys. Wind is also responsible for erosion and drying of soil and plants. When quinoa is cultivated in deserts and hot areas, high temperatures can cause flowers to abort and the death of pollen (Jacobsen *et al.*, 2003a). Fortunately, the genetic variability of quinoa makes it possible to select cultivars with greater tolerance to each of these environmental factors.

1.3.3 Economic uses

Quinoa has diverse uses. It is considered as one of the best leaf protein concentrate sources and so has the potential as a protein substitute for food and fodder and in the pharmaceutical industry. The whole plant can also be used as green fodder for cattle, sheep, pigs, horses and poultry. Results have indicated that up to 150 g/kg unprocessed or dehulled quinoa seed could be included in broiler feed (Jacobsen *et al.*, 1997). This incorporation of quinoa in poultry feeds can greatly benefit the poultry industry. The seeds can be eaten as a rice replacement, as a hot breakfast cereal or can be boiled in water to make infant cereal food (Bhargava *et al.*, 2006a). Quinoa seeds can be ground and used as flour, or sprouted, and can even be popped like popcorn. In Peru and Bolivia, quinoa flakes, tortillas, pancakes and puffed grains are produced commercially (Popenoe *et al.*, 1989). Quinoa flour in combination with wheat flour or corn meal is used in making biscuits, bread and processed food (Bhargava *et al.*, 2006a). There are numerous recipes for about 100 preparations, including tamales, huancaína sauce, leaf salad, pickled quinoa ears, soups and casseroles, stews, torrejas, pastries, sweets and desserts, and soft and fermented hot and cold beverages, as well as breads, biscuits and pancakes, which contain 15–20% quinoa flour. The flour has good gelation property, water absorption capacity, emulsion capacity and stability (Oshodi *et al.*, 1999). The high water absorptivity may be used in the formulation of some foods such as sausages, dough, processed cheese, soups and baked products (Oshodi *et al.*, 1999). Quantitative analysis of the sugar content and chemical composition of seed flour of quinoa has shown that it has a high proportion of D-xylose (120 mg/100 g), and maltose (101 mg/100 g), and a low content of glucose (19 mg/100 g) and fructose (19.6 mg/100 g) (Ogungbenle, 2003). Thus, quinoa could be effectively utilized in the beverage industry for the preparation of malted drink formulations. It can be fermented to make beer, or used to feed livestock (Galwey, 1989). Solid-state fermentation of quinoa with *Rhizopus oligosporus* Saito provides a good-quality tempeh (Valencia-Chamorro, 2003). Quinoa milk, a high quality and nutritive product, may have the potential for consumption as milk or as an ingredient of milky products (Jacobsen *et al.*, 2003b). This tasty and healthy product is of particular importance for people who are unable to digest casein or animal lactose.

Quinoa starch can be used for specialized industrial applications because of its small granules and high viscosity (Galwey *et al.*, 1990). Starches having small-sized granules could serve as dusting starches in cosmetics and rubber tyre mould release agents (Bhargava *et al.*, 2006a). Quinoa starch also has potential for utilization as biodegradable fillers in low-density polyethylene (LDPE) films (Ahamed *et al.*, 1996a). However, this aspect needs more investigation for effective utilization in the food, pharmaceutical and textile industries. Because of its mechanical properties, quinoa starch can be utilized in the manufacture of carrier bags, where tensile strength is important. Studies on freeze–thaw stability of quinoa starch have shown that its paste is resistant to retrogradation, suggesting applications in frozen and emulsion type food products (Ahamed *et al.*, 1996b; Bhargava *et al.*, 2006a). Another potential use of the plant could be in cloth dyeing and food preparation because of the presence of betalains, a natural colorant (Jacobsen *et al.*, 2003b).

Quinoa has been evaluated as a food with excellent nutritional characteristics by the National Research Council and the National Aeronautics and Space Administration (NASA) (Schlick and Bubenheim, 1996). The plant is being considered as a potential crop for NASA's Controlled Ecological Life Support System (CELSS), which aims to use plants to remove carbon dioxide from the atmosphere and generate food, oxygen and water for the crew of long-term space missions (Schlick and Bubenheim, 1996).

1.3.4 Medicinal importance

The use of quinoa for medicinal purposes has also been reported (Mujica, 1994). The plant is reportedly used in inflammation, as an analgesic and as a disinfectant of the urinary tract. It is also used in fractures and internal haemorrhaging and as an insect repellent (Mujica, 1994). The presence of glycine betaine, trigonelline and their derivatives has been reported in the plant (Jancurova *et al.*, 2009). In humans, glycine betaine can be readily absorbed through dietary intake or endogenously synthesized in the liver through choline catabolism. The concentration of glycine betaine in human blood plasma is highly regulated. Its concentrations are lower in patients with renal disease, and its urinary excretion is elevated in patients with diabetes mellitus (Dini *et al.*, 2006). Glycine betaine intake can lower plasma homocysteine levels in patients suffering from homocystinuria, and in chronic renal failure patients with hyperhomocysteinemia, as well as in healthy subjects (Tang *et al.*, 2002; Jancurova *et al.*, 2009). Recently, the cell wall polysaccharides of quinoa seeds (arabinan and arabinan-rich pectic polysaccharides) showed gastroprotective activity on ethanol-induced acute gastric lesions in rats (Cordeiro *et al.*, 2012). These reports can open new avenues for use of quinoa as a medicinal crop.

Dietary flavonoids are thought to have health benefits, possibly due to antioxidant and anti-inflammatory properties (Hirose *et al.*, 2010). Quinoa seeds are the most effective foodstuff as a source of flavonoids among cereals and pseudo-cereals. Recent studies have identified large amounts of flavonoid conjugates in quinoa seeds, such as kaempferol and quercetin oligomeric glycosides (Zhu *et al.*, 2001; Dini *et al.*, 2004; Hirose *et al.*, 2010). Flavonoids, one of the typical polyphenols in vegetables, fruits and tea, can prevent degenerative diseases such as coronary heart disease (Arts and Hollman, 2005), atherosclerosis (Kurosawa *et al.*, 2005), cancers (Rice-Evans and Packer, 1998), diabetes and Alzheimer's disease (Youdim *et al.*, 2004), through antioxidative action and/or the modulation of several protein functions (Hirose *et al.*, 2010). Quinoa also contains appreciable amount of vitamin E (Repo-Carrasco *et al.*, 2003). This is important since this vitamin acts as an antioxidant at the cell membrane level, protecting the fatty acids of the cell membranes against damage caused by free radicals.

The highly nutritious quinoa flour could be used to supplement protein-deficient wheat flour, commonly used for human consumption, in regions where protein deficiency occurs. Quinoa can be recommended for maturity-onset diabetes patients because of its low fructose and glucose. Quinoa flour can be used as a substitute for wheat flour in the production of bread for celiac consumers, with substitutions in small amounts having shown a positive effect on the quality of the breads (Park *et al.*, 2005). One study showed increase in the level of insulin-like growth factor-1 (IGF-1) in the plasma of children who consumed a supplementary portion of an infant food prepared by drum-drying a pre-cooked slurry of quinoa flour (Ruales *et al.*, 2002).

1.4 Concluding Remarks

Quinoa's ability to produce grains high in protein under ecologically extreme conditions makes it important for the diversification of future agricultural systems, not just in mountainous regions, but also in the plains (Bhargava *et al.*, 2006a). The high nutritional quality and multiple uses in food products makes quinoa seed ideal for utilization by the food industry. Other potential uses of quinoa include: a flow improver in starch flour products, fillers in the plastic industry, anti-offset and dusting powders, and a complementary protein for improving the amino acid balance of human and animal foods (Bhargava *et al.*, 2006a). Efforts should be directed to evolving edible varieties with high-quality components, better yield, large seed size and low saponin content. Making quinoa more popular would require dissemination of information about the crop among farmers as well as consumers, proper marketing and efficient post-harvest technologies. Quinoa has the potential to shed its underutilized status and become an important industrial and food crop of the 21st century.

References

Abugoch, L.E. (2009) Quinoa (*Chenopodium quinoa* Willd.): composition, chemistry, nutritional, and functional properties. *Advances in Food Nutrition Research* 58, 1–31.

Adolf, V.I., Shabala, S.N., Andersen, M.N., Razzaghi, F. and Jacobsen, S.-E. (2012) Varietal differences of quinoa's tolerance to saline conditions. *Plant and Soil* 357, 117–129.

Ahamed, N.T., Singhal, R.S., Kulkarni, P.R., Kale, D.D. and Pal, M. (1996a) Studies on *Chenopodium quinoa* and *Amaranthus paniculatas* starch as biodegradable fillers in LDPE films. *Carbohydrate Polymers* 31, 157–160.

Ahamed, N.T., Singhal, R.S., Kulkarni, P.R. and Pal, M. (1996b) Physicochemical and functional properties of *Chenopodium quinoa* starch. *Carbohydrate Polymers* 31, 99–103.

Andersen, S.D., Rasmussen, L., Jensen, C.R., Mogensen, V.O., Andersen, M.N. and Jacobsen, S.-E. (1996) Leaf water relations and gas exchange of field grown *Chenopodium quinoa* Willd. during drought. In: Stolen, O., Pithan, K. and Hill, J. (eds) *Small Grain Cereals and Pseudocereals*. Workshop at KVL, Copenhagen, Denmark.

Ando, H., Chen, Y., Tang, H., Shimizu, M., Watanabe, K. and Miysunaga, T. (2002) Food components in fractions of quinoa seed. *Food Science and Technology Research* 8, 80–84.

Anthony, K., Haq, N. and Cilliers, B. (eds) (1995) Genetic resources and utilization of underutilized crops in southern and eastern Africa. Proceedings of Symposium held at the Institute for Tropical and Subtropical Crops, Nelspruit, South Africa, August 1995. Food and Agriculture Organization, International Centre for Underutilised Crops and Commonwealth Science Council.

Arts, I.C.W. and Hollman, P.C.H. (2005) Polyphenols and disease risk in epidemiologic studies. *American Journal of Clinical Nutrition* 81, 317S.

Bermejo, J.E.H. and León, J. (1994) Neglected crops:1492 from a different perspective. *FAO Plant Production and Protection Series No. 26*. FAO, Rome, Italy.

Bhargava, A., Shukla, S., Katiyar, R.S. and Ohri, D. (2003) Selection parameters for genetic improvement in *Chenopodium* grain on sodic soil. *Journal of Applied Horticulture* 5, 45–48.

Bhargava, A., Shukla, S. and Ohri, D. (2006a) *Chenopodium quinoa*: an Indian perspective. *Industrial Crops and Products* 23, 73–87.

Bhargava, A., Shukla, S., Dixit, B.S., Bannerji, R. and Ohri, D. (2006b) Variability and genotype x cutting interactions for different nutritional components in *C. album* L. *Horticultural Science* 33, 29–38.

Bhargava, A., Shukla, S. and Ohri, D. (2007) Evaluation of foliage yield and leaf quality traits in *Chenopodium* spp. in multiyear trials. *Euphytica* 153, 199–213.

Bhargava, A., Shukla, S., Srivastava, J., Singh, N. and Ohri, D. (2008) *Chenopodium*: a prospective plant for phytoextraction. *Acta Physiologia Plantarum* 30, 111–120.

Bhargava, A., Fuentes, F.F., Bhargava, M. and Srivastava, S. (2012) Approaches for enhanced phytoextraction of heavy metals. *Journal of Environmental Management* 105, 103–120.

Bois, J.F., Winkel, T., Lhomme, J.P., Raffaillac, J.P. and Rocheteau, A. (2006) Response of some Andean cultivars of quinoa (*Chenopodium quinoa* Willd.) to temperature: effects on germination, phenology, growth and freezing. *European Journal of Agronomy* 25, 299–308.

Canahua, M.A. (1977) Observaciones del comportamiento de quinoa a la sequia. In: *Primer Congreso Internacional sobre cultivos Andinos*, Universidad Nacional San Cristobal de Huamanga, Instituto Interamericano de Ciencias Agricolas, Ayacucho, Peru, pp. 390–392.

Cardozo, A. and Tapia, M.E. (1979) Valor nutritivo. In: Tapia, M.E. (ed.) *Quinua y Kaniwa. Cultivos Andinos*. Serie Libros y Materiales Educativos. Instituto Interamericano de Ciencias Agricolas, Bogota, Colombia, pp. 149–192.

Casini P. (2002) Possibilita di introdurre la quinoa negli ambienti mediterranei. *Informatore Agrario* 58, 29–32.

Cordeiro, L.M.C., Reinhardt, V. and Baggio, C. (2012) Arabinan and arabinan-rich pectic polysaccharides from quinoa (*Chenopodium quinoa*) seeds: structure and gastroprotective activity. *Food Chemistry* 130, 937–944.

Cusack, D. (1984) Quinoa: grain of the Incas. *Ecologist* 14, 21–31.

De Bruin, A. (1964) Investigation of the food value of quinoa and canihua seed. *Journal of Food Science* 29, 872–876.

Delatorre-Herrera, J. and Pinto, M. (2009) Importance of ionic and osmotic components of salt stress on the germination of four quinua (*Chenopodium quinoa* Willd.) selections. *Chilean Journal of Agricultural Research* 69, 477–485.

Dini, I., Tenore, G.C. and Dini, A. (2004) Phenolic constituents of Kancolla seeds. *Food Chemistry* 84, 163–168.

Dini, I., Tenore, G.C., Trimarco, E. and Dini, A. (2006) Two novel betaine derivatives from Kancolla seeds (Chenopodiaceae). *Food Chemistry* 98, 209–213.

Dostalek, J. (1987) Influence of the mode of pollination on offspring of some species of the genus *Chenopodium*. *Preslia* 59, 263–269.

FAO (1998) *The State of the World's Plant Genetic Resources for Food and Agriculture*. Food and Agricultural Organization, Rome, Italy.

Fuentes, F.F. and Bhargava, A. (2011) Morphological analysis of quinoa germplasm grown under lowland desert conditions. *Journal of Agronomy and Crop Science* 197, 124–134.

Galwey, N.W. (1989) Exploited plants: quinoa. *Biologist* 36, 267–274.

Galwey, N.W., Leakey, C.L.A., Price, K.R. and Fenwick, G.R. (1990) Chemical composition and nutritional characteristics of quinoa (*Chenopodium quinoa* Willd.). *Food Science and Nutrition* 42F, 245–261.

Garcia, M. (2003) Agroclimatic study and drought resistance analysis of quinoa for an irrigation strategy in the Bolivian Altiplano. Dissertationes de Agricultura, PhD dissertation, Faculty of Applied Biological Sciences, K.U. Leuven, Belgium.

Garcia, M., Raes, D. and Jacobsen, S.-E. (2003) Evapotransporation analysis and irrigation requirements of quinoa (*Chenopodium quinoa*) in the Bolivian highlands. *Agricultural Water Management* 60, 119–134.

Giusti, L. (1970) El genero *Chenopodium* en Argentina 1: Numeros de cromosomas. *Darwiniana* 16, 98–105.

González, J.A., Roldán, A., Gallardo, M., Escudero, T. and Prado, F.E. (1989) Quantitative determination of chemical compounds with nutritional value from Inca crops: *Chenopodium quinoa* ('quinoa'). *Plant Foods for Human Nutrition* 39, 331–337.

González, J.A., Gallardo, M., Hilal, M., Rosa, M. and Prado, F.E. (2009) Physiological responses of quinoa (*Chenopodium quinoa* Willd.) to drought and waterlogging stresses: dry matter partitioning. *Botanical Studies* 50, 35–42.

González, J.A., Bruno, M., Valoy, M. and Prado, F.E. (2011) Genotypic variation of gas exchange parameters and leaf stable carbon and nitrogen isotopes in ten quinoa cultivars grown under drought. *Journal of Agronomy and Crop Science* 197, 81–93.

Gupta, K. and Wagle, D.S. (1988) Nutritional and antinutritional factors of green leafy vegetables. *Journal of Agricultural and Food Chemistry* 36, 472–474.

Hammer, K., Heller, J. and Engels, J. (2001) Monograph on underutilized and neglected crops. *Genetic Resources and Crop Evolution* 48, 3–5.

Hariadi, Y., Marandon, K., Tian, Y., Jacobsen, S.-E. and Shabala, S. (2011) Ionic and osmotic relations in quinoa (*Chenopodium quinoa* Willd.) plants grown at various salinity levels. *Journal of Experimental Botany* 62, 185–193.

Heiser, C.B. and Nelson, D.C. (1974) On the origin of cultivated Chenopods (*Chenopodium*). *Genetics* 78, 503–505.

Hirose, Y., Fujita, T., Ishii, T. and Ueno, N. (2010) Antioxidative properties and flavonoid composition of *Chenopodium quinoa* seeds cultivated in Japan. *Food Chemistry* 119, 1300–1306.

IAEA (2004) Genetic improvement of under-utilized and neglected crops in low income food deficit countries through irradiation and related techniques. *Proceedings of a Final Research Coordination Meeting organized by the Joint FAO/IAEA Division of Nuclear Techniques in Food and Agriculture*, Pretoria, South Africa, 19–23 May 2003, International Atomic Energy Agency, Vienna, Austria.

Jacobsen, E.E., Skadhauge, B. and Jacobsen, S.-E. (1997) Effect of dietary inclusion of quinoa on broiler growth performance. *Animal Feed Science and Technology* 65, 5–14.

Jacobsen, S.-E. (2003) The worldwide potential for quinoa (*Chenopodium quinoa* Willd.). *Food Reviews International* 19, 167–177.

Jacobsen, S.-E. (2007) Quinoa's world potential. In: Ochatt, S. and Jain, S.M. (eds) *Breeding of Neglected and Under-utilized Crops, Spices and Herbs*. Science Publishers, Enfield, UK, pp. 109–122.

Jacobsen, S.-E., Quispe, H. and Mujica, A. (2001) Quinoa: an alternative crop for saline soils in the Andes. In: *Scientists and Farmer–Partners in Research for the 21st Century*. CIP Program Report 1999–2000, pp. 403–408.

Jacobsen, S.-E., Mujica, A. and Jensen, C.R. (2003a) The resistance of quinoa (*Chenopodium quinoa* Willd.) to adverse abiotic factors. *Food Reviews International* 19, 99–109.

Jacobsen, S.-E., Mujica, A. and Ortiz, R. (2003b) The global potential of quinoa and other Andean crops. *Food Reviews International* 19, 139–148.

Jacobsen, S.-E., Monteros, C., Christiansen, J.L., Bravo, L.A., Corcuera, L.J. and Mujica, A. (2005) Plant responses of quinoa (*Chenopodium quinoa* Willd.) to frost at various phonological stages. *European Journal of Agronomy* 22, 131–139.

Jacobsen, S.-E., Monteros, C., Corcuera, L.J., Bravo, L.A., Christiansen, J.L. and Mujica, A. (2007) Frost resistance mechanisms in quinoa (*Chenopodium quinoa* Willd.). *European Journal of Agronomy* 26, 471–475.

Jacobsen, S.-E., Liu, F. and Jensen, C.R. (2009) Does root-sourced ABA play a role for regulation of stomata under drought in quinoa (*Chenopodium quinoa* Willd.). *Scientia Horticulturae* 122, 281–287.

Jancurova, M., Minarovicova, L. and Dandar, A. (2009) Quinoa: a review. *Czech Journal of Food Science* 27, 71–79.

Jellen, E.N., Kolano, B.A., Sederberg, M.C., Bonifacio, A. and Maughan, P.J. (2011) *Chenopodium*. In: Kole, C. (ed.) *Wild Crop Relatives: Genomic and Breeding Resources, Legume Crops and Forages*. Springer, Berlin, Germany, pp. 35–61.

Jensen, C.R., Jacobsen, S.-E., Andersen, M.N., Nunez, N., Andersen, S.D., Rasmussen, L. and Mogensen, V.O. (2000) Leaf gas exchange and water relation characteristics of field quinoa (*Chenopodium quinoa* Willd.) during soil drying. *European Journal of Agronomy* 13, 11–25.

Karyotis, T., Iliadis, C., Noulas, C. and Mitsibonas, T. (2003) Preliminary research on seed production and nutrient content for certain quinoa varieties in a saline-sodic soil. *Journal of Agronomy and Crop Science* 189, 402–408.

Konishi, Y., Hirano, S., Tsuboi, H. and Wada, M. (2004) Distribution of minerals in quinoa (*Chenopodium quinoa* Willd.) seeds. *Bioscience Biotechnology and Biochemistry* 68, 231–234.

Koziol, M.J. (1990) Composicion quimica. In: Wahli, C. (ed.) *Quinua, hacia su cultivo commercial*. Latinreco S.A., Casilla 17-110-6053, Quito, Ecuador, pp. 137–159.

Koziol, M.J. (1992) Chemical composition and nutritional value of quinoa (*Chenopodium quinoa* Willd.). *Journal of Food Composition and Analysis* 5, 35–68.

Kunkel, G. (1984) *Plants for Human Consumption*. Koeltz Scientific Books, Koenigstein, Germany.

Kurosawa, T., Itoh, F., Nozaki, A., Nakano, Y., Katsuda, S. and Osakabe, N. (2005) Suppressive effects of cacao liquor polyphenols (CLP) on LDL oxidation and the development of atherosclerosis in Kurosawa and Kusanagi-hypercholesterolemic rabbits. *Atherosclerosis* 179, 237–246.

León, J. (1964) *Plantas Alimenticias Andinas*. Boletín Técnico No. 6, IICA Zona Andina, Lima, Peru.

Maughan, P.J., Bonifacio, A., Coleman, C.E., Jellen, E.N., Stevens, M.R. and Fairbanks, D.J. (2007) Quinoa (*Chenopodium quinoa*). In: Kole, C. (ed.) *Genome Mapping and Molecular Breeding in Plants, Volume 3 Pulses, Sugar and Tuber Crops*. Springer, Berlin, Germany, pp. 147–158.

Mayes, S., Massawe, F.J., Alderson, P.G., Roberts, J.A., Azam-Ali, S.N. and Hermann, M. (2011) The potential of underutilized crops to improve security of food production. *Journal of Experimental Botany* 63, 1075–1079.

Moerman, D. (1998) *Native American Ethnobotany*. Timber Press, Portland, Oregon.

Mujica, A. (1994) Andean grains and legumes. In: Hernando-Bermujo, J.E. and Leon, J. (eds) *Neglected Crops: 1492 from a Different Perspective*. Vol. 26, FAO, Rome, Italy, pp. 131–148.

Mujica, A., Jacobsen, S.-E., Izquierdo, J. and Marathee, J.P. (2001) *Resultados de la Prueba Americana y Europes de la Quinua*. FAO, Rome, Italy/UNA-Puno, CIP, Peru.

Ogungbenle, H.N. (2003) Nutritional evaluation and functional properties of quinoa (*Chenopodium quinoa*) flour. *International Journal of Food Science and Nutrition* 54, 153–158.

Oshodi, A., Ogungbenle, H. and Oladimeji, M. (1999) Chemical composition, nutritionally valuable minerals and functional properties of benniseed, pearl millet and quinoa flours. *International Journal of Food Science and Nutrition* 50, 325–331.

Padulosi, S. and Hoeschle-Zeledon, I. (2004) Underutilized plant species: what are they? *Leisa* 5–6.

Park, S.H., Chung, O.K. and Seib, P.A. (2005) Effects of varying weight ratios of large and small wheat starch granules on experimental straight-dough bread. *Cereal Chemistry* 82, 166–172.

Partap, T. (1990) Exploiting underexploited crop plants of mountain agriculture: chenopods. In: Riley, K.W., Mateo, N., Hawtin, G.C. and Yadav, R.P. (eds) *Mountain Agriculture and Crop Genetic Resources*. Oxford and IBH, New Delhi, pp. 165–183.

Partap, T., Joshi, B.D. and Galwey, N.W. (1998) *Promoting the Conservation and Use of Underutilized and Neglected Crops*. International Plant Genetic Resources Institute, Rome, Italy.

Popenoe, H., King, S.R., Leon, J. and Kalinowski, L.S. (1989) Lost crops of the Incas. In: Vietmeyer, N.D. (ed.) *Little Known Plants of the Andes with Promise for Worldwide Cultivation*. National Academy Press, Washington, DC.

Prado, F.E., Boero, C., Gallardo, M. and Gonzalez, J.A. (2000) Effect of NaCl on germination, growth, and soluble sugar content in *Chenopodium quinoa* (Willd.) seeds. *Botanical Bulletin of Academia Sinica* 41, 27–34.

Prakash, D. and Pal, M. (1991) Nutritional and anti nutritional composition of vegetable and grain amaranth leaves. *Journal of the Sciences of Food and Agriculture* 57, 573–583.

Prakash, D. and Pal, M. (1998) *Chenopodium*: seed protein, fractionation and amino acid composition. *International Journal of Food Science and Nutrition* 49, 271–275.

Prakash, D., Nath, P. and Pal, M. (1993) Composition, variation of nutritional content in leaves, seed protein, fat and fatty acid profile of *Chenopodium* species. *Journal of the Sciences of Food and Agriculture* 62, 203–205.

Praznik, W., Mundigler, N., Kogler, A., Pelzl, B. and Huber, A. (1999) Molecular background of technological properties of selected starches. *Starch* 51, 197–211.

Prego, I., Maldonado, S. and Otegui, M. (1998) Seed structure and localization of reserves in *Chenopodium quinoa*. *Annals of Botany* 82, 481–488.

Prescott-Allen, R. and Prescott-Allen, C. (1990) How many plants feed the world? *Conservation Biology* 4, 365–374.

Pulgar-Vidal, J. (1954) La quinua o suba, alimento básico de los Chibchas. *Economía Colomiana* 1, 549–560.

Ranhotra, G., Gelroth, J., Glaser, B., Lorenz, K. and Johnson, D. (1993) Composition and protein nutritional quality of quinoa. *Cereal Chemistry* 70, 303–305.

Rea, J., Tapia, M. and Mujica, S.A. (1979) Practicas agronomicas. In: Tapia, M.E. (ed.) *Quinua y Kaniwa. Cultivos Andinos. Serie Libros y Materiales Educativos*. Vol. 49. Instituto Interamericano de Ciencias Agricolas, Bogota, Colombia, pp. 83–120.

Repo-Carrasco, R., Espinoza, C. and Jacobsen, S.-E. (2003) Nutritional value and use of the Andean crops quinoa (*Chenopodium quinoa*) and kañiwa (*Chenopodium pallidicaule*). *Food Reviews International* 19, 179–189.

Rice-Evans, C.A. and Packer, L. (1998) *Flavonoids in Health and Disease*. Marcel Dekker, New York.

Risi, J. and Galwey, N.W. (1984) The *Chenopodium* grains of the Andes: Inca crops for modern agriculture. *Advances in Applied Biology* 10, 145–216.

Risi, J. and Galwey, N.W. (1989) The pattern of genetic diversity in the Andean grain crop quinoa (*Chenopodium quinoa* Willd.). II Multivariate methods. *Euphytica* 41, 135–145.

Rosa, M., Hilal, M., González, J.A. and Prado, F.E. (2009) Low temperature effect on enzyme activities involved in sucrose-starch partitioning in salt-stressed and salt acclimated cotyledons of quinoa (*Chenopodium quinoa* Willd.) seedlings. *Plant Physiology and Biochemistry* 47, 300–307.

Ruales, J. and Nair, B.M. (1992) Nutritional quality of the protein in quinoa (*Chenopodium quinoa* Willd) seeds. *Plant Foods for Human Nutrition* 42, 1–12.

Ruales, J. and Nair, B.M. (1994a) Effect of processing on in vitro digestibility of protein and starch in quinoa seeds. *International Journal of Food Science and Technology* 29, 449–456.

Ruales, J. and Nair, B. (1994b) Properties of starch and dietary fibre in raw and processed quinoa (*Chenopodium quinoa* Willd.) seeds. *Plant Foods for Human Nutrition* 45, 223–246.

Ruales, J., Grijalva, Y., Jaramillo, P.L. and Nair, B.M. (2002) The nutritional quality of an infant food from quinoa and its effect on the plasma level of insulin-like growth factor-I (IGF-I) in undernourished children. *International Journal of Food Science and Nutrition* 53, 143–154.

Ruffino, A.M.C., Rosa, M., Hilal, M., González, J.A. and Prado, F.E. (2010) The role of cotyledon metabolism in the establishment of quinoa (*Chenopodium quinoa*) seedlings growing under salinity. *Plant and Soil* 326, 213–224.

Schlick, G. and Bubenheim, D.L. (1996) Quinoa: candidate crop for NASA's controlled ecological life support systems. In: Janick, J. (ed.) *Progress in New Crops*. ASHS Press, Arlington, Virginia.

Shukla, S., Pandey, V., Pachauri, G., Dixit, B.S., Bannerji, R. and Singh, S.P. (2003) Nutritional contents of different foliage cuttings of vegetable amaranth. *Plant Foods for Human Nutrition* 58, 1–8.

Tanaka, T. (1976) *Tanaka's Cyclopaedia of Edible Plants of the World*. Keigaku Publishing, Tokyo, Japan.

Tang, H., Watanabe, K. and Mitsunaga, T. (2002) Characterization of storage starches from quinoa, barley and adzuki seeds. *Carbohydrate Polymers* 49, 13–22.

Tapia, M. (1982) *The Environment, Crops and Agricultural Systems in the Andes and Southern Peru*. IICA, San Jose, Costa Rica.

USDA (2005) National Nutrient Database for Standard Reference, Release 18. Nutrient Data Laboratory, United States Department of Agriculture, Beltsville, Maryland.

Vacher, J.J. (1998) Responses of two main Andean crops, quinoa (*Chenopodium quinoa* Willd.) and papa amarga (*Solanum juzepczukii* Buk.) to drought on the Bolivian Altiplano: significance of local adaptation. *Agriculture Ecosystems and Environment* 68, 99–108.

Valencia-Chamorro, S.A. (2003) Quinoa. In: Caballero, B. (ed.) *Encyclopedia of Food Science and Nutrition*. Vol. 8. Academic Press, Amsterdam, The Netherlands, pp. 4895–4902.

Vega-Gálvez, A., Miranda, M., Vergara, J., Uribe, E., Puente, L. and Martínez, E.A. (2010) Nutrition facts and functional potential of quinoa (*Chenopodium quinoa* Willd.), an ancient Andean grain: a review. *Journal of Sciences of Food and Agriculture* 90, 2541–2547.

Vietmeyer, N.D. (1986) Lesser-known plants of potential use in agriculture and forestry. *Science* 232, 1379–1384.

Williams, J.T. and Harper, J.L. (1965) Seed polymorphism and germination. I. The influence of nitrates and low temperatures on the germination of *Chenopodium album*. *Weed Research* 5, 141–150.

Wilson, C., Read, J.J. and Abo-Kassem, E. (2002) Effect of mixed-salt salinity on growth and ion relations of a quinoa and a wheat variety. *Journal of Plant Nutrition* 25, 2689–2704.

Wilson, H.D. (1980) Artificial hybridization among species of *Chenopodium* sect. *Chenopodium*. *Systematic Botany* 5, 253–263.

Wilson, H.D. (1990) Quinua and relatives (*Chenopodium* sect. *Chenopodium* subsect. Cellulata). *Economic Botany* 44, 92–110.

Youdim, K.A., Shukitt-Hale, B. and Joseph, J.A. (2004) Flavonoids and the brain: interactions at the blood–brain barrier and their physiological effects on the central nervous system. *Free Radical Biology and Medicine* 37, 1683–1693.

Zhu, N., Sheng, S., Li, D., Lavoie, E.J., Karwe, M.V. and Rosen, R.T. (2001) Antioxidative flavonoid glycosides from quinoa seeds (*Chenopodium quinoa* Willd.). *Journal of Food Lipids* 8, 37–44.

2 Historical Perspectives and Domestication

D~IDIER~ B~AZILE~, F~RANCISCO~ F~UENTES~ AND Á~NGEL~ M~UJICA~

2.1 Introduction

Biodiversity is a key global concern of the international community. The increasing species extinction rate is alarming for the future wellbeing of societies. Attention to biodiversity is important because people around the world are managing this biological diversity for plant cultivation, pastoral activities, forests and many other occupations. Good practices in protecting biodiversity are beneficial for the society in order to provide food, fuel and shelter as well as protect the biota and habitats for future generations (Jackson *et al.*, 2007). Smallholder farmers are the guardians of both the biodiversity surrounding them and the knowledge to manage it (Altieri, 1987; Chevassus-au-Louis and Bazile, 2008).

Agroecosystems occupy 30% of the Earth's surface (Altieri, 1991). There are varied and changing ways in which small farmers, in particular in the developing world, use genetic resources for agriculture. But the dynamic diversity of small-scale farmers has limited literature and most of the information comes from Africa (Bazile and Weltzien, 2008).

Latin America, specifically the Highlands of the Andes, is one of the 'hotspots' of world biodiversity. This region has been used for thousands of years and has supported a large population in interaction with its agroecosystem. Importantly, quinoa (*Chenopodium quinoa* Willd.) has been cultivated by farmers in this region for more than 5000 years (Bazile and Negrete, 2009). A wild relative to quinoa exists in this region as parent of the cultivated form, along with other wild forms that could have participated in the evolution process.

In this chapter, we consider the domestication process at the origins of agriculture to explain the link between human settlements and agriculture development. With a particular focus on quinoa, we show the world importance of the genus *Chenopodium* (Chenopodiaceae), which has 250 species. We relate the complex process of the creation of quinoa from different wild parents and then present a new typology to describe the state of the actual diversification and utilization of the crop. Finally, we discuss how the importance of the biodiversity of quinoa could be related to other agricultures in the world and ask for rules to preserve farmers' rights.

2.2   General Process of Plant Domestication

Information on crop evolution is vital in the current effort to understand and conserve biodiversity, and provides a basis for the improvement of plant species. Plant domestication has changed considerably over the course of human history. The adaptation of plants to agriculture was vital to the shift from hunter–gatherer to agrarian societies. It is generally considered that crops were domesticated 10,000 years ago in diverse places called 'centers of origins' (Harlan, 1971).

From the perspective of population–environment systems, we need to move away from the notion of the individual, which is the term used by naturalists to describe each element or living thing in front of them. Individuals are replaced by the concept of population, a fundamental component of ecological systems. A population is a set of individuals of the same species that coexist in the same given environment. The concept of population is particularly interesting (Tilman, 1996) because it is considered to be a system characterized by different state variables. These state variables include:

- number (or density);
- spatial distribution;
- age structure;
- genetic class structure (gene frequency);
- social organization.

With the concept of population, it is easy to apply indicators of demographic processes (birth rate, mortality, emigration, immigration) that give a population its dynamics. Because these processes depend on both individual and environmental properties, we speak of the population–environment system. The concept of species diversity is based on the fact that an individual organism's variable features are recorded in its genetic heritage. The set of characteristics and behaviours of living things, known as the phenotype, studied by naturalists when they are working in the field depends first on their genetic structure, or genotype. Therefore, it is the set of genes and the genetic modifications that take place on the genes and chromosomes during DNA replication that determines species diversity (Collins and Qualset, 1998; Jarvis *et al.*, 2007).

With this short introduction, we investigate the relation between variability and evolution, and its uses for domestication. In this way, we will try to understand why agrobiodiversity is a human creation. We will focus here on the dynamic character of biodiversity and trace the development of cultivated diversity in different agroecosystems. At the first level of biodiversity, genetic diversity is a source of adaptability enabling farmers to respond to changes in the environment and, by allowing farmers to make selections, permits the production of new varieties that respond to new needs. It is thus due to genetic diversity that evolution within a species is possible and that farmers are able to match ecotypes and cultivars, corresponding to specific environments at a local scale (Bazile *et al.*, 2008).

The first important point is that hunter-gatherers around the world possessed a thorough knowledge of plants – their survival depended on it (Diamond, 2002). They knew which plants growing near their camps may be harvested, how to transform and process bitter or poisonous plants, and had knowledge on the range of medicinal or alimentary uses of these plants. Early farmers quickly understood that there was no point sowing or maintaining plants that already grew in the wild close to their camps. This is the reason we now believe that the beginning of agriculture involved secondary plants that could not be found easily.

Hunter-gatherer societies disappeared during the Neolithic period, although gathering, hunting and fishing practices have continued to this day. Over time new activities – ones that essentially were linked to a different food strategy – developed, with agriculture being one of them. The birth of agriculture is entwined with the search for new food products to support demographic growth (Cauvin, 2008). This effort also included developing techniques that allowed the consumption of these products, such as grinding and cooking. Yet, at the same time, people also continued to feed themselves through gathering, particularly wild cereals (wheat and barley in the Middle East, rice in Asia, millet and sorghum in Africa, maize in the Americas) (Wood and Lenné, 1999; Kaihura and Stocking, 2003).

The shift from gathering to cultivation involved a new way of thinking that was radically different from the past, requiring precise knowledge regarding the selection of seeds, when to sow, how to prepare the land into fields, rotate and distribute species, fertilize with manure, irrigate, and store (granaries, silos) and cook products. This is why there were numerous intermediary stages in agriculture, particularly in the protection of useful plant species and the destruction of harmful species. Thus, we speak of selection, conscious or not, of a certain number of plant types.

We can understand how this domestication took place if we consider, for example, wild cereals, which reproduce more easily when their grains detach easily from the ear. Farmers, however, need grains that remain on a solid ear and stem in order to be able to harvest the maximum amount in the shortest period of time. The same is true for vegetable plants, whose role was essential in the beginning of agricultural practices. The selection of desired characteristics, which was made almost automatically, was certainly at the origins of agriculture. The three essential conditions for the birth of agriculture were:

- people were settled into villages;
- they knew how to sow and harvest; and
- they were specialized in the gathering of species that later would be domesticated.

The original societies in Highland of the Andes existed in this way, with settlements around Lake Titicaca and a situation emerged that helped develop agriculture from wild species (Maxted *et al.*, 2012).

The initial stage of domestication was often determined entirely by unconscious selection. In fact, the phenotypic changes associated with domestication are likely to have arisen via unconscious selection occurring from automatic practices during harvest or unintentional practices that participated in the process of domestication. Generally, it is observed that the phenotypic changes associated with adaptation under domestication are substantial, and they are illustrative of the process and effects of natural selection combined with changes produced by human activities (Lenné and Wood, 2011).

Human societies have common features that explain the permanent domestication process. Farmers are looking for larger fruit or grain, reduced branching, gigantism, the loss or reduction of seed dispersal, the loss of seed dormancy, synchronized seed maturation, an increase in grain size, larger inflorescences, changes in photoperiod sensitivity, and the loss or reduction of toxic compounds. The phenotypic changes associated with domestication could be separated in two parts: characteristics such as colour or fruit size that were probably desired by humans and other traits resulting from unconscious selection that would have been difficult for early cultivators to notice or that would have changed without any direct effort. Finally, there is often a natural counterpart in the agroecosystems. Conscious or unconscious selection is not limited to the visible part of phenotypes. And much of the adaptation under domestication may have involved physiological or developmental changes corresponding to the new edaphic, photosynthetic, hydrological and competitive regimes associated with cultivation (Brookfield, 2001).

To summarize how farmers create diversity, there are three essential points to keep in mind:

1. Farmers domesticate wild plants. That is to say, they seek to adapt wild species to agricultural production. These species are ones that they had once obtained by gathering in the wild.
2. Farmers add to diversity by adapting crops to new ecosystems or changing needs. This may be in terms of human consumption or for other uses such as animal feed.
3. Farmers continuously discover new crops to cultivate, which means that diversity in agriculture is not fixed but is in a constant state of evolution.

2.3 Importance of the Genus *Chenopodium* and Domestication of *C. quinoa*

An important point is that plant genetic resources have been collected and exchanged for over 10,000 years, and more than 5000 years for quinoas (Planella *et al.*, 2011). As the practice of agriculture spread along with human migrations, genes, genotypes and crop populations were carried by people around the world. What is interesting to note is that each group of farmers who settled in a specific location continued to improve their cultivars in order to suit the specific requirements of their farming practices and the ecology of the environment in which they chose to settle and work. Farmers never stopped developing and improving plants, both in their places of origin and in locations very far away, and as populations became more settled, they began to cultivate the large number of species that are found today.

Agriculture has always been based on access to and exchange of plant genetic resources. Farmers give each other material in order to be able to sow from year to year. Agriculture was never based on the exclusive principles observed today with the extension of property rights over the living world. Through these free exchanges, people traded plants, local landraces and seeds. Through their travels, they brought back exotic species to cultivate alongside their usual plants. Farmers introduced exogenous material to avoid declines in productivity and a degeneration of variety cultivars in their fields.

Before discussing *Chenopodium quinoa* Willd., we need to consider all the species of the genus *Chenopodium* (Chenopodiaceae), which includes about 250 species that are mostly colonizing herbaceous annuals occupying large areas in the Americas, Asia and Europe, though some are also suffrutescent and arborescent perennials.

Nowadays, three important species of *Chenopodium* are in cultivation as food plants: *C. pallidicaule* Aellen and *C. quinoa* Willd. in South America and *C. nuttalliae* Safford in Mexico. Other species of the genus are known to have been important wild food sources in North America. Wild species were also used in Europe for food. This wide use of chenopods for food is not surprising because most species produce large numbers of seeds, which have a high protein content, and the leaves can also be used for human consumption (Risi and Galwey, 1984, 1989).

The economically important species of the genus *Chenopodium* are:

- *C. quinoa* ($2n = 36$) used as a grain crop;
- *C. pallidicaule* ($2n = 18$), *C. berlandieri* subsp. *nuttalliae* ($2n = 36$) used for both grain and vegetable; and
- *C. album* ($2n = 18, 36, 54$) mainly used as a leafy vegetable and foliage crop, though some Himalayan types are also cultivated for grain.

Using *Chenopodium* seeds for human consumption is not unique to the Andean region:

- *C. berlandieri* ssp. *nuttalliae*, a species similar to *C. quinoa*, is largely consumed in Mexico for its tender leaves and inflorescences.
- In the Himalayas in India, Nepal, Bhutan and China, farmers cultivate a kind of chenopod (classified as *C. album*) at altitudes of 1500–3000 m.
- *C. album* is a widespread weed and was part of the human diet in Europe according to prehistoric human remains found in Tollund (Denmark) and Cheshire (UK).

All of these examples show the importance of the genus *Chenopodium* and the need for research on the links between species. Phylogenetic relationships between cultivated and related wild taxa have been studied on the basis of allozyme studies, crossability and DNA structuration, but the complex and great morphological, ecological and chromosomal diversity found within the genus complex needs further studies to settle the taxonomic problems.

Quinoa, as a food grain, has been recognized for centuries as an important food crop in the high Andes of South America. Cultivated chenopods, especially *C. quinoa*, are gaining importance for their outstanding protein quality and high content of a range of minerals and vitamins. The genetic diversity of *C. quinoa*, with its salt and drought tolerance, offers a wide adaptation for many difficult environments. The very name quinoa in the Quechua and Aymara languages means 'Mother Grain'. Within South and Central America, two closely related species, Canihua (*C. pallidicaule*) and Huazontle (*C. nuttalliae*), are also utilized for food. The descendants of the Inca Empire, 8–10 million Quechua and Aymara Indians, still use quinoa as an important component of their diet.

Quinoa has been cultivated for more than 5000 years in the Andes. Some data report that it was probably domesticated by ancient civilizations at different times and in different geographic zones, including Peru (5000 BC), Chile (3000 BC) and Bolivia (750 BC) (Kadereit *et al.*, 2003). Nowadays, its existence is explained by the transmission of seeds by the Incas to other Chilean aboriginal groups living in distinct agroecological contexts, from the Chilean Altiplano (17°S) to Chiloé Island (42°S) and beyond (47°S, Puerto Rio Tranquilo). However, during the Spanish conquest this crop was strongly discouraged because of its cultural importance and because it was considered a sacred crop by the indigenous people (Ruas *et al.*, 1999).

2.4 Current Insights into the Evolution of Genetic Diversity in Quinoa

Ancient farmers along the Andes of South America took the first steps in domesticating quinoa from its wild or weedy forms in a domestication/cultivation period of approximately 5000 years. In this context, quinoa has been subjected to diverse selection processes for desirable traits for its cultivation and consumption by people belonging to different cultures and territories in South America, such as Chibchas, Andaki, Inganos in southern Colombia; Aymara and Quechua in areas of Peru, Bolivia and northern Chile; Diaguitas and Calchaquies in northern Argentina; and Mapuches in southern Chile (Mujica, 2004). This process, conducted in multiple local landraces, led to the loss of many allele combinations that were disadvantageous to farmers, such as dehiscent seeds (seed shattering) and thick seed coats, as well as the gain of useful ones such as increased seed size and environmental adaptations. Similarly, modern breeding procedures have also continued this process by crossing the best phenotypes to increase yields and agronomic performance and, thereby, narrowing the genetic diversity of quinoa in a wide sense. Nevertheless, the diversity can still be observed in the fields in a wide array of colours in plants and seeds, and differences in the types of branching and panicles, as well as having differences in grain productivity, abiotic stress tolerance and disease resistance (Fuentes and Bhargava, 2011; Ruiz-Carrasco *et al.*, 2011) (Fig. 2.1).

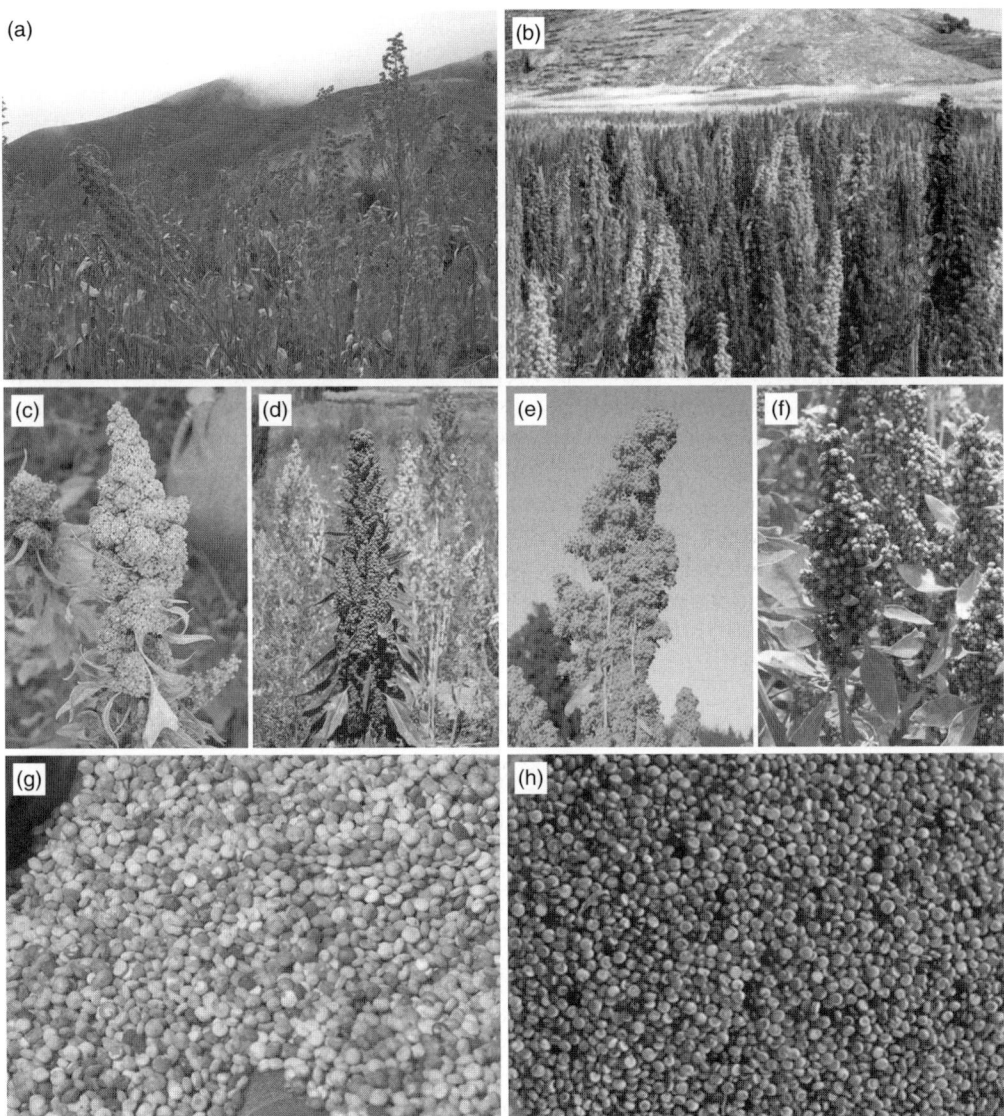

Fig. 2.1. Morphological variation in *C. quinoa*: (a) Lax panicle of quinoa in Salta, Argentina. (b) Compact panicle and colour variation of quinoa in Peru. (c) Glomerulate panicle of quinoa in Bolivia. (d) Amaranthiform panicle of quinoa in Salta, Argentina. (e) Glomerulate panicle of yellow quinoa in Temuco, Chile. (f) Semi-amaranthiform panicle of quinoa in Pichincha, Ecuador. (g) Representative quinoa seeds from highlands Chile (north). (h) Representative quinoa seeds from coastal/lowland Chile (south).

Quinoa diversity, at a continental scale, has been associated with five main ecotypes: Highlands (Peru and Bolivia), Inter-Andean valleys (Colombia, Ecuador and Peru), Salares (salt lakes; Bolivia, Chile and Argentina), Yungas (Bolivia) and Coastal/Lowlands (Chile); each of these are associated with sub-centres of diversity that originated around Lake Titicaca (Risi and Galwey, 1984). In the beginning, Gandarillas (1979) and Wilson (1988) identified the southern highlands of Bolivia as the genetic diversity centre for quinoa. Subsequently, Christensen *et al.* (2007), using SSR (simple sequence repeats) markers, identified the genetic diversity centre in the Altiplano area between Peru and Bolivia (Central Andean highlands). Germplasm from Ecuador and Argentina in the same study showed limited diversity, indicating that the most probable point of introduction for Ecuadorian accessions was the Altiplano (Peru–Bolivia), while for Argentina the original introduction could have been from the Chilean highlands and coastal zone (south of Chile) (Fig. 2.2). In addition, Christensen *et al.* (2007) highlighted the differences between coastal/lowlands accessions from Chile and those from the northern highlands of Peru, confirming the hypothesis proposed by Wilson (1988) that quinoas from Chile show more similarity with quinoas from the southern Altiplano of Bolivia. Nevertheless, Fuentes *et al.* (2009), while assessing genetic diversity on Chilean germplasm using SSR markers, reported for first time that Chilean coastal/lowland germplasm was much more genetically diverse than was previously believed. This finding was consistent with a cross-pollination system in the coastal/lowland quinoa fields with weed populations of *C. album* and/or *C. hircinum* and agrees well with the difficulty experienced by coastal/lowland quinoa breeders to obtain pure new cultivars in south-central Chile (I. and E. von Baer, personal communication). Taken together, the recent genetic-based analyses are consistent with the idea that quinoa itself has existed until now as two distinct germplasm pools: Andean highland quinoa with its associated weedy complex (ajara or ashpa quinoa, *C. quinoa* ssp. *milleanum* Aellen, also referred to as *C. quinoa* var. *melanospermum* Hunziker) and kinwa among the Mapuche people of the central and southern Chilean coastal/lowlands, representing in addition a second centre of major quinoa diversity (Jellen *et al.*, 2011). Interestingly, the weedy *C. hircinum* from lowland Argentina can be mentioned as a third distinct germplasm pool, which may represent remnants of archaic quinoa cultivation in that part of South America (Wilson, 1990).

When quinoa was originally classified in 1797, it was assumed to be an exclusively domesticated species in a section of the genus from the New World. In 1917 other important cultivated tetraploid chenopods were discovered in Central America (Wilson and Heiser, 1979). These plants were originally classified by Safford as *C. nuttalliae*, consisting of three different cultivars: huazontle, red chia and quelite. From its original classification, these plants have been reclassified several times, including a period in which they were considered conspecific with quinoa. These species are part of the complex of *C. berlandieri*, commonly known as *C. berlandieri* var. *nuttalliae* (Wilson and Heiser, 1979), including some extinct subspecies (subsp. *jonesianum* for example) that have been identified in several archaeological sites in the east of North America (Smith and Funk, 1985; Smith and Yarnell, 2009). Although it is widely accepted that these species share a common gene pool with quinoa, several studies have pointed out that it is likely that they were grown independently (Heiser and Nelson, 1974). Since systematic genetic research on quinoa began at the end of 1970s, it was believed that quinoa originated in South America

Fig. 2.2. Model of biodiversity dynamics of quinoa associated with the five ecotypes in the Andes: (a) Inter-Andean Valley (Ecuador–Colombia). (b) Yungas (Bolivia). (c) Highlands (Peru–Bolivia–Argentina). (d) Salares (northern Chile–southern Bolivia). (e) Coastal/lowland (southern Chile). Arrows show most likely seed migration routes, as inferred from genetic similitude and from ancestral people's interactions and cultural exchanges. [Reprinted from Fuentes *et al.* (2012), with permission from Cambridge University Press.]

from diploid descendants from the highlands such as *C. pallidicaule* Aellen (Kañawa), *C. petiolare* Kunth and *C. carnasolum* Moq., as well as from tetraploid weed species from South America such as *C. hircinum* Schard, or *C. quinoa* var. *melanospermum* (Mujica and Jacobsen, 2000). An alternative hypothesis, originally raised by Wilson and Heiser (1979), is that quinoa descended from the tetraploid *C. berlandieri* in North America.

However, when the Mexican complex of *C. berlandieri* was described, it was considered conspecific with quinoa. Thus, the prevalent paradigm is whether *C. quinoa* comes from early tetraploids of *C. berlandieri*; most probably *C. berlandieri* var. *zschackei*, considering that domesticated Mexican tetraploids descend from *C. berlandieri* var. *sinuatum*. This idea has been supported by diverse studies based on morphological, experimental crosses, isozymes and genetic analysis (Heiser and Nelson, 1974; Wilson and Heiser, 1979; Wilson, 1980; Walters, 1988; Maughan *et al.*, 2006). If this hypothesis is correct, it could be indicated that the tetraploid origin of *C. quinoa* is in North America, from an ancestor similar to *C. berlandieri* var. *zschackei*. If so, it is possible that this wild North American tetraploid progenitor travelled to Mexico and South America through human migration or by long-distance bird dispersals, probably as *C. hircinum*, and was subsequently domesticated as quinoa (Wilson, 1990).

Archaeobotanical studies based on patterns of seed morphology and frequencies of *C. quinoa* and its associated weedy complex have shown interesting perspectives to support Wilson's hypothesis. Studies conducted by Bruno and Whitehead (2003) have shed light on some processes contributing to the development of agricultural systems between 1500 BC and AD 100 in the southern Lake Titicaca Basin (Bolivia). The results of this study suggest that during the Early Formative period, farmers maintained small gardens where both the crop and weed species were grown and harvested. However, around 800 BC there was a drastic decrease in frequency of weedy seeds compared with quinoa seeds, revealing a significant change in crop management and use. This suggests that in the Middle Formative period farmers became more meticulous cultivators of quinoa, perhaps through weeding, careful seed selection and creating formal fields for cultivation. Although there have been an increasing number of *Chenopodium* studies, more information is still needed to accurately construct the probable ancestor of quinoa and the correct phylogeny of its genus, specifically those relations among quinoa and wild relatives and *C. berlandieri*, as well as with spontaneous hybrids between *C. quinoa* and wild relatives under crop conditions.

2.5 Current Typology to Describe the Diversification and Utilization of the Quinoas in South America

From many generations of farmers' selection, quinoa today presents high variability and genetic diversity that allows it to adapt to different ecological environments (valleys, highlands, salt flats, etc.). It can tolerate different relative humidity conditions (from 40% to 88%), a large range of altitudes (from sea level up to 4000 m) and a wide range of temperatures (from −8°C to 38°C). This shows its high adaptation to climate change and its potential for agricultural development in other parts of the world. But, in order to understand where the quinoas could be used in the future, we need to better understand the diversity of the actual contexts of cultivation in its area of origin.

Different types of quinoa exist in the Andean region whose characteristics vary from one agroecological zone to another. They differ in behaviour, phenology, morphology, cultivation technology, resistance to biotic and abiotic factors and utilization. There are eight quinoa typologies to describe its diversification and utilization according to its agroecological zone: the Altiplano (northern Andean highlands); the shore of salt lakes (southern Andean highlands); the inter-Andean valleys; arid zones and dry conditions (eastern Andean highlands); high altitudes and cool climates; coastal regions and near the sea; jungle and tropical zones; high rainfall and humidity zones. Wild relatives of quinoa have both nutritional and medicinal uses.

2.5.1 Quinoas of the Altiplano (northern Andean highlands)

This zone, lying near Lake Titicaca and the northern highlands, represents a relatively temperate and watered region. Annual average precipitation is about 600 mm, which determines various possibilities for diversifying the cropping system. Quinoa is associated with other crops such as potato, barley, oats, beans and various tubers. Quinoa from this region is mainly produced for home consumption or local markets. The associated crops that are present in the lake region are often rare and do not appear in the rest of the Bolivian Highland because precipitation decreases sharply to the south. In the northern and central highlands, soils are relatively rich in water and minerals; quinoa seed is broadcast or sown in rows, unlike the southern highlands where it is grown in pockets to optimize water resources.

In this zone, the quinoas are small plants of different colours adapted to the shores of Lake Titicaca, and have a variable content of saponin and a vegetative period of 6 months. They produce small to medium grains, have less resistance to cold and drought, and can sometimes grow in saline soils. They are generally sown in *aynokas*, an ancestral system of management and utilization (Mujica and Jacobsen, 2000), or in a mixture of varieties. Plants in the highlands are little branched and have a unique panicle, with abundant foliage. In this part of the northern highlands, quinoas are moderately resistant to mildew (*Peronospora farinosa* Fr.), and are attacked by young plants cutters (*Feltia experta* Walker), Kona Kona (*Eurisacca quinoae* Povolny), and by birds on the shore of Lake Titicaca. When there is a moisture deficit, the lower leaves turn yellow and drop off. These quinoas of the northern highlands are usually called *Jiura*. They normally require an annual rainfall of between 700 and 800 mm, and can grow at an altitude of 3850 m. Examples of names for theses quinoas include Kancolla, Blanca de July, Chullpi and Pasankalla in Puno, Peru and La Paz, Bolivia.

2.5.2 Quinoas from the Salares (southern highlands)

In the highlands, quinoa can grow up to an altitude of 4200 m, its extreme limit of cultivation (Bazile *et al.*, 2011). In the southern Altiplano with the Salar de

Uyuni at its centre, there are all the conditions of a typical cold desert, perched up to 3600 m above sea level and surrounded by volcanoes. Rainfall averages 350 mm on the north part of the Salar, but rarely exceeds 150 mm on the south shore, with over 200 days of frost per year.

Quinoa plots are frequently observed on the flanks of the volcanoes (Tunupa, for example) in this region. Quinoa is traditionally cultivated in the middle of lava blocks, to which different local varieties are adapted.

Another traditional agricultural landscape of quinoa is plots scattered on the slopes, where frosts are less frequent than on the plains and lowlands, which are left to pasture. In the south of the Salar, with only 150 mm of rainfall per year, quinoa fields are located in the lowlands. Throughout the region of the Salar (north and south), the plots are cultivated only every two years to allow the necessary soil water accumulation for the crop cycle. Here, quinoa is really the only plant cultivated because of the extreme conditions of this cold high-altitude desert. But in this area mechanization has revolutionized the production system. As a result, quinoa has been able to respond to international market demand, initially from Peru in the 1980s and North America and Europe from the 1990s. The slopes are inaccessible to tractors, so mechanization has been used on the plains. Farmers have converted their pastures to vast monocultures.

Plants of the southern highlands are large, branched, have different colours, large grain size (2.2–2.9 mm) (Bertero *et al.*, 2004) and high saponin content. They are drought resistant, adapted to saline and sandy soils of the shore of salt lakes, and have specific mechanisms of absorption, partition and excretion of salts (Mujica *et al.*, 2010a, 2010b). Traditional technology involves planting in holes, using wide plant spacing and sowing in soil reclaimed from Thola (*Parastrephia quadrangularis* (Meyen) Cab). Plants are susceptible to *Eurisacca quinoae* Povolny and *Peronospora farinosa* Fr., which are adapted to high, dry and cold conditions.

The quinoas of this region are called 'Quinoa Real' in the international market. They correspond to a group of landraces from this specific desert region with low annual rainfall (150–350 mm) and high altitude (3800–4200 m). Although quinoas have the same commercial appellation, it is possible to identify under them a high diversity, such as Pandela, Utusaya, Toledo, Achachino, which are well adapted to grow in the Uyuni salt flats: Salinas de Garci Mendoza, Coipasa, Llica in Bolivia, Colchane in Chile and Quebrada del Toro in Argentina.

2.5.3 Quinoas from Inter Andean Valleys

In the Andean valleys, quinoa can be grown in the more fertile soils and warmer climate. However, this is determined by the moisture level, which affects production by increasing sensitivity to parasitic attacks. So quinoa is generally produced on small and medium plots, mostly for home consumption.

Plants are tall and branched, with large leaves, producing large to small grains of different colours, and have a long vegetative period. The green

leaves are used as a leafy vegetable. Plants are susceptible to mildew and can
have high or low saponin content. In the Andean valleys, quinoa is not cultivated as a sole crop, but is usually associated with maize, beans, and potatoes and other tubers (Mujica, 2009). These types grow at altitudes of
2500–3200 m in areas having an annual rainfall of 800–900 mm. They are
usually called 'Quinua' and grouped by their genetic and phenotypic characteristics (Medina *et al.*, 2004).

Examples of this varietal diversity include Amarilla de Marangani, Blanca
de Junín, Acostambo, Roja Coporaque and Nariño. They are located in Peru
in the Valle del Vilcanota, Cusco; Valle del Mantaro, Huancayo; Callejon de
Huaylas, Ancash; Valle del Colca, Arequipa; Huancabamba, Piura; Cajabamba,
Cajamarca; and also Chimborazo in Ecuador and Cochabamba in Bolivia.

2.5.4 Quinoas from arid zones and dry conditions (western highlands)

Quinoas of this region are small with a short vegetative period because of the
environmental conditions (only two rainy months), and show morphological,
physiological, anatomical, biochemical and phenological modifications to withstand drought stress (Mujica *et al.*, 2010c). They grow in areas above 3900 m
with an annual rainfall of 150–350 mm. Plants have small leaves, nyctinastic
movements according to daylight hours, various colours, high betacyanin and
calcium oxalate content, and a deep and highly branched roots structure. The
leaves and inflorescences are used as food. They produce small to medium
grains, and have high saponin content. They are planted as a sole crop, generally as mixtures of local varieties. The cultivation technology under these rainfall conditions uses more seed for planting. Farmers usually plant this quinoa
with camelid and sheep manure. Seeds are sown just after the first rain in these
areas in Peru and at the end of winter in dry conditions in Chile. Under certain
conditions roots and stems are used for *llipta* or *llucta* (preparation of the
pungent ashes of the quinoa).

These quinoas are also called 'quinua' with variety names such as
Antahuara, Ucha, Ccoyto, Roja Ayauchana. They are located in Condoriri,
Puno; Cangallo, Ayacucho; Acobamba, Huancavelica in Peru; Colchane and
San Pedro de Atacama in Chile; Pisiga Choque in Bolivia.

2.5.5 Quinoas from high altitudes and cool climate

Plants are small with vivid colours like yellow, red or purple in the plants and
grains. Seeds may be grey (Coitos) or black (Ayaras). The plant has small and
compact glomerular panicles, with accentuated mechanisms of leaf rolling and
decumbent panicles, especially at night (nyctinastic movements). This is a
form of defence against cold, protecting the panicle and primordia, bending
down or protecting young leaves with adult leaves. The bitter grain, which
is high in protein, forms quite early and the plants exhibit cold resistance with
mechanisms for tolerating frost (Jacobsen *et al.*, 2007) and long and strong winds.

They are grown above 4000 m in areas like Canchas and Qochas (Mujica, 2011), which have annual rainfall of 800 mm and short, sunny days. These types are resistant to ultraviolet radiation and have intense and varied pigmentation.

Examples of landraces include Huariponcho, Witulla, Kellu, Kancolla Roja, grown in Macusani, Nuñoa, Laraqueri in Puno, Peru and Carhuamayo, Ondores, Junín, Peru.

2.5.6 Quinoas from the coastal regions and near the sea

This group of quinoas show characteristics specific to their environment, which includes an average annual rainfall of about 500–650 mm spread over 4–5 months and a high evapotranspiration index (Bazile *et al.*, 2010; Núñez *et al.*, 2010). The plants are adapted to salty and sandy soils. Small-scale farmers maintain this quinoa in marginal conditions along with other crops such as cereals. In the south of Chile, Mapuche women conserve a high diversity of landraces in home gardens, associated with cultural uses (Aleman *et al.*, 2010; Thomet *et al.*, 2010).

These quinoas have medium branched plants, with glomerular panicles, and are salt-tolerant with small leaves. All of them have small and hard grain, usually protected by a perigonium that strongly adheres to the grain. The plants are resistant to excessive moisture, some of them growing in areas with over 2000 mm of annual rainfall, such as the Precordillera. In the southern region of Chile, these quinoas are called Kinwa or Dawe (Sepúlveda *et al.*, 2004). These types have various uses as food, for example, in preparing beverages, and the grain flour is sometimes cooked in water or with soup.

Examples of landraces include: Quinoa blanca in the central zone of Chile; Kinwa mapuche, Lito, Faro, Islunga in Temuco and Valdivia, Chile. Currently Altiplano and valley varieties have adapted perfectly to the Peruvian coast, using drip and sprinkler irrigation, obtaining yields of 7.5 t/ha.

2.5.7 Quinoas from jungle and tropical zones

This zone has tall, highly branched quinoas with a long vegetative period, large leaves typical of chenopods, bright and intense colours, large and loose panicles, usually amarantiform (loose), and small grains. They are resistant to mildew and excessive moisture, and can even grow in flooded soils. Quinoa plantings in this zone are associated with other plants such as maize, cassava, fruit trees and potatoes. The plants grow at altitudes between 800 and 1800 m that have an annual rainfall of over 1500 mm, and are heat and evapotranspiration resistant.

Example of landraces include: Tupiza, A. Marangani in the Yungas of Bolivia, and Sandia, Puno, Ambo-Huánuco, Lares-Cusco, Marcapata, Cusco in Peru.

2.5.8   Quinoa from high rainfall and humidity zones

These quinoas are tall, highly branched, large panicled, small grained, high yielding and have a long vegetative period. The plants have a wide and thick root system that can grow in poorly drained soils. They are resistant to lodging and heavy rainfall (2000–3000 mm), with the lower leaves dropping when there is excessive moisture in the soil. These types are resistant to mildew but are strongly attacked by snails and slugs, and chewing and leaf miner larvae. The plants of this zone are consumed mainly as a leafy vegetable.

Example of landraces include: Tupiza, Nariño, Sogamoso, Tunkahuan in El Dorado, Sogamoso, Colombia; Mérida in Venezuela; Tupiza in Bolivia and Amazonas, Peru.

In this category, the quinoas from the Andes are short-day type, while those from coastal and southern Chile require a long photoperiod.

2.5.9   Wild relatives of quinoa

Quinoa is a sympatric plant, because areas of distribution and expansion are always accompanied by their wild relatives, which cross over and maintain variability. In each of the types of quinoa, specific wild relatives can be found (Mujica, 2010). The most common species are *C. carnosolum* Moq. (Chocca chiwa) and *C. quinoa* ssp. *melanospermum* Hunz. (ayara, ajara or aara) in the Andean area of Titicaca; *C. petiolare* Kunth., *C. hircinum* Schard. (Jatacco), *C. insisum* and *C. ambrosioides* in the inter-Andean valleys; *C. carnosolum*, *C. hircinum* and *C. petiolare* (aaras, ajaras) in the salares; *C. petiolare* and *C. hircinum* in the dry, arid and high zones; *C. pallidicaule* Aellen, *C. hircinum* and *C. quinoa* ssp. *melanospermum* in the high and cold areas (having valuable genes for resistance to drought and cold) (Mujica *et al.*, 2008); *C. ambrosioides*, *C. quinoa* ssp. *melanospermum* and other introduced species such as *C. album* (Hierba de gallinazo o cenizo) in the coastal areas (Mujica and Jacobsen, 2006); *C. ambrosioides* L. (Paicco) and *C. insisum* (Asna paicco or Arka paicco) in the Yungas and tropical zones (used to control gastrointestinal amoebas); *C. carnosolum* and *C. quinoa* ssp. *melanospermum* in the humid zones and high rainfall area (consumed by the Andean people as a vegetable and for medicinal uses) (Mujica, 2007).

There are other quinoas in Central America, in the central valley of Mexico, called Huatzontle, used for their inflorescences and leaves. These are medium-sized plants, with a high saponin content, and are not consumed as grains. Plants with green and yellow colours and medium-sized grains correspond to *C. berlandieri* ssp. *nuttalliae* or *C. nuttalliae*, and its wild relative is *C. graveolens*. The dish prepared with the inflorescences is called Capeado de huatzonthe.

2.6 Concluding Remarks

The domestication process for the genus *Chenopodium*, and for the species quinoa in particular, took place in various independent or linked areas over the same period or through migrations of people that conferred an adaptation to new ecological environments. So, current landraces are closely connected with specific geographical locations leading to the generation of distinct genotypes within the same species. These adaptations to specific agroecological regions generated five main ecotypes of the crop, which are associated with diversity sub-centres corresponding to the geographical regions and each of these groups displays high variability under specific agricultural practices. Considering these farmers' practices under agro-meteorological constraints, a new typology for quinoa could separate eight typologies and wild relatives.

Through this high agrobiodiversity and wide ecology, the adaptation of Andean quinoa offers great potential to bring into production underutilized areas such as dry and salty fields. It confers the potential for these agricultural systems to adapt to climate change. Considerations about the importance of quinoa for subsistence agriculture and small-scale farms under low input systems are needed to implement new agricultural systems all around the world with quinoa species with respect to Andean local communities. Scenarios for the future diffusion of quinoa to newer areas should integrate the Farmers' Property Rights and the Nagoya Protocol (attached to the Convention on Biological Diversity) that offers a framework for Access and Benefices Sharing.

The actual diffusion of quinoa across all the continents (North America, Europe, Asia and Africa) has occurred in diverse ways and has different objectives. But an international network that primarily includes researchers and farmers could provide an opportunity for better characterization and understanding of this species.

The biogeography of quinoa, an ancestral and highly nutritional crop, provides a global foresight of this underutilized crop in world agriculture, and also shows its broad geographic extension. Several aspects linked to its high genetic diversity and plasticity demonstrate that quinoa could become one of the most important crops of the South American Andes and could extend its area of cultivation in other contexts in the world giving, due respect to the farmers' rights for the local communities from the areas of domestication.

Acknowledgements

The authors wish to express their appreciation to farmers who cared for their seeds for telling us their stories, and also to projects that made possible funding of reported research activities BRG08, IMAS (ANR07 BDIV 016-01) and IRSES (PIRSES-GA-2008-230862). We also gratefully acknowledge Dr Eric Jellen (BYU, USA), Dr Daniel Bertero (UBA, Argentina) and Ingrid von Baer (AgroGen, Chile) for their images of quinoa grown in Bolivia, Argentina and Chile, respectively.

References

Aleman, J., Thomet, M., Bazile, D. and Pham, J.L. (2010) Central role of nodal farmers in seed exchanges for biodiversity dynamics: example of 'curadoras' for the quinoa conservation in Mapuche communities in south Chile. In: Coudel, E., Devautour, H., Soulard, C. and Hubert, B. (eds) *International Symposium ISDA 2010. Innovation and Sustainable Development in Agriculture and Food: Abstracts and Papers.* CIRAD, Montpellier. http:// hal.archives-ouvertes.fr/hal-00530950/fr/

Altieri, M.A. (1987) Peasant agriculture and the conservation of crop and wild plant resources. *Conservation Biology* 1, 49–58.

Altieri, M. (1991) How best can we use biodiversity in agroecosystems? *Outlook on Agriculture* 20, 5–23.

Bazile, D. and Negrete, J. (2009) Quínoa y biodiversidad: cuáles son los desafíos regionales? *Revista Geografica de Valparaíso* 42, 1–141.

Bazile, D. and Weltzien, E. (2008) Agrobiodiversités: numéro spécial. *Cahiers Agricultures* 17, 73–256.

Bazile, D., Dembélé, S., Soumaré, M. and Dembele, D. (2008) Utilisation de la diversité variétale du sorgho pour valoriser la diversité des sols au Mali. *Cahiers Agricultures* 17, 86–94.

Bazile, D., Olguin Manzano, P.A., Nuñez, L., Croce, P., Alarcon, G., Lagos, J., Parra, F., Peredo, P. and Negrete Sepulveda, J. (2010) Differenciación territorial asociada a la quinua en el secano costero de la sexta región, Chile: consideraciones sobre las práticas y representaciones sociales para un desarrollo sostenible. *Anales de la Sociedad Chilena de Ciencias Geograficas* 103–109.

Bazile, D., Carrié, C., Vidal, A. and Negrete Sepulveda, J. (2011) Modélisation des dynamiques spatiales liées à la culture du quinoa dans le Nord chilien. [Modelisation of spatial dynamics linked to the cultivation of quinoa in northern Chile.] *Mappemonde* (102) (article 11204), 14. http://mappemonde.mgm.fr/num30/articles/art11204.html

Bertero, D., De la Vega, A., Correa, G., Jacobsen, S.-E. and Mujica, A. (2004) Genotype and genotype-by-environment interaction effect for grain yield and grain size of quinoa (*Chenopodium quinoa* Willd.) as revealed by pattern analysis of international multi-environment trials. *Field Crop Research* 89, 299–318.

Brookfield, H. (2001) *Exploring Agrobiodiversity.* Columbia University Press, West Sussex, UK.

Bruno, M. and Whitehead, W.T. (2003) *Chenopodium* cultivation and formative period agriculture at Chiripa, Bolivia. *Latin American Antiquity* 14, 339–355.

Cauvin, J. (2008) *The Birth of Gods and the Origins of Agriculture.* Cambridge University Press, Cambridge.

Chevassus-au-Louis, B. and Bazile, D. (2008) Cultiver la diversité. *Cahiers Agricultures* 17, 77–78.

Christensen, S.A., Pratt, D.B., Pratt, C., Stevens, M.R., Jellen, E.N., Coleman, C.E., Fairbanks, D.J., Bonifacio, A. and Maughan, P.J. (2007) Assessment of genetic diversity in the USDA and CIP-FAO international nursery collections of quinoa (*Chenopodium quinoa* Willd.) using microsatellite markers. *Plant Genetic Resources* 5, 82–95.

Collins, W.W. and Qualset, C.O. (1999) *Biodiversity in Agroecosystems.* CRC Press LLC, Boca Raton, Florida.

Diamond, J. (2002) Evolution, consequences and future of plant and animal domestication. *Nature* 418, 700–707.

Fuentes, F. and Bhargava, A. (2011) Morphological analysis of quinoa germplasm grown under lowland desert conditions. *Journal of Agronomy and Crop Science* 197, 124–134.

Fuentes, F.F., Martínez, E.A., Hinrichsen, P.V., Jellen, E.N. and Maughan, P.J. (2009) Assessment of genetic diversity patterns in Chilean quinoa (*Chenopodium quinoa* Willd.) germplasm using multiplex fluorescent microsatellite markers. *Conservation Genetics* 10, 369–377.

Fuentes, F., Bazile, D., Bhargava, A. and Martínez, E.A. (2012) Implications of farmers' seed exchanges for on-farm conservation of quinoa, as revealed by its genetic diversity in Chile. *The Journal of Agricultural Science* 150, 702–716.

Gandarillas, H. (1979) Genetica y origen. In: Tapia, M.E. (ed.) *Quinoa y Kaniwa*. Instituto Interamericano de Ciencias Agricolas, Bogota, Colombia, pp. 45–64.

Harlan, J.R. (1971) Agricultural origins: centers and noncenters. *Science* 174, 468–474.

Heiser, C.B. and Nelson, C.D. (1974) On the origin of cultivated chenopods (*Chenopodium*). *Genetics* 78, 503–505.

Jackson, L.E., Pascual, U. and Hodgkin, T. (2007) Biodiversity in agricultural landscapes: investing without losing interest. *Agriculture, Ecosystems and Environment* 121, 196–210.

Jacobsen, S.-E., Monteros, C., Corcuera, L.J., Bravo, L.A., Christiansen, J.L. and Mujica, A. (2007) Frost resistance mechanisms in quinoa (*Chenopodium quinoa* Willd.). *European Journal of Agronomy* 26, 471–475.

Jarvis, D.I., Padoch, C. and Cooper, H.D. (2007) *Managing Biodiversity in Agricultural Ecosystems*. Columbia University Press, New York.

Jellen, E.N., Kolano, B.A., Sederberg, M.C., Bonifacio, A. and Maughan, P.J. (2011) *Chenopodium*. In: Kole, C. (ed.) *Wild Crop Relatives: Genomic and Breeding Resources*. Springer, Berlin, Germany, pp. 35–61.

Kadereit, G., Borsch, T., Welsing, K. and Freitag, H. (2003) Phylogeny of *Amaranthaceae* and *Chenopodiaceae* and the evolution of C4 photosynthesis. *International Journal of Plant Science* 164, 959–986.

Kaihura, F. and Stocking, M. (2003) *Agricultural Biodiversity in Smallholder Farms of East Africa*. The United Nations University, Tokyo, Japan.

Lenné, J.M. and Wood, D. (2011) *Agrobiodiversity Management for Food Security*. CAB International, Wallingford, UK.

Maughan, P.J., Kolano, B.A., Maluszynska, J., Coles, N.D., Bonifacio, A., Rojas, J., Coleman, C.E., Stevens, M.R., Fairbanks, D.J., Perkinson, S.E. and Jellen, E.N. (2006) Molecular and cytological characterization of ribosomal RNA genes in *Chenopodium quinoa* and *Chenopodium berlandieri*. *Genome* 49, 825–839.

Maxted, N., Ford-Lloyd, B.V., Kell, S.P., Iriondo, J.M. and Dulloo, M.E. (2012) *Crop Wild Relative Conservation and Use*. CAB International, Wallingford, UK.

Medina, W., Janiak, A., Szarejko, I., Mujica, A. and Jacobsen, S.-E. (2004) Análisis de relaciones genéticas entre variedades de quinua (*Chenopodium quinoa* Willd.) utilizando la técnica de ALFP (Amplified Fragment Length Polymorphism). In: *Libro de Resúmenes XI Congreso Internacional de cultivos andinos*. Cochabamba, Bolivia.

Mujica, A. (2004) La quínoa Indígena, características e historia. In: Sepúlveda, J., Thomet, M.I., Palazuelos, F. and Mujica, A. (eds) *La Kinwa Mapuche, Recuperación de un Cultivo para la Alimentación*. Fundación para la Innovación Agraria, Ministerio de Agricultura Temuco, Chile, pp. 22–42.

Mujica, A. (2007) Usos etnofarmacobotanicos de la quinua (*Chenopodium quinoa* Willd.) y parientes silvestres en el altiplano Peruano. In: *Avances de la Farmacobotánica en Latinoamérica (2004–2007). IX Simposio Argentino y XII Simposio Latinoamericano de Farmacobotánica*, 4–6 July, Tucumán, Argentina, pp. 42–43.

Mujica, A. (2009) Rol de las comunidades indígenas en el desarrollo y uso sustentable de los recursos fitogenéticos en los países de América Latina y el Caribe: Caso Quinua (*Chenopodium quinoa* Willd.). In: *Proceeding: VII Simposio de Recursos Genéticos para América Latina y el Caribe*, 28–30 October, Pucón, Chile, pp. 107–108.

Mujica, A. (2010) Determinación de la distribución e identificación de parientes silvestres y variedades nativas de cañigua (*Chenopodium pallidicaule* Aellen) y quinua (*Chenopodium quinoa* Willd.) para su conservación *ex situ* e *in situ* en el altiplano Peruano. In: *Anales del Encuentro Científico Internacional 2010 INVIERNO, ECI 2010i*, 2–6 August 2010, Lima, Peru, pp. 1–20.

Mujica, A. (2011) Conocimientos y prácticas tradicionales indígenas en los Andes para la adaptación y disminución de los impactos del cambio climático. In: *Compilación de Resúmenes Workshop: Indigenous Peoples, Marginalized Populations and Climate Change: Vulnerability, Adaptation and Traditional Knowledge*, 19–21 July 2011, IPMPCC, Mexico, D.F.

Mujica, A. and Jacobsen, S.-E. (2000) Agrobiodiversidad de las aynokas de quinua (*Chenopodium quinoa* Willd.) y la seguridad alimentaria. *Seminario Agrobiodiversidad en la Región Andina y Amazónica* 151–156.

Mujica, A. and Jacobsen, S.-E. (2006) La quinua (*Chenopodium quinoa* Willd.) y sus parientes silvestres. In: Moraes, M., Ollgaard, B., Kvist, L.P., Borchsenius, F. and Balslev, H. (eds) *Botánica Económica de los Andes*. Universidad Mayor de San Andrés, La Paz, Bolivia, pp. 449–457.

Mujica, A., Viñas, O., Mamani, F., Dela Torre, J. and Jacobsen, S.-E. (2008) Conservación *in situ* de parientes silvestres de quinua (*Chenopodium quinoa* Willd.) con genes de resistencia a factores abióticos adversos en el altiplano Peruano-Boliviano-Chileno. In: *Memorias del 13° Congreso Latinoamericano de Genética y VI Congreso Peruano de Genética, Recursos Genéticos Latinoamericanos: Vida para la vida*, May 2008, ALAG, SPG, Lima, Peru.

Mujica, A., Chura, E., Ruiz, E. and Martínez, R. (2010a) Mecanismos de resistencia a sequía de la quinua (*Chenopodium quinoa* Willd.). In: *Proceedings Primer Congreso Peruano de Mejoramiento Genético de Plantas y Biotecnología Agrícola*, 17–19 May, UNALM, EPG, Lima, Peru, pp. 111–114.

Mujica, A., Chura, E., Ruiz, E., Rossel, J. and Pocco, M. (2010b) Mecanismos de resistencia a sales y selección de variedades de quinua (*Chenopodium quinoa* Willd.) resistentes a salinidad. In: *Anales XII Congreso Nacional de las Ciencias del Suelo y V Congreso Internacional de las Ciencias del Suelo*, Arequipa, Peru, 11–15 October 2010, pp. 187–189.

Mujica, A., Rossel, J., Chura, E., Ruiz, E., Martínez, R., Cutipa, S. and Gomel, Z. (2010c) Saberes y conocimientos de comunidades originarias sobre conservación de diversidad y variabilidad de quinua (*Chenopodium quinoa* Willd.) en el altiplano Peruano. In: *Memoria-Resúmenes. II Congreso Mundial de la Quinua*, 16–19 March, UTO, Oruro, Bolivia.

Núñez Carrasco, L., Bazile, D., Chia, E., Hocdé, H., Negrete Sepúlveda, J. and Martínez, E.A. (2010) Representaciones sociales acerca de la conservación de la biodiversidad en el caso de productores tradicionales de *Chenopodium quinoa* Willd del secano costero en las regionses de O'Higgins y el Maule. *Anales de la Sociedad Chilena de Ciencias Geograficas* 181–187.

Planella, M.T., Scherson, R. and McRostie, V. (2011) Sitio el Plomo y nuevos registros de cultigenos iniciales en cazadores del arcaico IV en alto Maipo, Chile central. [New evidence on the use of initial cultigens by the hunter-gatherer groups of the archaic IV period at El Plomo, Alto Maipo, Central Chile.] *Chungara, Revista de Antropologia Chilena* 43, 189–202.

Risi, J.C. and Galwey, N.W. (1984) The *Chenopodium* grains of the Andes: Inca crops for modern agriculture. *Advances in Applied Biology* 10, 145–216.

Risi, J.C. and Galwey, N.W. (1989) The pattern of genetic diversity in the Andean grain crop quinoa: association between characteristics. *Euphytica* 41, 147–162.

Ruas, P., Bonifacio, A., Ruas, C., Fairbanks, D. and Andersen, W. (1999) Genetic relationship among 19 accessions of six species of *Chenopodium* L., by randomly amplified polymorphic DNA fragments (RAPD). *Euphytica* 105, 25–32.

Ruiz-Carrasco, K., Antognoni, F., Coulibaly, A.K., Lizardi, S., Covarrubias, A., Martínez, E.A., Molina-Montenegro, M.A., Biondi, S. and Zurita-Silva, A. (2011) Variation in salinity tolerance of four lowland genotypes of quinoa (*Chenopodium quinoa* Willd.) as assessed by growth, physiological traits, and sodium transporter gene expression. *Plant Physiology and Biochemistry* 49, 1333–1341.

Sepúlveda, J., Thomet, M., Palazuelos, P. and Mujica, A. (2004) *La Kinwa Mapuche: Recuperación de un Cultivo Para la Alimentación*. Fundación para la Innovación Agraria, CET, CETSUR, Temuco, Chile.

Smith, B.D. and Yarnell, R.A. (2009) Initial formation of an indigenous crop complex in eastern North America at 3800 B.P. *Proceedings of the National Academy of Sciences (USA)* 106, 6561–6566.

Smith, B.G. and Funk, V.A. (1985) A newly described subfossil cultivar of *Chenopodium* (*Chenopodiaceae*). *Phytologia* 57, 445–448.

Thomet, M., Aleman, J., Bazile, D. and Pham, J.L. (2010) Impactos de la redefinición del concepto de *Trafkintü* sobre la diversidad de variedades de quínoa cultivadas por agricultores mapuches en cuatro comunas de la región de la Araucanía del sur de Chile. *Anales de la Sociedad Chilena de Ciencias Geograficas* 244–249.

Tilman, D. (1996) Biodiversity: population versus ecosystem stability. *Ecology* 77, 350–363.

Walters, T.W. (1988) Relationship between isozymic and morphologic variations in the diploids *Chenopodium fremontii*, *C. neomexicanum*, *C. palmeri*, and *C. watsonii*. *American Journal of Botany* 75, 97–105.

Wilson, H.D. (1980) Artificial hybridization among species of *Chenopodium* sect. *Chenopodium*. *Systematic Botany* 5, 253–263.

Wilson, H.D. (1988) Quinoa biosystematics I: domesticated populations. *Economic Botany* 42, 461–477.

Wilson, H.D. (1990) Quinua and relatives (*Chenopodium* sect. *Chenopodium* subsect. Cellulata). *Economic Botany* 44, 92–110.

Wilson, H.D. and Heiser, C.B. (1979) The origin and evolutionary relationships of 'Huauzontle' (*Chenopodium nuttalliae* Safford), domesticated chenopod of Mexico. *American Journal of Botany* 66, 198–206.

Wood, D. and Lenné, J.M. (1999) *Agrobiodiversity: Characterization and Management*. CAB International, Wallingford, UK.

3 Distribution

3.1 Introduction

A number of species of chenopod are cultivated and consumed in various forms in different parts of the world (Bhargava *et al.*, 2007a). Cañihua (*Chenopodium pallidicaule*) has been domesticated in the Andes for grain and forage, but it is not as productive or as highly domesticated as quinoa (*C. quinoa* Willd.). Cañihua is adapted to high-altitude environments above 4000 m because of its extreme frost tolerance and is still cultivated in these areas (Galwey, 1995). Huazontle (*C. berlandieri* subsp. *nuttalliae*) was domesticated in Mexico, where it was cultivated both as grain and as potherb, but it is now cultivated mainly for its leaves and immature inflorescences (Bhargava *et al.*, 2005). A domesticated form of *C. album* is cultivated in the Himalayas as a seed grain and potherb (Partap and Kapoor, 1985; Bhargava *et al.*, 2007b), while foliage of other species like *C. giganteum*, *C. murale* and *C. bushianum* is also consumed as food and fodder, but to a lesser extent (Kunkel, 1984; Partap *et al.*, 1998; Bhargava *et al.*, 2008). Quinoa has been cultivated in the Andean region for thousands of years as a grain crop suited to diverse climatic conditions.

3.2 The Andean Region and Quinoa

The Andean region covers 2 million square kilometres and extends from southern Venezuela to northern Argentina and Chile, and includes Colombia, Ecuador, Peru and Bolivia (Izquierdo and Roca, 1998). This ecoregion is one of the world's most fragile and least understood environments and has been severely damaged by soil erosion, deforestation, overgrazing, contamination by mining wastes and poor water management (National Research Council, 1989). In terms of biodiversity, the Andean ecoregion is very rich: more than a third of the world's foods are indigenous to the region. Apart from their ability to grow at high altitudes under extremely harsh temperature and water stress conditions, these crops have a wide range and mix of desirable characteristics: high protein, vitamin and starch content, high yield and important medicinal properties (Izquierdo and Roca, 1998).

Quinoa has been an important food grain source in the Andean region since 3000 BC and occupied a place of prominence in the Incan empire next only to maize (Tapia, 1982; Cusack, 1984). However, following the Spanish conquest of the region in 1532, other crops, such as potato, faba beans, oats and barley, relegated quinoa to the background (Galwey, 1995; Bhargava *et al.*, 2006). During the colonial period the cultivation of quinoa was discouraged, possibly because of its honoured position in Incan society and religion (Risi and Galwey, 1989a). Quinoa's religious significance for the Incas made it a less attractive crop to the Spanish than the potato. The status of quinoa as 'Mother Grain' and the 'Grain of the Gods' put it in direct conflict with the Catholic religion promoted by the Spanish Conquistadors. Their religious leaders therefore discouraged its production and consumption in the newly conquered territories. The trend continued until the sporadic failure of the green revolution in the Andes and widespread destruction of other crops by droughts. This once again brought native crops, like quinoa, to the forefront because its yields were less affected in severe conditions (Cusack, 1984; Bhargava *et al.*, 2006). During the 1980s a market for quinoa was established in Europe and North America, mainly in the health-food sector. The market demand was met partly by imports from South America and partly by development of quinoa in 'new regions' outside its centre of origin.

3.3　Quinoa in its Native Region

Willdenow was the first to botanically describe quinoa as a species native to South America, whose centre of origin, according to Buskasov, was in Bolivia and Peru (Cárdenas, 1944). However, during the last decade, Chile, Ecuador, Argentina and Colombia have started extensive cultivation and research projects on quinoa, such as SICA (Agricultural Census and Information System) of the Agricultural Ministry of Ecuador; Quinuacoche CANOE Program promoted by the Latin American Foundation in Colombia; Provincial Congress for Quinoa promoted by the Chamber of Deputies of Salta, Argentina; Program of Encouragement for Business Design and Innovation promoted by the Euro Chile Foundation (Taboada *et al.*, 2011).

Quinoa is seen as an oligocentric species with a broad centre of origin and multiple diversification. The Andean altiplano, a high plains region encompassing Lake Titicaca and extending approximately 800 km from north to south, lies mostly between 3500 and 4300 m above sea level. The shores of Lake Titicaca in the Andean region are considered to be the area of greatest genetic diversity and variation of the crop (Mujica, 1992). Quinoa is distributed throughout the Andean region, from Colombia (Pasto) to northern Argentina (Jujuy and Salta) and Chile (Antofagasta), where a group of quinoas have been found at sea level (Lescano, 1994).

In fact, quinoa cultivation occurred in areas where it is absent today. Most notable of these are northern Colombia near Bogota, and Cordoba Province in eastern Argentina. The major areas of current quinoa cultivation (Fig. 3.1) appear to extend southward from extreme southern Colombia through Ecuador, Peru and Bolivia, with extensions into the Chilean altiplano (eastern Tarapaca) and northern Argentina (Jujuy and Salta) (Wilson, 1990). According to Rojas (1998), the geographical distribution of quinoa in the region extends from 5°N in southern Colombia to 43°S in the Xth Region of Chile. Its altitudinal distribution ranges from sea level in Chile to 4000 m in the altiplano of Bolivia and Peru.

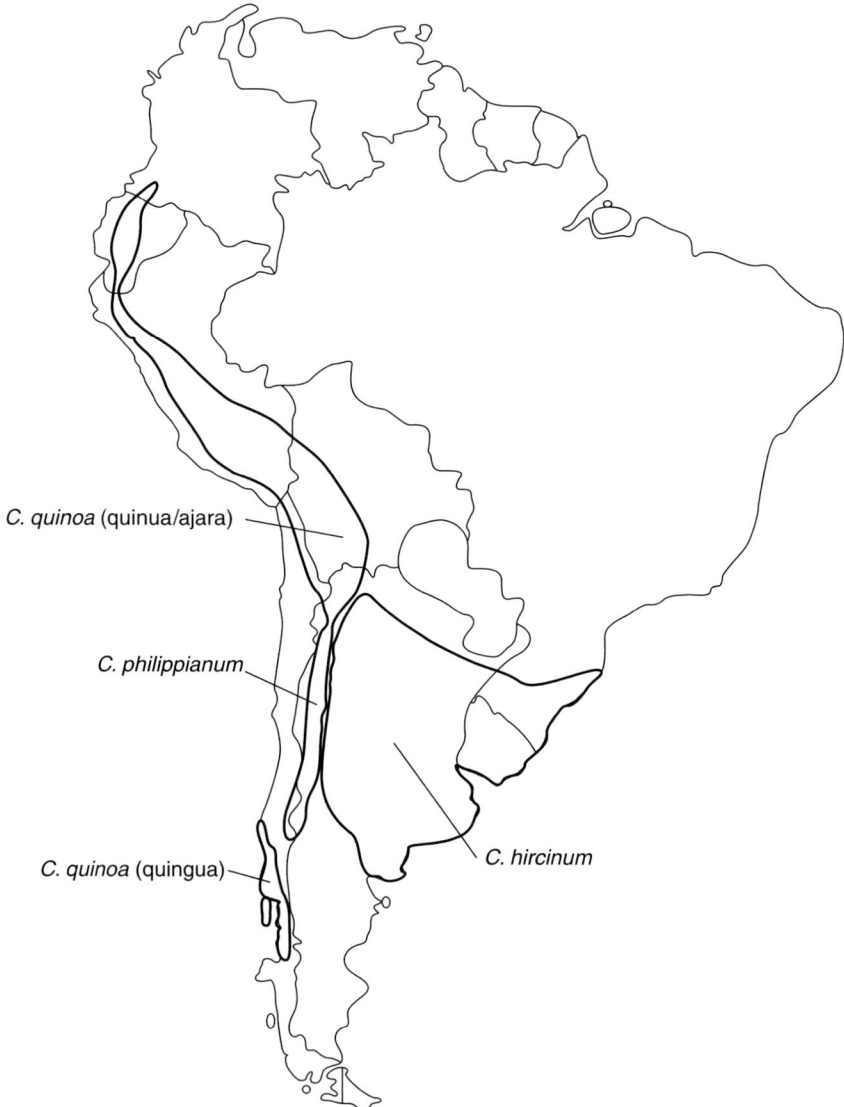

Fig. 3.1. Distribution of quinoa and other members of subsect. Cellulata in South America. [Reprinted from Wilson *et al.* (1990), with permission from Springer.]

3.3.1 Bolivia

Historically, the best quality quinoa has been produced in Bolivia. The high-quality quinoa from Bolivia is sold in European, North American and Asian markets at high prices. Apart from being a major producer, Bolivia is also the world's number one exporter of quinoa and fulfils almost 42% of the quinoa demand in the international market (Antonio, 2011).

The Bolivian altiplano is a vast basin lying between the Royal Cordillera and the Western Cordillera (Vacher, 1998). It consists of a series of plateaus covering an area of 70,000 km², whose altitudes vary from 3700 to 4100 m, and a secondary mountain range whose summits reach up to 5000 m (Geerts *et al.*, 2006). More than 35,000 hectares of quinoa have been cultivated throughout the Bolivian altiplano. The main cultivation areas are in La Paz, in the provinces of Aroma and Gualberto Villaroel, in Oruro, in the region of Salinas de Garci Mendoza in the province of Ladislao Cabrera, and in Potosí, which includes the region of Llica in the Province of Daniel Campos and the Province of Nor López, one of the areas producing high-quality quinoa (Rojas *et al.*, 2004).

3.3.2 Peru

The agroecological zone on the Peruvian altiplano is complex and large, containing tremendous variation in agricultural production determined by physical factors (Aguilar and Jacobsen, 2003). Three major agroecological zones are identified in Peru, namely Puna, Suni and Circunlacustre, which are further subdivided into smaller areas called 'ayonoqas'. Peru has a good range of genetic variability in quinoa and cultivation occurs under environmental conditions that change from year to year. Puno is the main quinoa-producing region (75% of total production), followed by Huancayo (10%) and Cusco (5%) (Mujica *et al.*, 2003). Other important quinoa-producing areas include Cajamarca, Callejón de Huayllas, Valle del Mantaro and Andahuayllas.

3.3.3 Chile

The long Chilean mainland territory (3000 km) extends between 18°S to 43°S. Three different biogeographic regions of quinoa are recognized in Chile, namely, the north altiplano, the centre and the southern region (Bazile and Negrete, 2009; Fuentes *et al.*, 2009). The quinoa crop in north Chile is cultivated primarily by indigenous Aymara Indians in the northern Chilean altiplano, and the cultivation extends to the south-central zone of Chile in a fragmented pattern (Fuentes *et al.*, 2009). The lowland landraces are cultivated on small family farms in the southern regions of Chile from elevations of 1000 m to near sea level, practices inherited from Pehuenche's ancient cultures at 34–36°S and from the Mapuches at 40°S (Martínez *et al.*, 2007). The crop is also cultivated at high elevations above 3500 m in the Chilean altiplano (Isluga and Iquique) (Martínez *et al.*, 2009).

3.3.4   Ecuador

Quinoa is grown mainly in the areas of Imbabura, Chimborazo, Cotopaxi, Pichincha, Carchi, Tungurahua, Loja, Latacunga, Ambato and Cuenca.

3.3.5   Argentina

In Argentina, the crop is grown in isolation in Jujuy and Salta. It is also cultivated in the Calchaquíes valleys of Tucumán (Gallardo and Gonzalez, 1992).

3.3.6   Colombia

The crop is grown in the area of Ipiales, Puesres, Contadero, Cordova, San Juan, Mocondino and Pasto.

3.4   The European Experiment

The main breeding goals in Europe were adaptation to local climatic conditions, improvement of the agronomic performance as a seed crop with respect to early maturity, uniformity, seed yield and seed weight, as well as qualitative research on starch, protein and saponin content in the seeds (Risi and Galwey, 1989b; Jacobsen and Stølen, 1993; Limburg and Mastebroek, 1997; Mastebroek and Limburg, 1997; Mastebroek et al., 2002). Research in Sweden and Denmark concentrated on breeding for fodder quality (Carlsson, 1980; Haaber, 1991).

Quinoa was introduced to England in the 1970s and to Denmark shortly after. In 1993, a project titled 'Quinoa: a multipurpose crop for EC's agricultural diversification' was initiated in the European Union (Jacobsen, 2003). The project envisaged setting up laboratories in Scotland and France, and field trials in England, Denmark, the Netherlands and Italy. Quinoa has been cultivated in Denmark on a small-scale and experimental basis since the late 1980s. Trials in Denmark have demonstrated seed yields of 2–3 t/ha with 12–16% protein and 6–8% oil contents. Seed yields, however, have varied considerably between years and locations, because the establishment of the crop, weed control, harvest and post-harvest techniques have not yet been optimized. Quinoa's experiment in Polish conditions seems to be satisfactory (Gęsiński and Kwiatkowska, 1999a, 1999b; Gęsiński, 2000, 2001). Field trials of 24 quinoa cultivars were conducted in north-west Poland in 1998 (Gęsiński, 2001). Some cultivars showed good potential for cultivation under the climatic and soil conditions in Poland despite being sown 2 months after the supposed optimal date. Field examinations of American and European cultivars in 1999–2001 at Chrzastowo (Poland) have revealed that the European cultivar had a short compact inflorescence with a short flowering period, reaching 120 cm as compared to the American cultivar that showed a slower growth, loose long inflorescence and long period of flowering (Gęsiński, 2006). The study concluded that adequate growth, the course of flowering and a stable yield of the European cultivar make the group ideal for cultivation under Polish conditions.

In Italy and Greece, Danish quinoa has given the best yield. In Greece the suitability of grain quinoa for agriculture has been experimentally tested since 1995 (Karyotis *et al.*, 1996, Iliadis *et al.*, 1997). Field experiments were carried out to test the adaptation of quinoa to Greek soils in loamy, well-drained soils classified as 'Inceptisol' and in heavy textured soils, classified as 'Vertrisol'. The results showed that a wide range of soils seem to be suitable for cultivation of quinoa. Heavy soils could be used for cultivation of the crop, even at pH values as high as 7.5. Quinoa has also been evaluated in Sweden as a source for leaf protein concentrate (Carlsson, 1980; Carlsson *et al.*, 1984). The suitability of quinoa for southern Italy was evaluated in Vitulazio, Italy in a 2-year field trial (2006–2007) under rain-fed conditions (Pulvento *et al.*, 2010). Two quinoa genotypes, KVLQ520Y and Regalona Baer, were compared for yield and two sowing dates. The results showed that April was the best sowing time for quinoa in the typical Mediterranean region. Of the two genotypes, Regalona Baer recorded better growth and productivity, apparently being more tolerant to abiotic stress (high temperatures associated with water stress). Chemical analyses reveal the potential of quinoa seed as a valuable ingredient in the preparation of cereal foods, having improved nutritional characteristics.

3.5 North America

In the USA, quinoa represents one of the relatively few successful introductions of a new food plant. Quinoa has been cultivated in the USA since the early 1980s and commercially produced since the mid-1980s in the Colorado Rockies, especially in the San Luis Valley. The first commercial crop was produced in Colorado in 1987 on five farms after a processing facility was provided by Pillsbury Company to remove saponins from the pericarp (Oelke *et al.*, 1992). About a dozen ecotypes were selected from various regions of South America that had characteristics matching the high, semi-arid mountain valleys and plains of Colorado (Cusack, 1984). Most of the quinoa varieties seemed to be well adapted to high altitudes and cool temperatures. Seed yield at 2000 m above sea level in Colorado was 1000 kg/ha. The North American Quinoa Producers Association was formed in 1988 and a small processing plant was started for the crop produced in the area. Production has also been attempted in California, New Mexico, Oregon and Washington.

In other parts of the US, quinoa has been successfully grown at high elevations; at lower elevations considerable pollen sterility and poor seed set has been observed (Bhardwaj *et al.*, 1996). The quinoa cultivars adapted for production at elevations above 2300 m in Colorado failed to produce viable seeds at Moscow, Idaho (Kephart *et al.*, 1990). In North Dakota, good stands of the crop were obtained at three southern locations, but serious insect problems were encountered in these areas (Berti and Schneiter, 1993). In 1992–93, quinoa was grown near Blacksburg and at the Northern Piedmont Agricultural Experiment Station in Orange County (Bhardwaj *et al.*, 1996). The average seed yield at the Blacksburg location was 2804 kg/ha. However, in 1992 seed set did not occur, probably due to the warm weather. In 2010, 44 varieties of quinoa representing a broad diversity of regions and environmental conditions were grown in three locations in Washington State (Port Townsend, Olympia and Pullman). Data were collected for traits like emergence, aphid resistance, plant height, lodging and grain maturity. Of these 44 varieties, 11 superior varieties were chosen for inclusion in larger scale variety × nitrogen fertility trials in Pullman in 2011. These 11 varieties were also evaluated in the greenhouse under three different nitrogen regimes. The results of these detailed trials are awaited.

In the late 1980s/early 1990s, some farmers in Alberta successfully grew quinoa. Research at the Crop Diversification Centre South in Brooks, Alberta has demonstrated that quinoa can be successfully grown in Southern Alberta (AAFRD, 2005). The north-eastern Saskatchewan region in Canada resembles the Andean region in climate and is most suitable for cultivation of quinoa. The Northern Quinoa Company (NQC) is a specialty grain and food processing company located at Kamsack, Saskatchewan, Canada. In Saskatchewan, yields are said to be highly variable, and can be up to 2000 lb per acre. In this area the average yield over the 2000–2004 season was said to be in the range 750–1250 lb per acre. In terms of quality, the quinoa crop in Saskatchewan is somewhat smaller and darker in colour than the South American exports. Quinoa is currently grown in Saskatchewan and Manitoba, about 80% grown organically. Approximately 1600 acres of quinoa was grown in Saskatchewan in 2005, primarily to supply the NQC in Kamsack (AAFRD, 2005). Quinoa could successfully be adapted to the Canadian Prairies because it was known that the crop has also done well in field trials in Northern Michigan, where it is not the elevation, but the temperature that is critical to the proper development of the quinoa plant.

3.6 Quinoa and Africa

Quinoa has been field tested in Kenya and the initial results indicate a high seed yield, comparable to that in the Andean region (Mujica *et al.*, 2001). All the quinoa cultivars matured in the Kenyan conditions, although the growth period was shorter (65–98 days). In Kenya, seed yields up to 9 t/ha and biomass yields up to 15 t/ha have been obtained. The late cultivars from Colombia and the inter-Andean valleys gave the highest yield. This is important for Kenya, since it is a primary rural economy relying heavily on the agricultural sector for economic growth. Increased crop diversification is important for improving food security, and quinoa can be considered a promising option for introduction to Kenya and other African countries having similar agroclimatic conditions (Jacobsen, 2003).

A partnership between the Danish Company Eghøjgaard and the Egyptian Natural Oil Company (NATOIL) was formed in 2007 that aimed to develop quinoa in Egypt. The requirements of temperate temperatures during flowering, short day-length to flower and produce seeds for most genotypes mean that quinoa is suitable for the Egyptian winter climate. Adaptation and selection work is underway through a DANIDA-supported project to select the best genotype with respect to length of growing period, plant height, flowering time quality and many other characters. Other properties of quinoa for its successful development in Egypt are its drought resistance, high water use efficiency, good growth in poor soils and salt tolerance. In Ismalia, the salinity in the irrigation water is 3000 ppm, which would make most other crops suffer severely. In Egypt, the fertile farmland area along the Nile and in the delta is under pressure from infrastructure and urban development. Thus new farmland is often poor reclaimed desert land with salinity problems. Quinoa can therefore play a key role in food production in reclaimed desert land. Quinoa was put in formal field trials in the Sinai Peninsula, with 13 varieties and strains being tested in the deserts of South Sinai governorate (near Nuwaiba city) (Shams, 2011) and its introduction in the desert lands proved to be a success.

3.7 Experiments in Asia

The experimental introduction of quinoa in Asia has been quite impressive, with the crop showing good adaptation and abundant yield in the Indian subcontinent. In Pakistan, the crop has been experimentally cultivated in Faisalabad, Chakwal and Bahawalpur. Quinoa was introduced in Pakistan in 2007 in the central Punjab to lessen people's dependence on conventional foods (Munir *et al.*, 2012). Continuous experimentation has demonstrated that quinoa accessions showed variation in their seed yield with respect to change in the growing environments, with some accessions showing good stability in the new environment. Under uncertain and unpredictable climatic conditions, quinoa gave promising seed yields, proving its domestication a success in the Pakistani conditions. The accessions of Danish origin with short stature set seed in the shortest time, whereas the Chilean sea-level accessions gave most viable seed with a medium duration life cycle. A fiscal balance sheet displaying coefficient of profitability has shown that quinoa has the potential to be a new cash crop in the region and could be a sound choice for farmers who have small land holdings (Munir *et al.*, 2012). Thus quinoa could be an important new crop for Pakistan, providing highly nutritive and versatile food products for the population and a new raw material for industry. In particular, it could be cultivated in many of the marginal environments afflicted by drought or salinity stress, which currently suffer from very low productivity (Jacobsen *et al.*, 2002). The crop has a promising future in the northern areas, where conventional agriculture is difficult due to loss of fertile soil and non-availability of suitable crops to improve the agricultural economy. Quinoa's adaptation to severe winters would help alleviate poverty in those areas. It can also help improve food production in the dry western mountains of Balochistan, where the degraded land and declining ground water resources severely hamper production of many crops.

India, located between 8° and 38°N and 68° and 93.5°E, exhibits enormous diversity for agroclimatic regions and edapho-climatic conditions (Bhargava *et al.*, 2006). Research on quinoa has been underway since the early 1990s at the National Botanical Research Institute (NBRI), Lucknow, in an area situated in the heart of the Indo-Gangetic Plain. The Indo-Gangetic Plain, a region covering a large area of India, Pakistan, Nepal and Bangladesh, is characterized by fertile soils and an abundant supply of water (Aggarwal *et al.*, 2004). The research at NBRI increased greatly in the year 2000, when extensive field trials were performed in cooperation with many departments, namely genetics and plant breeding, lipid chemistry, plant pathology, experimental taxonomy and biomass biology. Trials in the Indo-Gangetic Plain (120 m above sea level) have shown that the crop can be successfully cultivated in this region, with many cultivars giving high grain yields (Bhargava *et al.*, 2007b). A thorough assessment of yield potential of quinoa germplasm lines of diverse origin showed that 41% of the lines were high yielding. The results reflected greater adaptability of the Chilean and US germplasm lines to North Indian agroclimatic conditions (Bhargava *et al.*, 2007b). Thus quinoa could serve as an alternative winter crop for the North Indian Plains and other subtropical regions having similar agroclimatic and edaphic conditions (Bhargava *et al.*, 2007b). This is important for India because a large proportion of the population has little access to a protein-rich diet, rice and wheat being the principal food crops. Quinoa's highly proteinaceous grain can help to make diets more balanced in this region and can play an important role in combating 'silent hunger' prevalent among poor populations who have little access to a protein-rich diet. Quinoa can be termed 'underutilized', especially for India, since in spite of its wide adaptability and nutritional superiority, its commercial potential has remained untapped. Quinoa can play a major role in the future diversification of agricultural systems in India, not only in the Himalayan region, but also in the North Indian Plains.

Vietnam is another country where quinoa yield trials were held and the crop has shown potential for future studies. The growth period of quinoa was short (87–96 days) and similar to that observed in Kenya. Seed yields up to 1125–1685 kg/ha and biomass yields up to 9 t/ha were obtained. Quinoa has also been field tested in Japan in the climatic conditions of Southern Kanto District (Yamashita *et al.*, 2007). The maximum yield was obtained in the Sea-level type variety sown in March to May, followed by the Valley type variety sown in July.

3.8 Concluding Remarks

The distribution of quinoa in different agroclimatic regions and its successful cultivation in various parts of the world make us believe that this crop is highly adaptable and could be used to diversify agriculture in new regions. However, detailed trials are needed before successfully recommending this crop in newer areas. Also, the cultivators need to be informed of the immense economic potential of quinoa, so that they may be able to readily accept quinoa over previously cultivated traditional crops.

References

AAFRD (2005) *Quinoa: The Next Cinderella Crop for Alberta?* Alberta Agriculture, Food and Rural Development, Lethbridge, Canada.

Aggarwal, P.K., Joshi, P.K., Ingram, J.S.I. and Gupta, R.K. (2004) Adapting food systems of the Indo-Gangetic plains to global environmental change: key information needs to improve policy formulation. *Environmental Science and Pollution* 7, 487–498.

Aguilar, P.C. and Jacobsen, S.-E. (2003) Cultivation of quinoa on the Peruvian altiplano. *Food Reviews International* 19, 31–41.

Antonio, K. (2011) *The Challenges of Developing a Sustainable Agroindustry in Bolivia: the Quinoa Market*. Duke University, Durham, North Carolina.

Bazile, D. and Negrete, J. (2009) Quínoa y biodiversidad: cuáles son los desafíos regionales? *Revista Geografica de Valparaíso* 42, 1–141.

Berti, M.T. and Schneiter, A.A. (1993) Preliminary agronomic evaluation of new crops for North Dakota. In: Janick, J. and Simon, J.E. (eds) *New Crops*. Wiley, New York, pp. 105–109.

Bhardwaj, H.L., Hankins, A., Mebrahtu, T., Mullins, J., Rangappa, M., Abaye, O. and Welbaum, G.E. (1996) Alternative crops research in Virginia. In: Janick, J. (ed.) *Progress in New Crops*. ASHS Press, Alexandria, Virginia, pp. 87–96.

Bhargava, A., Rana, T.S., Shukla, S. and Ohri, D. (2005) Seed protein electrophoresis of some cultivated and wild species of *Chenopodium* (Chenopodiaceae). *Biologia Plantarum* 49, 505–511.

Bhargava, A., Shukla, S. and Ohri, D. (2006) *Chenopodium quinoa*: an Indian perspective. *Industrial Crops and Products* 23, 73–87.

Bhargava, A., Shukla, S. and Ohri, D. (2007a) Evaluation of foliage yield and leaf quality traits in *Chenopodium* spp. in multiyear trials. *Euphytica* 153, 99–213.

Bhargava, A., Shukla, S. and Ohri, D. (2007b) Genetic variability and interrelationship among various morphological and quality traits in quinoa (*Chenopodium quinoa* Willd.). *Field Crops Research* 101, 104–116.

Bhargava, A., Shukla, S., Srivastava, J., Singh, N. and Ohri, D. (2008) Genetic diversity for mineral accumulation in the foliage of *Chenopodium* spp. *Scientia Horticulturae* 118, 338–346.

Cárdenas, M. (1944) Descripción preliminar de las variedades de *Chenopodium quinoa* de Bolivia. *Revista de Agricultura* (Universidad Mayor San Simón de Cochabamba, Bolivia) 2, 13–26.

Carlsson, R. (1980) Quantity and quality of leaf protein concentrates from *Atriplex hortensis* L., *Chenopodium quinoa* Willd. and *Amaranthus caudatus* L., grown in Southern Sweden. *Acta Agriculturae Scandinavica* 30, 418–426.

Carlsson, R., Hanczakowski, P. and Kaptur, T. (1984) The quality of the green fraction of leaf protein concentrate from *Chenopodium quinoa* Willd. grown at different levels of fertilizer nitrogen. *Animal Feed Science and Technology* 11, 239–245.

Cusack, D. (1984) Quinoa: grain of the Incas. *Ecologist* 14, 21–31.

Fuentes, F.F., Martínez, E.A., Hinrichsen, P.V., Jellen, E.N. and Maughan, P.J. (2009) Assessment of genetic diversity patterns in Chilean quinoa (*Chenopodium quinoa* Willd.) germplasm using multiplex fluorescent microsatellite markers. *Conservation Genetics* 10, 369–377.

Gallardo, M.G. and Gonzalez, J.A. (1992) Efecto de algunos factores ambientales sobre la germinación de *Chenopodium quinoa* W. y sus posibilidades de cultivo en algunas zonas de la Provincia de Tucumán (Argentina). *LILLOA* XXXVIII, 55–64.

Galwey, N.W. (1995) Quinoa and relatives. In: Smartt, J. and Simmonds, N.W. (eds) *Quinoa and Relatives*. Longman Scientific and Technical, Harlow, UK.

Geerts, S., Mamani, R.S., Garcia, M. and Raes, D. (2006) Response of quinoa (*Chenopodium quinoa* Willd.) to differential drought stress in the Bolivian Altiplano: towards a deficit irrigation strategy within a water scarce region. In: *Proceedings of the 1st International Symposium on Land and Water Management for Sustainable Irrigated Agriculture*, 4–8 April 2006, Cukurova University, Adana, Turkey.

Gęsiński, K. (2000) Potential for *Chenopodium quinoa* Willd acclimatisation in Poland. *Crop Development of the Cool and Wet Regions of Europe*. European Communities, Belgium.

Gęsiński, K. (2001) Test of quinoa (*Chenopodium quinoa* Willd.) in Poland. *Proecto Quinoa CIP-Danida*. Universidad Nacional Agraria, La Molina, Lima, Peru.

Gęsiński, K. (2006) Evaluation of growth and flowering of *Chenopodium quinoa* Willd. under Polish conditions. *Acta Agrobotanica* 59, 487–496.

Gęsiński, K. and Kwiatkowska, B. (1999a) Justification for the introduction of *Chenopodium quinoa* Willd. Part one. Cultivation potential, phenology and morphology. *Zeszyty Naukowe Rolnictwo* 44, 95–100.

Gęsiński, K. and Kwiatkowska, B. (1999b) Justification for the introduction of *Chenopodium quinoa* Willd. Part two. Yielding on light soil. *Naukowe Rolnictwo* 44, 101–105.

Haaber, J. (1991) *Chenopodium quinoa* Willd. as a green crop for the pelleting industry – the effect of heat treatment on the palatability in green pellets made of quinoa. *First European Symposium on Industrial Crops and Products*, Maastricht, The Netherlands.

Iliadis, C., Karyotis, T. and Mitsimponas, T. (1997) Research on quinoa (*Chenopodium quinoa*) and amaranth (*Amarantus caudatus*) in Greece. In: Ortiz, R. and Stolen, O. (eds) *Crop Development for the Cool and Wet Regions of Europe. Spelt and Quinoa*. COST 814, pp. 85–91.

Izquierdo, J. and Roca, W. (1998) Underutilized Andean food crops: status and prospects of plant biotechnology for the conservation and sustainable agricultural use of genetic resources. *Acta Horticulturae* 457, 157–172.

Jacobsen, S.-E. (2003) The worldwide potential for quinoa (*Chenopodium quinoa* Willd.). *Food Reviews International* 19, 167–177.

Jacobsen, S.-E. and Stølen, O. (1993) Quinoa: morphology, phenology and prospects for its production as a new crop in Europe. *European Journal of Agronomy* 2, 19–29.

Jacobsen, S.-E., Hollington, P.A. and Hussain, Z. (2002) Quinoa (*Chenopodium quinoa* Willd.), a potential new crop for Pakistan. In: Ahmad, R. and Malik, K.A. (eds) *Prospects for Saline Agriculture*. CAB International, Wallingford, UK.

Karyotis, T., Mitsimponas, T., Iliadis, C., Kapetanaki, G. and Haroulis, A. (1996) Adaptation of quinoa under Greek climatic conditions. In: Stolen, O., Bruhn, K., Pithan, K. and Hill, J. (eds) *Crop Development for the Cool and Wet Regions of Europe: Small Grain Cereals and Pseudo-cereals*. COST 814 European Commission, Luxembourg, pp. 133–137.

Kephart, K.D., Murray, G.A. and Auld, D.L. (1990) Alternate crops for dryland production systems in northern Idaho. In: Janick, J. and Simon, J.E. (eds) *Advances in New Crops*. Timber Press, Portland, Oregon, pp. 62–67.

Kunkel, G. (1984) *Plants for Human Consumption*. Koeltz Scientific Books, Koenigstein, Germany.

Lescano, J.L. (1994) *Mejoramiento y fisiologia de cultivos andinos. Cultivos andinos en el Peru*. Consejo Nacional Cientifico y Tecnico (CONCYTEC), Proyecto FEAS.

Limburg, H. and Mastebroek, H.D. (1997) Breeding high yielding lines of *Chenopodium quinoa* Willd. with saponin free seed. In: Stølen, O., Bruhn, K., Pithan, K. and Hill, J. (eds) *Small Grain Cereals and Pseudo-Cereals*. Proceedings of COST 814 workshop, 22–24 February 1996, Copenhagen, Denmark, pp. 103–114.

Martínez, E.A., Delatorre, J. and Von Baer, I. (2007) Quinoa: las potencialidades de un cultivo subutilizado en Chile. *Tierra Adentro (INIA)* 75, 24–27.

Martínez, E.A., Veas, E., Jorquera, C., San Martín, R. and Jara, P. (2009) Re-introduction of quinoa into arid Chile: cultivation of two lowland races under extremely low irrigation. *Journal of Agronomy and Crop Science* 195, 1–10.

Mastebroek, H.D. and Limburg, H. (1997) Breeding for harvest security in *Chenopodium quinoa*. In: Stølen, O., Bruhn K., Pithan, K. and Hill, J. (eds) *Small Grain Cereals and Pseudo-Cereals*. Proceedings COST 814 workshop, 22–24 February 1996, Copenhagen, Denmark, pp. 79–86.

Mastebroek, H.D., van Loo, E.N. and Dolstra, O. (2002) Combining ability for seed yield traits of *Chenopodium quinoa* breeding lines. *Euphytica* 125, 427–432.

Mujica, A. (1992) Granos y leguminosas andinas. In: Hernandez, J., Bermejo, J. and Leon, J. (eds) *Cultivos marginados: otra perspectiva de 1492*. Organización de la Naciones Unidas para la Agricultura y la Alimentación, FAO, Rome, pp. 129–146.

Mujica, A., Jacobsen, S.-E., Ezquierdo, J. and Marathee, J.P. (2001) *Resultados de la Prueba Americana y Europes de la Quinua*. FAO, Rome, Italy/UNA-Puno, CIP, Peru.

Mujica, A., Marca, S. and Jacobsen, S.-E. (2003) Current production and potential of quinoa (*Chenopodium quinoa* Willd.) in Peru. *Food Reviews International* 19, 149–154.

Munir, H., Sehar, S., Basra, S.M.A., Jacobsen, H.-J. and Rauf, S. (2012) Growing quinoa in Pakistan as a potential alternative for food security. In: *Resilience of Agricultural Systems Against Crises*, 19–21 September 2012, Göttingen-Kassel/Witzenhausen, Germany.

National Research Council (1989) *Lost Crops of the Incas: Little Known Plants of the Andes with Promise for Worldwide Cultivation*. National Academy Press, Washington, DC.

Oelke, E.A., Putnam, D.H., Teynor, T.M. and Oplinger, E.S. (1992) *Alternative Field Crops Manual*. Center for Alternative Plant and Animal Products, University of Wisconsin Cooperative Extension Service, Madison, Wisconsin.

Partap, T. and Kapoor, P. (1985) The Himalayan grain chenopods I. distribution and ethnobotany. *Agriculture Ecosystems and Environment* 14, 185–199.

Partap, T., Joshi, B.D. and Galwey, N.W. (1998) *Chenopods: Chenopodium spp. Promoting the Conservation and Use of Underutilized and Neglected Crops*. I.P.G.R.I., Rome, Italy.

Pulvento, C., Riccardi, M., Lavini, A., D'Andria, R., Iafelice, G. and Marconi, E. (2010) Field trial evaluation of two *Chenopodium quinoa* genotypes grown under rain-fed conditions in a typical Mediterranean environment in South Italy. *Journal of Agronomy and Crop Science* 196, 407–411.

Risi, J. and Galwey, N.W. (1989a) *Chenopodium* grains of the Andes: a crop for the temperate latitudes. In: Wickens, G.E., Haq, N. and Day, P. (eds) *New Crops for Food and Industry*. Chapman and Hall, New York.

Risi, J. and Galwey, N.W. (1989b) The pattern of genetic diversity in the Andean grain crop quinoa (*Chenopodium quinoa* Willd). I. Associations between characteristics. *Euphytica* 41, 147–162.

Rojas, W. (1998) Análisis de la diversidad genética del germoplasma de quinua (*Chenopodium quinoa* Willd.) de Bolivia, mediante métodos multivariados. MSc. thesis, Universidad Austral de Chile, Facultad de Ciencias Agrarias, Valdivia, Chile.

Rojas, W., Soto, J.-L. and Carrasco, E. (2004) *Study on the Social, Environmental and Economic Impacts of Quinoa Promotion in Bolivia*. PROINPA Foundation, La Paz, Bolivia.

Shams, A. (2011) Combat degradation in rain fed areas by introducing new drought tolerant crops in Egypt. *International Journal of Water Resources and Arid Environments* 1, 318–325.

Taboada, C., Mamani, A., Raes, D., Mathijs, E., García, M., Geerts, S. and Gilles, J. (2011) Farmers' willingness to adopt irrigation for quinoa in communities of the Central Altiplano of Bolivia. *Revista Latinoamericana de Desarrollo Económico* 16, 7–28.

Tapia, M. (1982) *The Environment, Crops and Agricultural Systems in the Andes and Southern Peru*. IICA, Lima, Peru.

Vacher, J.J. (1998) Responses of two main Andean crops, quinoa (*Chenopodium quinoa* Willd.) and papa amarga (*Solanum juzepczukii* Buk.) to drought on the Bolivian Altiplano: significance of local adaptation. *Agriculture, Ecosystems and Environment* 68, 99–108.

Wilson, H.D. (1990) Quinua and relatives (*Chenopodium* sect. *Chenopodium* subsect. *Cellulata*). *Economic Botany* 44, 92–110.

Yamashita, A., Isobe, K. and Ishii, R. (2007) Agronomic studies on quinoa cultivation in Japan. I. Determination of the proper seeding time in the southern Kanto district for good performance of the grain yield. *Japanese Journal of Crop Science* 76, 59–64.

4 Taxonomy

4.1 Introduction

Chenopodium, commonly known as pigweed or goosefoot, includes a wide array of species and is native to all the inhabited continents as well as far-flung archipelagos like Juan Fernandez, New Zealand and Hawaii (Jellen *et al.*, 2011). The genus *Chenopodium* sensu lato was formerly in the family Chenopodiaceae but phylogenetic revision has merged the Chenopodiaceae and Amaranthaceae under the name Amaranthaceae (APG, 1998). The species are mostly colonizing herbaceous annuals occupying large areas in the Americas, Asia and Europe, though some are also suffrutescent (sect. *Ambrina*) and arborescent perennials (sect. *Skottsbergia*) (Wilson, 1990; Bhargava *et al.*, 2005). Most of the annual members of the genus show a series of complex adaptive modifications associated with breeding system, dispersal and germination that allow occupation of disturbed areas in otherwise prime habitat (Wilson, 1990; Bhargava *et al.*, 2007). These ecological specializations represent a strong, positive preadaptive syndrome for exploitation of global disturbance associated with the relatively recent development and spread of agriculture (Wilson, 1990). However, as compared with other plants of dry environments, they lack typical adaptations like Kranz type leaf anatomy and the C_4 photosynthetic pathway, both frequent in other Chenopodiaceae (Carolin *et al.*, 1975; Jacobs, 2001). Members of *Chenopodium* are well known for the taxonomic difficulty created by phenotypic plasticity and lack of macroscopic morphological differences (Woodlot, 2001). Species of *Chenopodium* are often unidentifiable without mature fruits, even among closely related species. In addition, limited collecting and poor distributional data create a situation in which species may be misidentified or overlooked (Woodlot, 2001).

© Bhargava and Srivastava 2013. *Quinoa: Botany, Production and Uses*
(A. Bhargava and S. Srivastava)

4.2 Chenopodiaceae and Amaranthaceae

At the family level, Amaranthaceae and Chenopodiaceae (Caryophyllales) have long been treated as two closely related families (Brown, 1810; Bentham and Hooker, 1880; Baillon, 1887; Volkens, 1893; Ulbrich, 1934; Aellen, 1965; Behnke, 1976; Thorne, 1976; Carolin, 1983; Kühn *et al.*, 1993). The two families were largely treated as separate entities, although most authors admitted difficulty in identifying distinguishing characters. Studies based on the morphology, anatomy and phytochemistry of the two families have revealed a number of common characters like minute sessile flowers; cymose inflorescence; a pentamerous, imbricate, uniseriate perianth; a single whorl of epipetalous stamens; a single basal ovule; pantoporate pollen; sieve elements with P-type plastids but without a central protein crystalloid; occurrence of the betacyanins amaranthin and celosianin; and presence of 6,7-methylenedioxyflavonol and isoflavones (Hegnauer, 1964, 1989; Wohlpart and Mabry, 1968; Behnke, 1976; Natesh and Rau, 1984; Sandersson *et al.*, 1988; Hershkovitz, 1989; Rodman, 1990, 1994; Behnke and Mabry, 1994; Judd and Ferguson, 1999; Kadereit *et al.*, 2003). Molecular data also supported the combination of the two families into Amaranthaceae. Rettig *et al.* (1992) supported the placement of Amaranthaceae and Chenopodiaceae as a monophyletic lineage based on the nucleotide subunit data of ribulose 1,5-bisphosphate carboxylase/oxygenase (*rbc*L). Similar results were also reported by Downie *et al.* (1997) by studying the phylogeny based on the sequence of partial chloroplast DNA ORF 2280 homolog. It was concluded that phylogeny lacked the separation between two families (Singh, 2010).

The independent status of Chenopodiaceae and Amaranthaceae families ceased to exist in 1998 when APG I (1998) combined the two species on the basis of the similarities in morphological and molecular data. According to the APG II (2003), the plants formerly treated as the family Chenopodiaceae were categorized under the family Amaranthaceae, making Amaranthaceae a large family with 160 genera and 2400 species (Singh, 2010). Both families were merged into one family on the assumption that the Chenopodiaceae are paraphyletic in relation to Amaranthaceae (APG, 1998; Judd *et al.*, 1999). To clarify this, a phylogenetic analysis of Chenopodiaceae and Amaranthaceae was carried out using sequence variation of the chloroplast gene *rbc*L (Kadereit *et al.*, 2003). However, the relationship between Chenopodiaceae and Amaranthaceae was poorly resolved in the *rbc*L tree. Branches at the base of the Amaranthaceae/Chenopodiaceae lineage were short and largely collapsed in the strict consensus tree (Kadereit *et al.*, 2003).

4.3 Chenopod Classification and Quinoa

Different approaches have been followed for classifying chenopods. Bassett and Crompton (1982) worked on clarifying taxonomic circumscriptions and the nomenclature and distribution of taxa in *Chenopodium* across Canada, mainly based on characteristics such as seed testa and pericarp. Crawford and

Reynolds (1974) used phonetic characters to understand the association among narrow-leaved *Chenopodium* species. Pollen grain structure and cytological and seed characteristics are other parameters that have been studied to gain insight into the *Chenopodium* taxa (Dvořák, 1983).

Although a large number of different species and intraspecific taxa had been described, the most comprehensive and detailed classification system of chenopods was first published by Aellen and Just (1943), which was later expanded by Aellen (1960) to encompass 120 world-wide species. In this system, the subsectional distinctions were based on inflorescences and pericarps of different *Chenopodium* species. Domesticated species of *Chenopodium* were placed in the subsections Leiosperma and Cellulata of section *Chenopodium* (Aellen, 1960; Scott, 1978). Domesticated species of each subsection reflect a convergent phyletic response to human selection (Wilson, 1990). The subsection Cellulata contains members that are all tetraploids (2n = 4× = 36) namely quinoa (*C. quinoa*), quinoa Silvestre (*C. hircinum*) and huazontle (*C. berlandieri* subsp. *nuttalliae*). The other subsection Leiosperma includes the South American species cañihua (*C. pallidicaule*) (2n = 2× = 18) and the Eurasian species group *C. album* (2n = 6× = 54).

This classification was widespread until the late 1990s, when a new morphology-based system was devised by Mosyakin and Clemants (1996) and modified in 2002 and 2008. In this revised system, all the genera under Sections Ambrina, Botryoides, Orthosporum and Roubieva were placed under a new genus, *Dysphania*. Quinoa has been placed under Section Chenopodium, Subsection Favosa that also includes weedy South American *C. hircinum*, weedy ecotypes of North American *C. berlandieri* and at least four extinct or surviving domesticates of *C. berlandieri* (Mosyakin and Clemants, 1996).

A comprehensive picture of relationships of *Chenopodium* sensu lato was provided using Maximum parsimony and Bayesian analyses of the non-coding trnL-F (cpDNA) and nuclear ITS regions (Fuentes-Bazan *et al.*, 2012). The genus *Chenopodium* was found to be highly paraphyletic within Chenopodioideae, consisting of five major clades. The tribe Dysphanieae with three genera *Dysphania*, *Teloxys* and *Suckleya* (comprising the aromatic species of *Chenopodium* s.l.) formed one of the early branches in the tree of Chenopodioideae. The tribe Spinacieae included *Spinacia*, several species of *Chenopodium* and the genera *Monolepis* and *Scleroblitum*. The *C. rubrum* and the *C. murale* clades were newly discovered as distinct major lineages. The delimitation of *Chenopodium* to include Einadia and Rhagodia was suggested since these were part of the crown group composed of species of subg. *Chenopodium* that appeared sister to the Atripliceae. The tetraploid crops such as *C. berlandieri* subsp. *nuttalliae* and *C. quinoa* also belonged to *Chenopodium* sensu stricto.

4.3.1   Genetic interrelationships based on protein analysis

Phylogenetic relationships between cultivated and their related wild taxa have been investigated on the basis of allozyme studies (Wilson, 1976, 1981, 1988a,

1988b, 1988c, 1990; Wilson and Heiser, 1979; Walters, 1987), crossability (Wilson and Heiser, 1979; Wilson, 1980) and RAPD studies (Ruas *et al.*, 1999; Gangopadhyay *et al.*, 2002). Initially the taxonomic treatments of quinua and allied free-living populations were based on two perspectives. Domesticated populations were classified according to conspicuous, often agronomic characters found in field populations (inflorescence form, pigmentation), whereas free-living taxa were defined from herbarium specimens according to established key fruit characters of the genus (Wilson, 1990). This resulted in a well-defined but confusing array of lower taxa that were broadly circumscribed into two primary elements: *C. hircinum* and *C. quinoa*. Depending on fruit size, the free-living populations sympatric with *quinua* were placed into either *C. quinoa* subsp, *milleanum* (Aellen) Aellen (or var. *melanospermum* Hunziker) or Andean taxa of *C. hircinum* (Wilson, 1990).

Allozyme analysis applied to domesticated and wild collections of quinoa from the Andes has shown that wild and crop populations share a low level of allozyme variation with no distinction between sympatric domesticated and free-living populations (Wilson, 1988a, 1988b). On the other hand, free-living Andean populations share a wide range of variation in fruit shape and size with domesticated populations. Comparative analysis of allozyme and leaf blade data among 187 *Chenopodium* taxa separated free-living populations inhabiting eastern South America (*C. hircinum* s.str.) from a complex of domesticated and free-living populations occurring in the western highlands and along the Chilean coast (*C. quinoa* s.1.) (Wilson, 1990). Free-living populations of the Andes, known as 'ajara', showed a closer affinity to Andean 'quinua' than either does to domesticates, known locally as 'quingua', which is an outlier of the quinoa complex cultivated at low elevations on the Chilean coast. Quingua is thought to be an archaic domesticated element representing a phyletic base within the quinua complex, that would be expected to have phenetic or genetic linkage to free-living sister taxa that gave rise to the complex (Wilson, 1990). The results indicate that domestication in South America involved three main elements: *C. hircinum* sensu Schrader (not Aellen), the Andean crop (quinua)/ weed (ajara) complex and quingua.

Seed protein electrophoresis has proved to be a powerful tool in elucidating the origin and evolution of a number of cultivated and wild plants (Vladova *et al.*, 2000; El Naggar, 2001; Ghafoor *et al.*, 2002; Alvarez *et al.*, 2006; Stoilova *et al.*, 2006; Mirali *et al.*, 2007; Rout and Chrungoo, 2007; Yuzbasioglu *et al.*, 2008). In *Chenopodium*, although sodium dodecyl sulphate-polyacrylamide gel electrophoresis (SDS-PAGE) has been performed on other many other species of *Chenopodium* (Crawford, 1974; Crawford and Julian, 1976; Bera and Mukherjee, 1995), such studies involving quinoa are rare (Drzewiecki *et al.*, 2003; Bhargava *et al.*, 2005). Drzewiecki *et al.* (2003) examined genetic diversity and interrelationships of 11 species and cultivars of different angiospermic families using sodium dodecyl sulfate (SDS) seed protein markers. The protein was resolved into 41 bands for quinoa, 36 bands for soybean and 28–39 bands for amaranth. The similarity coefficients calculated on the basis of presence or absence of bands ranged from 0.08 to 0.97. The results did not support the Tachtadzjan hypothesis that Polygonales (buckwheat) and Caryophyllales (quinoa and amaranth) are closely related. It was also doubtful that amaranth and quinoa are closely related (similarity coefficients varied from 0.16 to 0.25). A comparison of protein bands of quinoa and amaranth led to the conclusion that both of them can be considered as phylogenetically distant taxa, dispelling the notion that these two are closely related.

Seed storage and analogous proteins have been present in the early progenitors of plants and are responsible for providing a reservoir of nitrogen and amino acids for the developing seedling on germination. Seed protein profiles of 40 cultivated and wild taxa of *Chenopodium* (involving 11 accessions of quinoa) were compared by SDS-PAGE (Bhargava *et al.*, 2005). Seventy-two unique polypeptide bands were identified in the taxa studied. The dendrogram divided all the taxa in two main groups, which included different subgroups and clusters. A remarkable feature of the study was the separation of *C. quinoa* accessions in two main groups of the dendrogram (Fig. 4.1). The first group comprised three accessions of quinoa, while the second group had eight accessions clustered together. The similarity within the first and second group ranged between 40.0 to 70.0 and 43.8 to 73.9%, respectively, while these two groups shared 2.9 to 14.3% similarity between them. Two accessions of *C. berlandieri* ssp. *nuttalliae* also belonged to the same subgroup in the dendrogram as *C. quinoa* and *C. bushianum*. Incidentally, all these belong to the subgroup *Chenopodium* sect. *Chenopodium* subsect. *Favosa* (Mosyakin and Clemants, 1996). The lone accession of *C. bushianum* showed very close similarity of 70.8 to 76.2% with two accessions of *C. quinoa*. The accessions of *C. berlandieri* ssp. *nuttalliae*, however, showed only 26.5 to 64.5% genetic similarity with eight *C. quinoa* accessions, which was in agreement with independent origin of *C. berlandieri* ssp. *nuttalliae* vis-à-vis *C. quinoa* as derived from morphological, biochemical and crossability studies (Wilson and Heiser, 1979). Other wild species studied are placed in the dendrogram more or less according to their taxonomic position. Thus, the seed protein data were congruent with taxonomic position, crossability relationships and other biochemical characters.

4.3.2   Molecular markers for phylogenetic analysis

Molecular marker techniques have also proved useful in the investigation of the origin and domestication of crop species, as well as providing information on evolutionary relationships (phylogenetic trees), and identifying geographical areas of admixture among populations of diverse genetic origins. The random amplified polymorphic DNA (RAPD) technique has proved to be quite effective for identifying genetic variation and interrelationships in several groups of plants (Kafkas and Perl-Treves, 2002; Lalhruaitluanga and Prasad, 2009; Schlag and McIntosh, 2012). Ruas *et al.* (1999) used the RAPD technique to identify genetic relationships in 19 accessions that included eight cultivated and two wild varieties of *C. quinoa*, three cultivated varieties of *C. nuttalliae*, two varieties of *C. pallidicaule*, two varieties of *C. ambrosioides*, and one variety each of *C. berlandieri* and *C. album*. A dendrogram was constructed using UPGMA from 399 DNA markers (Fig. 4.2). The molecular data clustered species

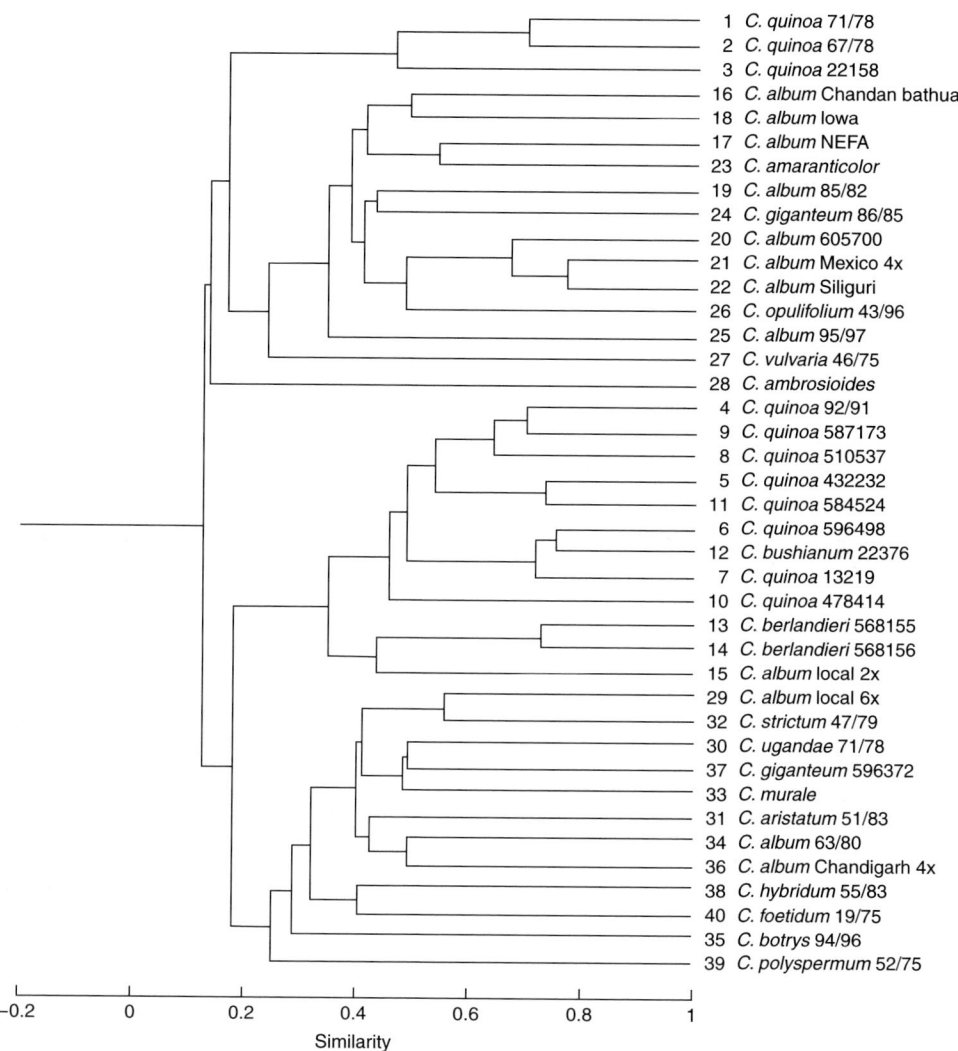

Fig. 4.1. Dendrogram (UPGMA) of 40 wild and cultivated taxa of *Chenopodium* based on similarity matrix of seed storage proteins. [Reprinted from Bhargava *et al.* (2005), with permission from Springer.]

and accessions into five different groups. Group 1 included three cultivated varieties of *C. nuttalliae*, Group 2 included eight cultivars and two wild varieties of *C. quinoa*, Group 3 included *C. berlandieri* and *C. album*, Group 4 included two accessions of *C. pallidicaule* and Group 5 included two accessions of *C. ambrosioides*. The polymorphic patterns generated by RAPD profiles showed different degrees of genetic relationship among the species studied. A low level of intraspecific variation was found within the accessions of *C. quinoa*, *C. nuttalliae* and *C. pallidicaule*. RAPD data also showed that *C. nuttalliae* and *C. quinoa* were located in different genetic groups, although

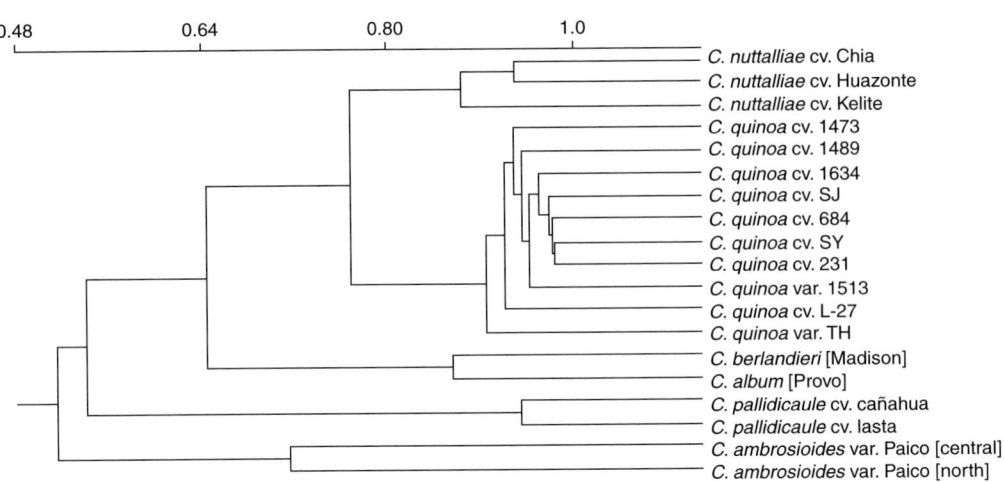

Fig. 4.2. Genetic relationship among 19 taxa of *Chenopodium* based on RAPD markers. The scale represents UPGMA similarity values. [Reprinted from Ruas *et al.* (1999), with permission from Springer.]

the similarity between the two species was about 80%. The proximity among some species, such as *C. quinoa* and *C. nuttalliae*, pointed towards the possibility of introgression of some favourable traits that are present in one species and lacking in others (Ruas *et al.*, 1999).

The genomic and cDNA sequences for two copies of the 11S gene at two different loci in quinoa were determined by Balzotti *et al.* (2008). Transcription initiation and termination sites, coding regions and introns were determined by comparing genomic to cDNA sequences. Gene expression and protein accumulation data were obtained using real time RT-PCR and separation by SDS-PAGE, respectively. Using the coding DNA sequence for the well-conserved 11S basic subunit, phylogenetic relationships were analysed between quinoa and homologous proteins for 50 different plant species. Two phylogenetic trees were constructed, one based on maximum parsimony and another on Bayesian analysis. The two quinoa 11S sequences formed a monophyletic group with that of *Amaranthus hypochondriacus* sequence. The 11S basic subunit showed over 74% sequence identity between amaranth and quinoa. Although both reconstructions show similar relationships between taxa and all monophyletic groups defined by the maximum parsimony tree were also supported by the Bayesian tree, the Bayesian consensus tree (Fig. 4.3) had better support, indicated by high posterior probability values, than the parsimony tree for unresolved portions of the tree including relationships between taxa from species of different families and orders (Balzotti *et al.*, 2008).

Genetic relationships in 55 taxa of *Chenopodium* comprising 14 species (*C. quinoa*, *C. berlandieri* ssp. *nuttalliae*, *C. album*, *C. giganteum*, *C. ugandae*, *C. opulifolium*, *C. ficifolium*, *C. vulvaria*, *C. pallidicaule*, *C. strictum*, *C. botrys*, *C. bushianum*, *C. murale*, *C. foetidum*) and *Amaranthus viridis* as an outgroup were studied using RAPD and directed amplification of minisatellite

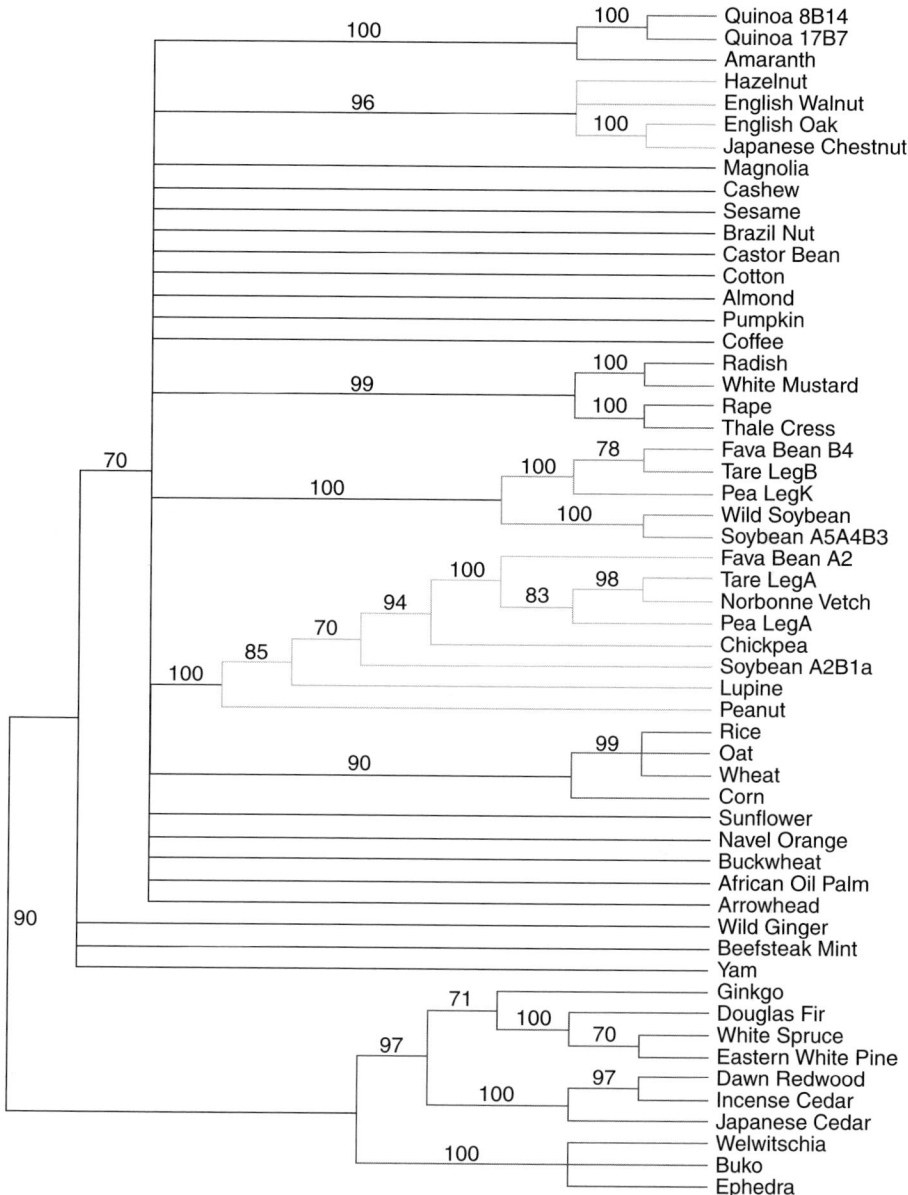

Fig. 4.3. Bayesian consensus tree reconstruction of coding DNA sequences of the legumin basic subunit. Unresolved portions of the tree represent posterior probabilities below 90. Unresolved portions of the tree represent bootstrap values under 70. [Reprinted from Balzotti *et al.* (2008).]

DNA (DAMD) markers (Rana *et al.*, 2010). A UPGMA dendrogram based on 242 DNA markers divided the taxa into two main clusters (Fig. 4.4). The first cluster joined all the accessions of quinoa (4×) with *C. berlandieri* ssp. *nuttalliae* (4×), one *C. album* (4×) and three *C. album* (2×). The second cluster

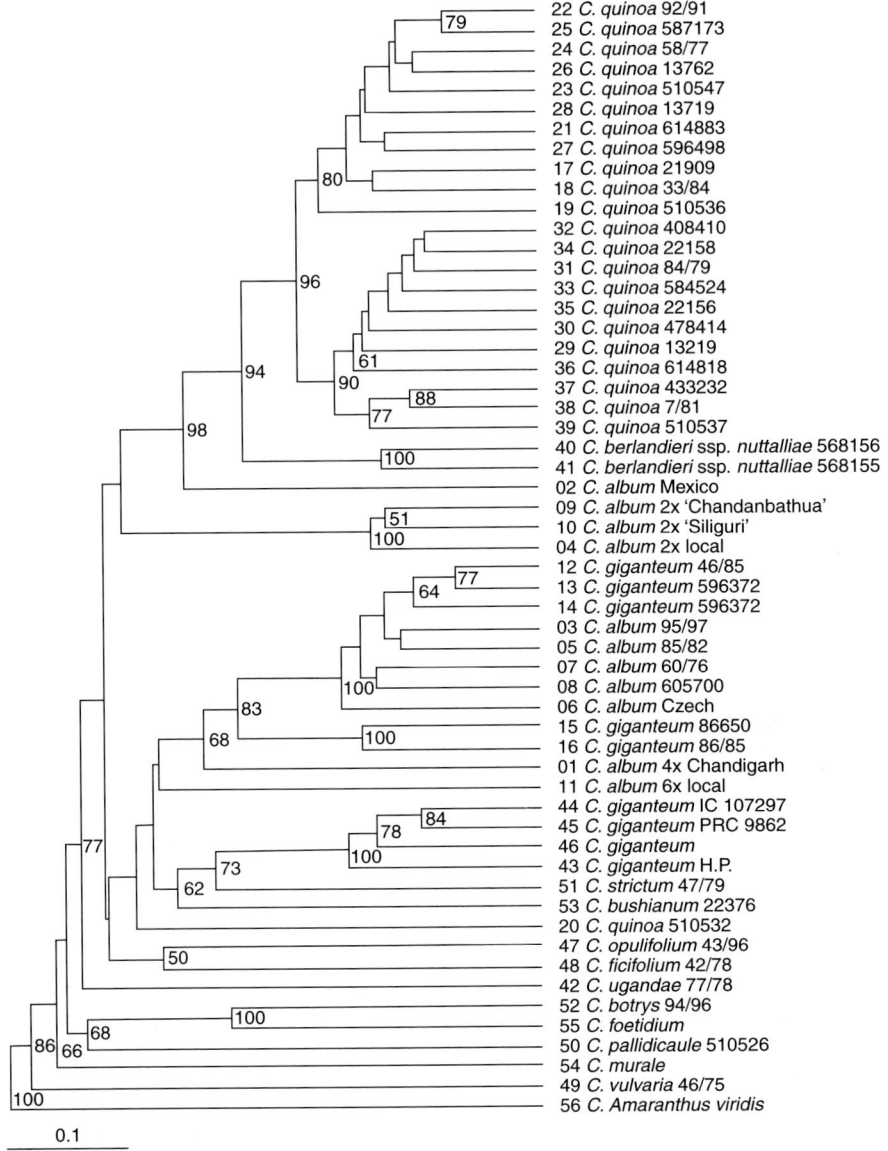

Fig. 4.4. Cluster analysis of the combined RAPD and RAMD data of 55 accessions of *Chenopodium* spp. [Reprinted from Rana *et al.* (2010), with permission from Current Science Association.]

included various taxa under *C. album* and *C. giganteum*, *C. strictum*, *C. bushianum*, *C. opulifolium* and *C. ficifolium*. Other species were present as separate branches in the tree (Fig. 4.4). The results support the possibility of some phyletic relationships between *C. quinoa* and *C. berlandieri* ssp. *nuttalliae*, which is in accordance with taxonomic grouping of these two species

(Rana *et al.*, 2010). The study showed that DNA markers are useful tools to assess inter and intra specific variation within wild and cultivated chenopods.

4.4 Concluding Remarks

A wide range of marker techniques is now available for genotyping plant genomes. Many of these modern techniques, along with conventional ones, have been successfully employed for deciphering the taxonomic intricacies and phylogenetic relationships of chenopods, including quinoa. Apart from quinoa, some of quinoa's close relatives have been domesticated, which could potentially be developed into productive cultivars, either as vegetable or as seed crops.

References

Aellen, P. (1960) *Chenopodium. Illustrierte Flora von Mitteleuropa*, 2nd edn, Vol. 3C. G Hegi, Hanser, Munich, Germany, pp. 533–657.

Aellen, P. (1965) Chenopodiaceae. In: Conert, H.J., Hamann, U., Schultze-Motel, W. and Wagenitz, G. (eds) *Gustav Hegi, Illustrierte Flora von Mitteleuropa*. Parey, Berlin/Hamburg, Germany, pp. 533–747.

Aellen, P. and Just, T. (1943) Key and synopsis of American species of the genus *Chenopodium* L. *American Midland Naturalist* 30, 47–76.

Alvarez, J.B., Moral, A. and Martin, L.M. (2006) Polymorphism and genetic diversity for the seed storage proteins in Spanish cultivated einkorn wheat (*Triticum monococcum* L. ssp. *monococcum*). *Genetic Resources and Crop Evolution* 53, 1061–1067.

APG (Angiosperm Phylogeny Group) (1998) An ordinal classification for the families of flowering plants. *Annals of the Missouri Botanical Garden* 85, 531–553.

APG (Angiosperm Phylogeny Group) (2003) An update of the angiosperm phylogeny group classification for the orders and families of flowering plants. APG II. *Botanical Journal of the Linnean Society* 141, 399–436.

Baillon, H.E. (1887) *Histoire des Plantes*, Vol. 5. Librairie Hachette, Paris.

Balzotti, M.R., Thornton, J.N., Maughan, P.J., McClellan, D.A., Stevens, M.R., Jellen, E.N., Fairbanks, D.J. and Coleman, C.E. (2008) Expression and evolutionary relationships of the *Chenopodium quinoa* 11S seed storage protein gene. *International Journal of Plant Sciences* 169, 281–291.

Bassett, I.J. and Crompton, C.W. (1982) The genus *Chenopodium* in Canada. *Canadian Journal of Botany* 60, 586–610.

Behnke, H.-D. (1976) Ultrastructure of sieve-element plastids in Caryophyllales (Centrospermae): evidence for the delimitation and classification of the order. *Plant Systematics and Evolution* 126, 31–54.

Behnke, H.-D. and Mabry, T.J. (1994) *Caryophyllales: Evolution and Systematics*. Springer, Berlin, Germany.

Bentham, G. and Hooker, J.D. (1880) *Genera Plantarum*, Vol. 3. Lovell Reeve, London.

Bera, B. and Mukherjee, K.K. (1987) Phenotypic variability in *Chenopodium album*. *Nucleus* 30, 50–53.

Bhargava, A., Rana, T.S., Shukla, S. and Ohri, D. (2005) Seed protein electrophoresis of some cultivated and wild species of *Chenopodium* (Chenopodiaceae). *Biologia Plantarum* 49, 505–511.

Bhargava, A., Shukla, S. and Ohri, D. (2007) Gynomonoecy in *Chenopodium quinoa* Willd. (Chenopodiaceae): variation in inflorescence and floral types in some accessions. *Biologia* 62, 19–23.

Brown, R. (1810) *Prodromus Florae Novae Hollandiae*, Vol. 1. Johnson, London.

Carolin, R.C. (1983) The trichomes of the Chenopodiaceae and the Amaranthaceae. *Botanische Jahrbucher fur Systematik, Pflanzengeschichte und Pflanzengeographie* 103, 451–466.

Carolin, R.C., Jacobs, S.W.L. and Vesk, M. (1975) Leaf structure in Chenopodiaceae. *Botanische Jahrbucher fur Systematik, Pflanzengeschichte und Pflanzengeographie* 95, 226–255.

Crawford, D.J. (1974) Variation in the seed proteins of *Chenopodium incanum* (Chenopodiaceae). *Bulletin of the Torrey Botanical Club* 101, 72–77.

Crawford, D.J. and Julian, E.A. (1976) Seed protein profiles in the narrow leaved species in *Chenopodium* in the western United States: taxonomic value and comparison with distribution of flavonoid compounds. *American Journal of Botany* 63, 302–308.

Crawford, D.J. and Reynolds, J.F. (1974) A numerical study of the common narrow-leaved taxa of *Chenopodium* occurring in the Western United States. *Brittonia* 26, 398–410.

Downie, S.R., Katz-Downie, D.S. and Cho, K.J. (1997) Relationship in the Caryophyllales as suggested by phylogenetic analyses of partial chloroplast DNA ORF2280 homolog sequences. *American Journal of Botany* 84, 253–273.

Drzewiecki, J., Delgado-Licon, E., Haruenket, R., Pawelzik, E., Martin-Belloso, O., Park, Y.S., Jung, S.T., Trakhtenberg, S. and Gorinstein, S. (2003) Identification and differences of total proteins and their soluble fractions in some pseudocereals based on electrophoretic patterns. *Journal of Agricultural and Food Chemistry* 51, 7798–7804.

Dvořák, F. (1983) A contribution to the study of *Chenopodium album* agg. *Folia Geobotanica et Phytotaxonomica Praha* 18, 29–43.

El Naggar, S.M. (2001) Implications of seed proteins in Brassicaceae systematics. *Biologia Plantarum* 44, 547–553.

Fuentes-Bazan, S., Mansion, G. and Borsch, T. (2012) Towards a species level tree of the globally diverse genus *Chenopodium* (Chenopodiaceae). *Molecular Phylogenetics and Evolution* 62, 359–374.

Gangopadhyay, G., Das, S. and Mukherjee, K.K. (2002) Speciation in *Chenopodium* in West Bengal, India. *Genetic Resources and Crop Evolution* 49, 503–510.

Ghafoor, A., Ahmad, Z., Qureshi, A.S. and Bashir, M. (2002) Genetic relationship in *Vigna mungo* (L.) Hepper and *V. radiata* (L.) R. Wilczek based on morphological traits and SDS-PAGE. *Euphytica* 123, 367–378.

Hegnauer, R. (1964) *Chemotaxonomie der Pflanzen*, Vol. 3. Birkhauser, Basel, Switzerland.

Hegnauer, R. (1989) *Chemotaxonomie der Pflanzen*, Vol. 8. Birkhauser, Basel, Switzerland.

Hershkovitz, M.A. (1989) Phylogenetic studies in Centrospermae: a brief appraisal. *Taxon* 38, 602–610.

Jacobs, S.W.L. (2001) Review of leaf anatomy and ultrastructure in Chenopodiaceae (Caryophyllales). *Journal of the Torrey Botanical Society* 128, 236–253.

Jellen, E.N., Kolano, B.A., Sederberg, M.C., Bonifacio, A. and Maughan, P.J. (2011) *Chenopodium*. In: Kole, C. (ed.) *Wild Crop Relatives: Genomic and Breeding Resources, Legume Crops and Forages*. Springer, Berlin, Germany, pp. 35–61.

Judd, W.S. and Ferguson, I.K. (1999) The genera of Chenopodiaceae in the southeastern United States. *Harvard Papers in Botany* 4, 365–416.

Judd, W.S., Campbell, C.S., Kellogg, E.A. and Stevens, P.F. (1999) *Plant Systematics: A Phylogenetic Approach*. Sinauer, Sunderland, Massachusetts.

Kadereit, G., Borsch, T., Weising, K. and Freitag, H. (2003) Phylogeny of Amaranthaceae, Chenopodiaceae and the evolution of C_4 photosynthesis. *International Journal of Plant Sciences* 164, 959–986.

Kafkas, S. and Perl-Treves, R. (2002) Interspecific relationships in *Pistacia* based on RAPD fingerprinting. *HortScience* 37, 168–171.

Kühn, U., Bittrich, V., Carolin, R., Freitag, H., Hedge, I.C., Uotila, P. and Wilson, P.G. (1993) Chenopodiaceae. In: Kubitzki, K. (ed.) *Families and Genera of Vascular Plants.* Springer, Berlin, Germany, pp. 253–281.

Lalhruaitluanga, H. and Prasad, M.N.V. (2009) Comparative results of RAPD and ISSR markers for genetic diversity assessment in *Melocanna baccifera* Roxb. growing in Mizoram state of India. *African Journal of Biotechnology* 8, 6053–6062.

Mirali, N., El-Khouri, S. and Rizq, F. (2007) Genetic diversity and relationships in some *Vicia* species as determined by SDS-PAGE of seed proteins. *Biologia Plantarum* 51, 660–666.

Mosyakin, S.L. and Clemants, S.E. (1996) New infrageneric taxa and combinations in *Chenopodium* L. (Chenopodiaceae). *Novon* 6, 398–403.

Mosyakin, S.L. and Clemants, S.E. (2002) New nomenclatural combinations in *Dysphania* R. Br. (Chenopodiaceae): taxa occurring in North America. *Ukrainian Botanical Journal* 59, 380–385.

Mosyakin, S.L. and Clemants, S.E. (2008) Further transfers of glandular pubescent species from *Chenopodium* subg. Ambrosia to Dysphania (Chenopodiaceae). *Journal of the Botanical Research Institute of Texas* 2, 425–431.

Natesh, S. and Rau, M.A. (1984) The embryo. In: Johri, B.M. (ed.) *Embryology of Angiosperms.* Springer, Berlin, Germany, pp. 377–443.

Rana, T.S., Narzary, D. and Ohri, D. (2010) Genetic diversity and relationships among some wild and cultivated species of *Chenopodium* L. (Amaranthaceae) using RAPD and DAMD methods. *Current Science* 98, 840–846.

Rettig, J.H., Wilson, H.D. and Manhart, J.R. (1992) Phylogeny of the Caryophyllales: gene sequence data. *Taxon* 41, 210–209.

Rodman, J.E. (1990) Centrospermae revisited. Part 1. *Taxon* 39, 383–393.

Rodman, J.E. (1994) Cladistic and phenetic studies. In: Behnke, H.D. and Mabry, T.J. (eds) *Caryophyllales: Evolution and Systematics.* Springer, Berlin, Germany, pp. 279–301.

Rout, A. and Chrungoo, N.K. (2007) Genetic variation and species relationships in Himalayan buckwheats as revealed by SDS-PAGE of endosperm proteins extracted from single seeds and RAPD based DNA fingerprints. *Genetic Resources and Crop Evolution* 54, 767–777.

Ruas, P.M., Bonifacio, A., Ruas, C.F., Fairbanks, D.J. and Anderson, W.R. (1999) Genetic relationships among 19 accessions of six species of *Chenopodium* L. by random amplified polymorphic DNA fragments (RAPD). *Euphytica* 105, 25–32.

Sandersson, S.C., Ge-Ling, C., McArthur, E.D. and Stutz, H.C. (1988) Evolutionary loss of flavonoids and other chemical characters in the Chenopodiaceae. *Biochemical Systematics and Ecology* 16, 143–149.

Schlag, E.M. and McIntosh, M.S. (2012) RAPD-based assessment of genetic relationships among and within American ginseng (*Panax quinquefolius* L.) populations and their implications for a future conservation strategy. *Genetic Resources and Crop Evolution* 59, 1553–1568.

Scott, A.J. (1978) A review of the classification of *Chenopodium* L. and related genera (Chenopodiaceae). *Botanische Jahrbucher fur Systematik, Pflanzengeschichte und Pflanzengeographie* 100, 205–220.

Singh, S. (2010) Understanding the weedy *Chenopodium* complex in the north central States. PhD. thesis, University of Illinois, Urbana-Champaign, Illinois.

Stoilova, T., Cholakova, N. and Markova, M. (2006) Variation in seed protein and isoenzyme patterns in *Cucurbita* cultivars. *Biologia Plantarum* 50, 450–452.

Thorne, R.F. (1976) A phylogenetic classification of the Angiospermae. *Evolutionary Biology* 9, 35–106.

Ulbrich, E. (1934) Chenopodiaceae. In: Engler, A. and Prantl, K. (eds) *Die naturlichen Pflanzenfamilien*, Vol. 16c. Engelmann, Leipzig, Germany, pp. 379–584.

Vladova, R., Pandeva, R. and Petcolicheva, K. (2000) Seed storage proteins in *Capsicum annum* cultivars. *Biologia Plantarum* 43, 291–295.

Volkens, G. (1893) Chenopodiaceae. In: Engler, A. and Prantl, K. (eds) *Die naturlichen Pflanzenfamilien*, Vol. 1a. Engelmann, Leipzig, Germany, pp. 36–91.

Walters, T.W. (1987) Electrophoretic evidence for the evolutionary relationship of the tetraploid *Chenopodium berlandieri* to its putative diploid progenitors. *Selbyana* 10, 36–55.

Wilson, H.D. (1976) Genetic control and distribution of leucine aminopeptidase in the cultivated chenopods (*Chenopodiaceae*) and related weed taxa. *Biochemical Genetics* 14, 913–919.

Wilson, H.D. (1980) Artificial hybridization among species of *Chenopodium* sect. *Chenopodium*. *Systematic Botany* 5, 253–263.

Wilson, H.D. (1981) Genetic variations among South American populations of tetraploid *Chenopodium* sect. *Chenopodium* subsect. *Cellulata*. *Systematic Botany* 6, 380–398.

Wilson, H.D. (1988a) Allozyme variation and morphological relationships of *Chenopodium hircinum* Schrader (s.lat.). *Systematic Botany* 13, 215–228.

Wilson, H.D. (1988b) Quinua biosystematics I. Domestic populations. *Economic Botany* 42, 461–477.

Wilson, H.D. (1988c) Quinua biosystematics II. Free living populations. *Economic Botany* 42, 478–494.

Wilson, H.D. (1990) Quinoa and relatives (*Chenopodium* sect. *Chenopodium* subsect. *Cellulata*). *Economic Botany* 44, 92–110.

Wilson, H.D. and Heiser, C.B. (1979) The origin and evolutionary relationships of 'Huazontle' (*Chenopodium nuttalliae* Safford) domesticated chenopod of Mexico. *American Journal of Botany* 66, 198–206.

Wohlpart, A. and Mabry, T.J. (1968) The distribution and phylogenetic significance of the betalains with respect to the Centrospermae. *Taxon* 17, 148–152.

Woodlot (2001) Identification and ecology of rare *Chenopodium* in Maine. Woodlot Alternatives Inc., Topsham, Maine.

Yuzbasioglu, E., Acik, L. and Ozcan, S. (2008) Seed protein diversity among lentil cultivars. *Biologia Plantarum* 52, 126–128.

5 Cytology and Genome Size

5.1 Introduction

The relation of chromosome architecture to the genetic function of an organism is an integral part of modern day cytogenetics (Gill *et al.*, 2008). Chromosome studies date back to as early as around 1600 with the invention of the compound microscope. Chromosomes were first observed by Karl Wilhelm von Nägeli in 1842. However, the term 'chromosome' was initially coined by Waldeyer in the year 1888. Since then numerous discoveries in the field of cytogenetics have contributed significantly to the knowledge of chromosome structure, function and behaviour (Gill *et al.*, 2008). Before the 1920s, cytological studies were carried out on tissues that were embedded in paraffin, sectioned and stained (Wilson, 1925; Darlington, 1937). During the 1920s and 1930s innovations were introduced that facilitated cytological and karyotypic analyses (Fedak and Kim, 2008). In 1921, Belling described a technique for studying meiosis in plant species involving the squashing of anthers. This method permitted the separation of pollen mother cells (PMCs) and facilitated the spreading of their chromosomes. By the early 1940s the squash technique, concomitant with appropriate modifications and pretreatment, completely replaced the method of microtome sectioning of tissues in chromosome studies using somatic and meiotic tissues of most plant species (Fedak and Kim, 2008). The modified squash technique was also used successfully in chromosome studies of insects, amphibians and other animals, with the exception of humans. The chromosome number in humans was first correctly determined to be $2n = 46$ by both Tjio and Levan, and Ford and Hamerton in 1956. The various techniques that have been used to this day for karyotyping hundreds of species using gross chromosomal morphological features have been detailed in La Cour (1947), Darlington and La Cour (1960), Sharma and Sharma (1965) and Haskell and Wills (1968).

5.2   Structural/Physical Organization of Plant Chromosomes

Chromosomal organization consists of a centromere (primary constriction) comprised of regions of condensed chromatin, flanked by pericentromeric regions rich in heterochromatin (Kellogg and Bennetzen, 2004) and telomeres that mark the ends of chromosomes. High-quality metaphase preparations containing high number of appropriate metaphase spread is a prerequisite for cytogenetic studies. Individual chromosomes can often be identified based on the differential condensation pattern at the prometaphase/metaphase stage (Fukui and Mukai, 1988). Based on morphological observation, quantitative chromosome maps called 'ideograms' can be constructed (Fukui *et al.*, 1998).

5.3   Cytological Techniques

Cytologists have devised cytological techniques to obtain precise information on chromosome numbers and structures, and to examine the mechanism of cell division in living species. The basic principles for handling mitotic and meiotic chromosomes of all plant species are similar and consist of collection of specimens, fixation and staining. However, cytological procedures are modified depending on crop species, objective of the experiments and the preference of the cytologists (Sharma and Sharma, 1965; Darlington and La Cour, 1960). The first step involves collection of actively growing roots, either from germinating seeds in a petri dish or from seedlings grown in pots or in the field. The root tips are then pretreated with ice-cold water or by treatment with chemicals like 8-hydroxyquinoline, colchicine, α-bromonaphthalene or paradichlorobenzene (Tjio and Levan, 1950; Tsuchiya, 1971; Palmer and Heer, 1973; Singh, 2002). Pretreatment stops the formation of spindles, increases the number of metaphase cells by arresting chromosomes at the metaphase plate, contracts the chromosome length with distinct constrictions, increases the viscosity of the cytoplasm and facilitates rapid penetration of the fixative by removing undesirable deposits in the tissues (Singh, 2002). The next step involves fixation of the material in fixatives like Carnoy's solution I, Carnoy's solution II and propionic acid alcohol solution (Swaminathan *et al.*, 1954; Sharma and Sharma, 1965). The function of the fixative is to fix or stop the cells at a desired stage of cell division without causing distortion, swelling or shrinkage of the chromosomes (Singh, 2002). The final step involves staining of chromosomes by stains like acetocarmine, Feulgen, giemsa or orcein, and observation of the stained chromosomes under a microscope.

5.4   Cytological Studies in Quinoa

Although much work has been done with regard to enumerating chromosome numbers in the genus *Chenopodium*, detailed karyotypic studies have been lacking until recently. The basic chromosome number in *Chenopodium* is reported to be $x = 8$ and $x = 9$ (Kawatani and Ohno, 1950, 1956). The number $x = 8$ is restricted to the section Ambrina (Uotila, 1973) that contains *C. ambrosioides* ($2n = 2x = 16$) as a representative number (Suzuka, 1950; Giusti, 1970). The number $x = 9$ is found in section Chenopodia, which has been further subdivided into three subsections, Cellulata, Leiosperma and Undata (Risi and Galwey, 1984). Early as well as recent cytological studies have established that *C. quinoa* and *C. berlandieri* have a chromosome number $2n = 4x = 36$ (Harbhajan, 1961; Palomino *et al.*, 1990, 2008; Bhargava, 2005; Bhargava *et al.*, 2006; Kolano *et al.*, 2012a) (Fig. 5.1). Gandarillas (1979) reported mixoploidy in *C. quinoa* with chromosome numbers of $2n = 18$, $2n = 27$, $2n = 36$ and $2n = 45$. Wang *et al.* (1993) studied the somatic chromosomes from root tips of nine taxa of five species of the genus *Chenopodium*. Two other wild species, *C. berlandieri* subsp. *berlandieri* and *C. hircinum*, two cultivars of *C. berlandieri* and three cultivars of *C. quinoa* were all reported to be tetraploid ($2n = 36$). Different subspecies within the same species had identical chromosome number. Mixoploidy was frequently observed in all the material studied.

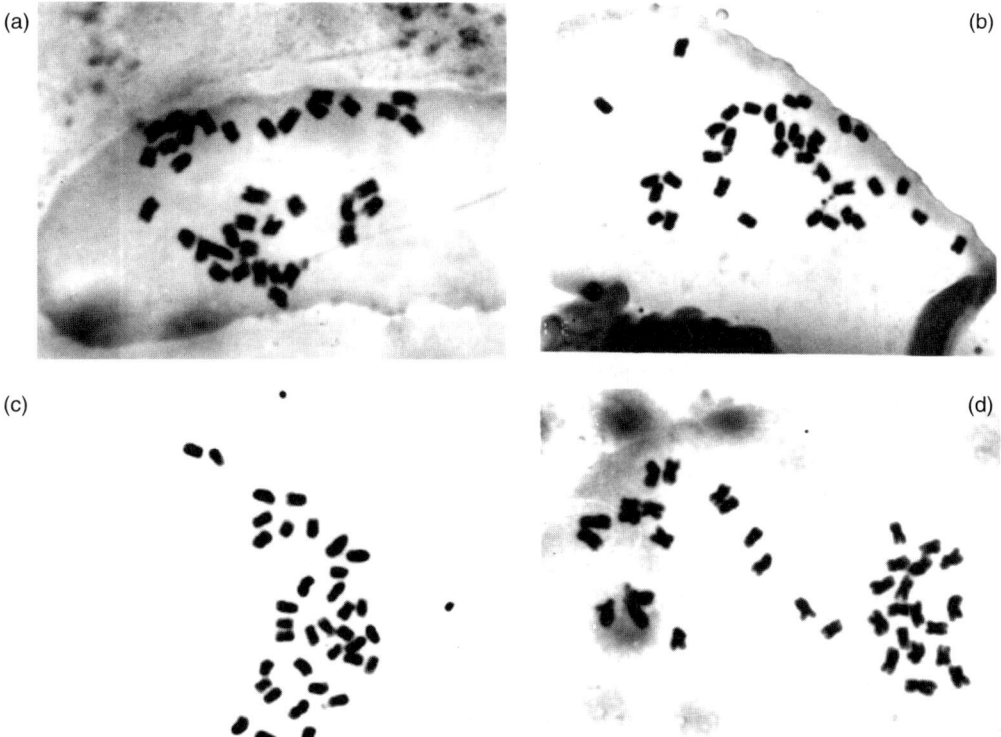

Fig. 5.1. Somatic chromosomes, $2n = 36$, $x = 9$ (a) *C. quinoa* PI 584524, (b) *C. quinoa* PI 596498, (c) *C. quinoa* CHEN 58/77, (d) *C. quinoa* PI 587173. [From Bhargava (2005).]

Although a few workers have discussed certain conspicuous karyotypic features in some species of the genus (Tanaka and Tanaka, 1980; Wang *et al.*, 1993; Kolano *et al.*, 2001), the complete cytogenetic status and detailed karyotyping in quinoa and its related species were lacking. Catacora (1977) performed karyotypic studies in some accessions of *C. quinoa* and found allopolyploidy in the species by measurement of chromosome arm length ratios. The 36 chromosomes of *C. quinoa* (2n = 36) were arranged into nine groups or four homologues on the basis of length and long-arm/short-arm ratio. Bhargava *et al.* (2006) were the first to undertake comprehensive classical cytogenetic studies in quinoa involving seven quinoa accessions. All the accessions of *C. quinoa* had broadly similar karyotypes and were divided into two groups based on the ratio between the longest and the shortest chromosomes in the complement which is <2.0 in 1a and >2.0 in 1b types of karyotypes (Figs 5.2 and 5.3; Table 5.1). The symmetry index (TF%) on the basis of arm ratios varied from 43.9% (PI 584524, most asymmetrical) to 47.4% (CHEN 58/77, most symmetrical) (Table 5.1). One satellite pair was reported in all the taxa, the position of which varied according to its comparative size in the complement (Figs 5.2 and 5.3; Table 5.1). The satellite pair was similar in all the accessions, being median (m) or median-submedian (msm), and had the satellite on the short arm. The first chromosome in different complements was either m or msm, with arm ratios varying between 1.18 (PI 510537) to 1.56 (CHEN 71/78), while 4th, 9th and 18th pairs were the most conserved in being median (M or m) (Figs 5.2 and 5.3; Table 5.1). The greatest variability was observed in 10th and 13th pair, with arm ratios ranging between 1.0 and 1.86 and 1.0 and 1.78 respectively (Figs 5.2 and 5.3; Table 5.1). *C. berlandieri* subsp. *nuttalliae* also had only one SAT pair, with the satellite on the short arm of 3rd pair (Figs 5.2 and 5.3; Table 5.1). The first pair was msm and the 18th pair was median point (M) and like most of the accessions of *C. quinoa* there was no sm pair in the complement. The symmetry index was 44.1% and the karyotype belonged to 1a class (Table 5.1).

The study carried out by Bhargava *et al.* (2006) clearly demonstrated that quinoa accessions showed minor though consistent differences in their karyotypes. This was expected because the species has a monophyletic origin from the Andean crop/weed system (Wilson, 1990). However, minor differences in karyotypes due to chromosomal alterations are being maintained in quinoa because of predominantly self-pollinating behaviour (Risi and Galwey, 1984) and are consistent with some degree of variability in morphological characters (Risi and Galwey, 1984; Wilson, 1988; Bhargava *et al.*, 2007), protein profiles (Bhargava *et al.*, 2005) and RAPD profiles (Ruas *et al.*, 1999). Only one satellite pair was observed, which was corroborated by studies on fluorescent in situ hybridization with 45S rDNA showing two sites of hybridization on two homologous chromosomes (Kolano *et al.*, 2001). Both these rDNA loci are transcriptionally active, which means that at least one such locus may have been lost (Kolano *et al.*, 2001). The karyotype of *C. berlandieri* subsp. *nuttalliae* did not show any distinct differences and was basically similar to those of different accessions of *C. quinoa*. This was clear from karyotype formula, symmetry index and one satellite pair of similar morphology as in *C. quinoa*.

A similar study of cytogenetic characterization by karyotyping was carried out by Palomino *et al.* (2008) in *C. quinoa* cultivar Barandales and six accessions

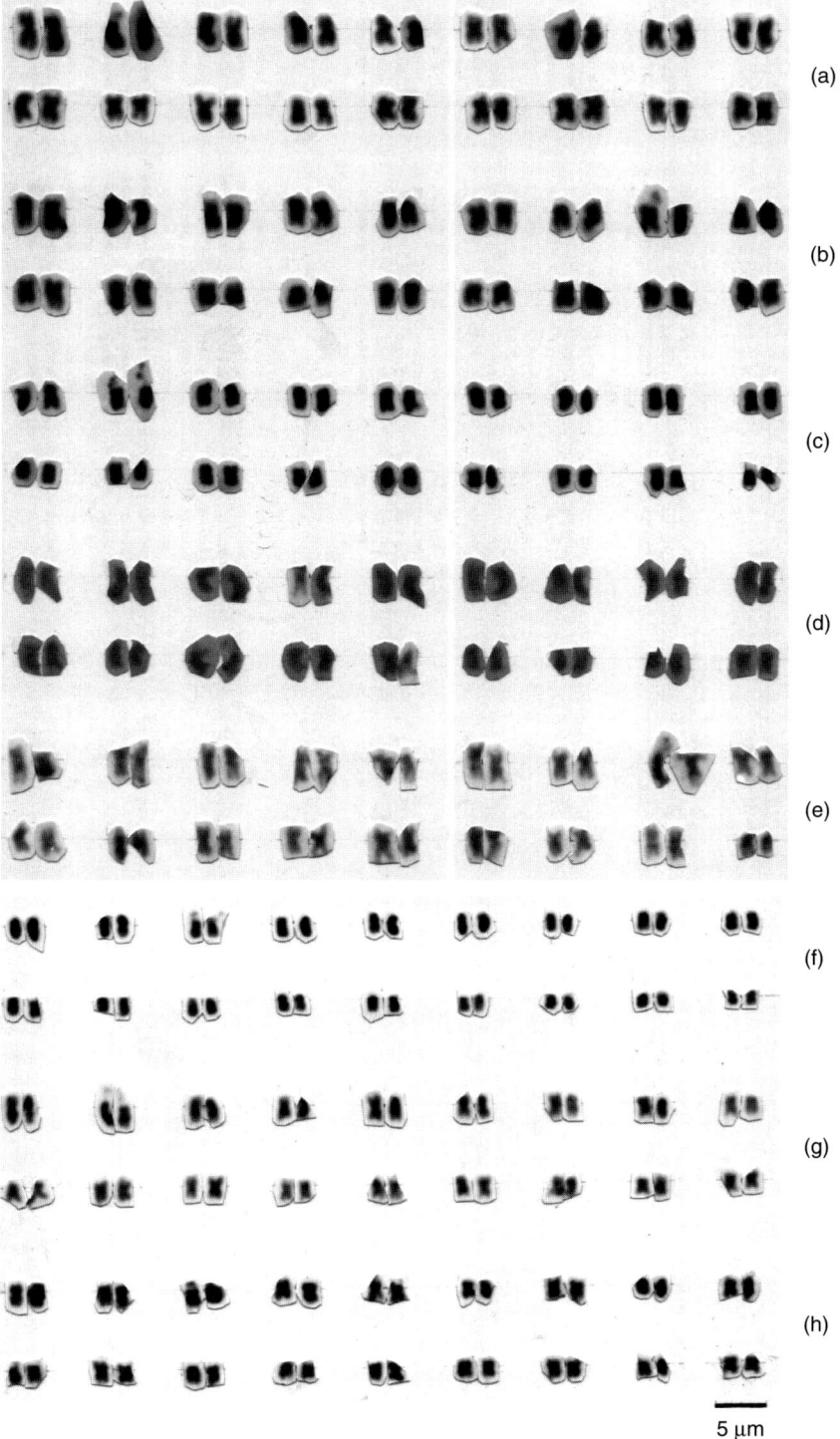

5 μm

Fig. 5.2. Karyotypes of (a) *C. quinoa* PI 587173, (b) *C. quinoa* PI 584524, (c) *C. quinoa* PI 596498, (d) *C. quinoa* PI 510537, (e) *C. quinoa* CHEN 71/78, (f) *C. quinoa* CHEN 58/77, (g) *C. quinoa* CHEN 33/84, (h) *C. berlandieri* subsp. *nuttalliae* PI 568156. [Reprinted from Bhargava *et al.* (2006), with permission from Springer.]

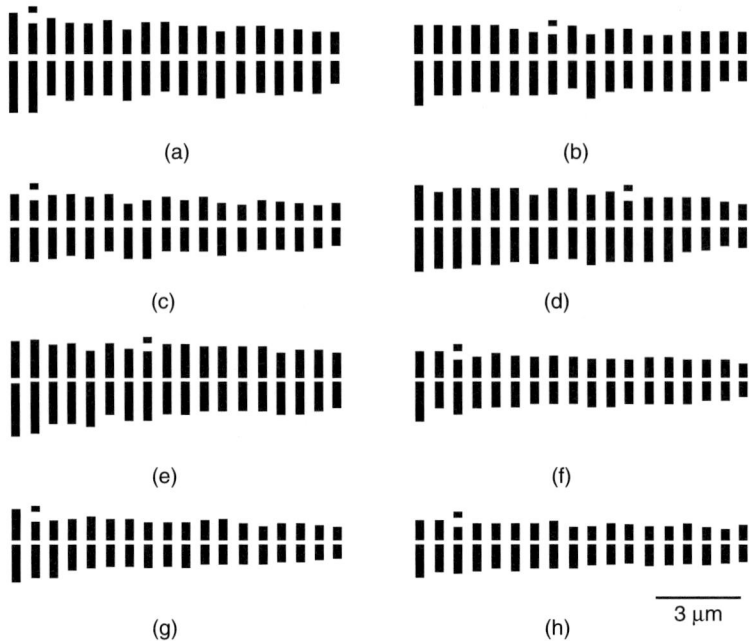

(a)   (b)

(c)   (d)

(e)   (f)

3 μm

(g)   (h)

Fig. 5.3. Idiograms of (a) *C. quinoa* PI 587173, (b) *C. quinoa* PI 584524, (c) *C. quinoa* PI 596498, (d) *C. quinoa* PI 510537, (e) *C. quinoa* CHEN 71/78, (f) *C. quinoa* CHEN 58/77, (g) *C. quinoa* CHEN 33/84, (h) *C. berlandieri* subsp. *nuttalliae* PI 568156. [Reprinted from Bhargava *et al.* (2006), with permission from Springer.]

of *C. berlandieri* subsp. *nuttalliae*. All karyotypes had nine groups of small metacentric chromosomes, supporting $x = 9$ (Fig. 5.4). Variation among varieties was evident in chromosome size, genome length (GL) and the position of satellites. Analysed plants exhibited two pairs of chromosomes with satellites. The first one was always located on chromosome pair number 1, while the position of the second satellite varied. *C. quinoa* cv. Barandales showed an index of asymmetry or TF = 43.80%, which was similar to TF% values of 43.9–47.4% reported for several collections of *C. quinoa* (Bhargava *et al.*, 2006).

5.5   Nuclear DNA Content

Deoxyribonucleic acid (DNA) is considered as the primary genetic material in most organisms. The DNA content per genome is usually constant, both between cells of an individual and between different individuals of the same species. The nuclei of cells with a single unreplicated copy of the species genome, such as newly formed products of normal meiosis (e.g. nuclei of tetrads), contain 1C DNA, while the nuclei of cells containing two fully replicated copies of the species genome, one from either parent, such as those at pachytene of meiosis, contain 4C DNA (Bennett and Leitch, 1995). DNA amounts in plants and other eukaryotes

Table 5.1. Karyotype arrangements in eight taxa of *Chenopodium* species. [Reprinted from Bhargava *et al.* (2006), with permission from Springer.]

Taxa	2n	No. of satellite pairs	Ratio of longest/ shortest x+SE	Maximum r-index x + SE	Symmetry index[a]	Karyotypic formula	Classification (Stebbins, 1958)
C. quinoa PI 587173	36	1(2)[b]	2.12 ± 0.04	1.58(7)[b] ± 0.05	44.7	4M + 9m + 5msm	1b
C. quinoa PI 584524	36	1(8)	1.67 ± 0.02	1.86(10) ± 0.02	43.9	6M + 4m + 6msm+2sm	1a
C. quinoa PI 596498	36	1(2)	1.64 ± 0.02	1.64(8) ± 0.04	44.7	4M + 6m + 8msm	1a
C. quinoa PI 510537	36	1(12)	2.11 ± 0.03	1.56(14) ± 0.04	46.2	5M + 8m + 5msm	1b
C. quinoa CHEN 71/78	36	1(8)	1.73 ± 0.06	1.68(5) ± 0.05	44.8	7M + 4m + 7msm	1a
C. quinoa CHEN 58/77	36	1(3)	2.13 ± 0.04	1.50(1) ± 0.02	47.4	10M + 7m + 1msm	1b
C. quinoa CHEN 33/84	36	1(2)	2.45 ± 0.02	1.64(3) ± 0.04	46.2	9M + 6m + 3msm	1b
C. berlandieri subsp. *nuttalliae* PI 568156	36	1(3)	1.63 + 0.03	1.58(3) + 0.02	44.1	4M + 6m + 8msm	1a

[a]Symmetry index (TF%) = (total sum of short arm length/total sum of chromosome length)× 100.
[b]Number within parentheses denotes the chromosome in order of decreasing size.

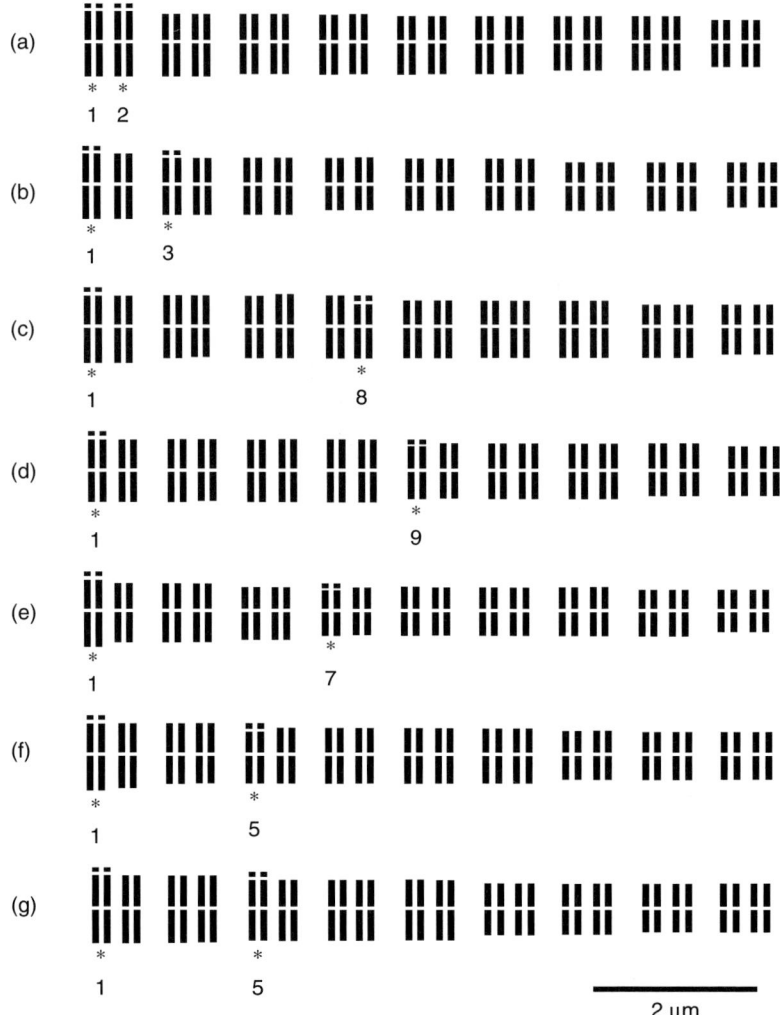

Fig. 5.4. Idiograms of (a) *C. berlandieri* subsp. *nuttalliae* cv. Chia roja 1-99, (b) *C. quinoa* cv. Barandales, (c) *C. berlandieri* subsp. *nuttalliae* cv. Quelite, (d) *C. berlandieri* subsp. *nuttalliae* cv. Huauzontle 8-01, (e) *C. berlandieri* subsp. *nuttalliae* cv. Huauzontle 7-01, (f) *C. berlandieri* subsp. *nuttalliae* cv. Chia roja 2-99, (g) *C. berlandieri* subsp. *nuttalliae* cv. Huauzontle 2-01. Numbers indicate chromosomes with satellites. [Reprinted from Palomino *et al.* (2008), with permission from Springer.]

are usually expressed in picograms (pg) or in megabase pairs of nucleotides (Mb) [1 pg = 10^{-12}; 1 Mb = 10^6 nucleotide base pairs; 1 pg = 965 Mb]. The studies of numerous and diverse consequences of large-scale variation in nuclear DNA amounts show that DNA C-value is a character of fundamental biological significance, which has considerable predictive value in a wide range of fields.

Nuclear DNA amounts are known for nearly 3000 species of angiosperms representing 1% of the global angiospermic flora and show over 600-fold variation in C-values (Ohri, 1998). Sizes of plant nuclear genomes vary over three orders of magnitude (from about 0.065 to 152.23 pg/1C) (Greilhuber *et al.*,

2006; Pellicer *et al.*, 2010). Moreover, wide variation is known to occur within many families and genera. This variation in genome size irrespective of organismic complexity has been termed the C-value paradox by Thomas (1971), where C stands for constancy of DNA amount of unreplicated haploid genome of an individual (Swift, 1950). The bulk of DNA responsible for such a wide variation has been termed as 'junk', 'selfish' or 'parasitic' (Ohno, 1972; Orgel and Crick, 1980). However, the nucleotypic theory envisages a functional and adaptive role for this DNA by showing that this non-informative DNA can produce various phenotypic effects at nuclear, cellular, tissue and organismic levels (Cavalier-Smith, 1985; Price, 1988).

Several methods have been used to estimate DNA C-values in angiosperms, notably chemical analysis, microdensitometry, reassociation kinetics and flow cytometry (Bennett and Leitch, 1995). Both chemical analysis and reassociation kinetics were tedious and cumbersome and were hardly used after the 1970s. Overall, 87.10% of the DNA C-values have been estimated using microdensitometry, which consequently has been by far the most widely used method (Bennett and Leitch, 1995). It is generally accepted that animal standards such as *Homo sapiens* (2C = 3.5 pg) and chicken erythrocytes (2C = 2.33 pg) may be used as calibration standards for angiosperm DNA C-values. However, Price *et al.* (1980) advised that a plant calibration standard is better for estimating genome size in plants. Several species with relatively stable genome size, whose DNA C-values are distributed at suitable intervals over the large range of genome sizes known for angiosperms, are needed as calibration standards. Use of the above-listed standard genotypes approaches is the ideal in practice because they were all estimated, directly or indirectly, against one calibration standard, namely *Allium cepa* cv. Ailsa Craig (Bennett and Leitch, 1995).

Interspecific variation in nuclear DNA is directly correlated with such organismic traits as minimum generation time (Bennett, 1972), growth in different latitudes and ecological conditions (Knight *et al.*, 2005), environmental variation (Kalendar *et al.*, 2000; Walker *et al.*, 2005), water relations in conifers (Ohri and Khoshoo, 1986a) and development (Bharathan, 1996). Therefore, a comparison of C-values explains the phylogenetic relationships and systematics of narrow taxonomic groups in a better way (Ricroch *et al.*, 2005; Garnatje *et al.*, 2006). Intraspecific variability in genome size at a constant basic chromosome number and type has been debatable. Studies in many plant species have shown striking variation, which was previously known to have occurred only between different species. Intraspecific variability in genome size has been reported for a number of plant species, such as *Zea mays* (Laurie and Bennett, 1985), *Cajanus cajan* (Ohri *et al.*, 1994) and *Vicia faba* (Ceccarelli *et al.*, 1995), among others. However, recent reports have disproved of intraspecific variation in genome size (Greilhuber, 1988; Greilhuber and Obermeyer, 1997).

Initial studies estimated the haploid genome size (1C value) of quinoa at between 1.005 and 1.596 pg (Kolano *et al.*, 2012b). However, determination based on Feulgen photometry (1.330–1.596 pg) has suggested a comparatively bigger genome than flow cytometry (1.005–1.500 pg) (Bennett and Smith, 1991; Stevens *et al.*, 2006; Bhargava *et al.*, 2007; Kolano *et al.*, 2009). Bhargava *et al.* (2007) studied the DNA content in 21 accessions of quinoa and two accessions of *C. berlandieri* subsp. *nuttalliae*, along with several other species of the genus *Chenopodium*. Actively growing shoot tips were fixed for

2 hours in 1:3 acetic:ethanol, rinsed in distilled water for 5 min and then hydrolysed in 5N HCl for 30 min at room temperature and stained in Feulgen solution
(pH 2.2) for 2 hours. The material was then given three 10-min washes in SO_2
water. Squash preparations were made in glycerol. Measurements were made
on a Vickers M86 scanning microdensitometer set at a wavelength of 565 nm.
Twenty-one accessions of *C. quinoa* showed only 1.02-fold variation in 4C
DNA amounts, ranging from 6.34 to 6.47 pg (Table 5.2). The two accessions
of *C. berlandieri* subsp. *nuttalliae* showed DNA content in the range 5.79–
5.90 pg. The average DNA amount of two accessions of *C. berlandieri*
subsp. *nuttalliae* was 8.31% less than the mean DNA amount of 21 accessions of *C. quinoa*. The significant differences in DNA amounts of *C. quinoa* and
C. berlandieri subsp. *nuttalliae* pointed to the fact that both of them evolved

Table 5.2. Chromosome number (2n), ploidy level, genome size and nuclear DNA contents
4C in picograms (pg) and 1C in megabase pairs (Mbp)[a] in 23 taxa of *Chenopodium*.
[Modified from Bhargava *et al.* (2007), with permission from Taylor & Francis.]

Taxa	2n	Ploidy level (x)	2C DNA amount/ploidy level (pg)	4C DNA amount ± SE (pg)	1C DNA amount (Mbp)[a]
C. berlandieri subsp. *nuttalliae* (Saff.) Wilson and Heiser PI 568156	36	4	0.74	5.90 ± 0.51	1445
C. berlandieri subsp. *nuttalliae* (Saff.) Wilson and Heiser PI 568155	36	4	0.72	5.79 ± 0.48	1418
C. quinoa Willd. PI 433232	36	4	0.80	6.40 ± 0.60	1568
C. quinoa Willd. PI 510536	36	4	0.79	6.35 ± 0.59	1556
C. quinoa Willd. PI 510547	36	4	0.79	6.34 ± 0.52	1553
C. quinoa Willd. Ames 21909	36	4	0.80	6.37 ± 0.61	1561
C. quinoa Willd. PI 478410	36	4	0.80	6.38 ± 0.60	1563
C. quinoa Willd. PI 510532	36	4	0.80	6.41 ± 0.54	1570
C. quinoa Willd. PI 614938	36	4	0.80	6.42 ± 0.26	1573
C. quinoa Willd. Ames 13762	36	4	0.80	6.42 ± 0.49	1573
C. quinoa Willd. PI 614883	36	4	0.80	6.37 ± 0.55	1561
C. quinoa Willd. PI 584524	36	4	0.79	6.34 ± 0.46	1553
C. quinoa Willd. PI 587173	36	4	0.80	6.37 ± 0.41	1561
C. quinoa Willd. CHEN 84/79	36	4	0.80	6.39 ± 0.59	1566
C. quinoa Willd. Ames 13719	36	4	0.81	6.45 ± 0.63	1580
C. quinoa Willd. CHEN 92/91	36	4	0.81	6.47 ± 0.62	1585
C. quinoa Willd. PI 478414	36	4	0.80	6.37 ± 0.53	1561
C. quinoa Willd. CHEN 33/84	36	4	0.80	6.36 ± 0.52	1558
C. quinoa Willd. PI 596498	36	4	0.80	6.38 ± 0.60	1563
C. quinoa Willd. CHEN 7/81	36	4	0.79	6.35 ± 0.61	1556
C. quinoa Willd. CHEN 58/77	36	4	0.80	6.36 ± 0.56	1558
C. quinoa Willd. PI 614881	36	4	0.80	6.39 ± 0.42	1566
C. quinoa Willd. Ames 22156	36	4	0.80	6.41 ± 0.60	1570

[a]1pg = 980 Mbp (Cavalier-Smith, 1985).

in widely separated geographical areas subsequent to their independent origin. This was in conformance with evidence like genetic complementation for light fruited condition (Heiser and Nelson, 1974), morphological and electrophoretic differences and low pollen stainability (3–4%) of F$_1$ between *C. berlandieri* subsp. *nuttalliae* and *C. quinoa/C. hircinum* (Andean Complex), thus suggesting an independent origin of both the cultigens (Wilson and Heiser, 1979).

The nuclear DNA content of *C. quinoa* cv. Barandales and six accessions of Huauzontle (*C. berlandieri* subsp. *nuttalliae*) were determined by Palomino *et al.* (2008) through flow cytometry. More than 90% of nuclei of the accessions analysed were in phase G0/G1, exhibiting nuclear 2C DNA content in peak 1 (channel 50) for *Lycopersicon esculentum* and in channel 75–78 for *Chenopodium* (Fig. 5.5). Nuclear 2C DNA content ranged from 2.96 pg in quinoa to 3.04 pg in Huauzontle 7-01, exhibiting a difference of 2.7% (Table 5.3). The average value of the seven cultivars studied was 2C DNA = 2.98 pg,

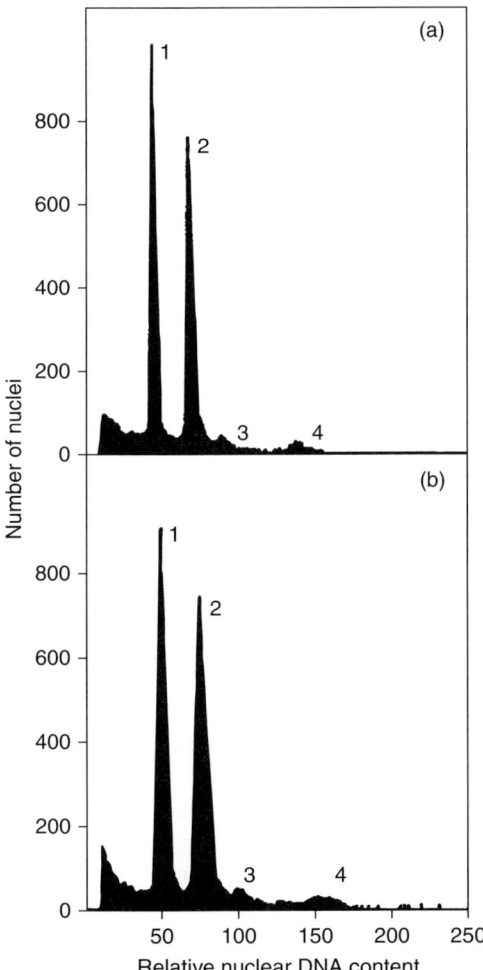

Fig. 5.5. Histograms of fluorescence intensity using flow cytometric analysis of nuclei isolated from *Chenopodium* taxa and *Lycopersicon esculentum* used as internal standard. Peaks 1 and 3 represent G1 and G2 nuclei of *L. esculentum*; peaks 2 and 4 represent G1 and G2 nuclei of (a) *C. quinoa* cv. Barandales and (b) *C. berlandieri* subsp. *nuttalliae* cv. Chia roja 1-99. [Reprinted from Palomino *et al.* (2008), with permission from Springer.]

Table 5.3. Nuclear DNA content and genome size of *C. quinoa* and *C. berlandieri* subsp. *nuttalliae*. [Reprinted from Palomino *et al.* (2008), with kind permission from Springer Science + Business Media B.V.]

Taxa	Accession number	2C DNA (pg)	Genome size 1C x Mbp
C. quinoa cv. Barandales	M-50	2.96	724
C. berlandieri subsp. *nuttalliae* cv. Huauzontle	8-01	2.98	729
C. berlandieri subsp. *nuttalliae* cv. Huauzontle	7-01	3.04	744
C. berlandieri subsp. *nuttalliae* cv. Huauzontle	2-01	2.97	726
C. berlandieri subsp. *nuttalliae* cv. Quelite	10-01	2.99	731
C. berlandieri subsp. *nuttalliae* cv. Chia roja	1-99	2.96	724
C. berlandieri subsp. *nuttalliae* cv. Chia roja	2-99	2.97	726

1 picogram (pg) = 978 megabase pairs (Mbp) (Dolezel *et al.*, 2003).

indicating a small genome size according to the range defined for angiosperms of 1.4–3.5 pg. Species with small genome size, such as quinoa, are considered to be more evolutionarily flexible, allowing them to colonize new and more diverse environments (Leitch *et al.*, 1998; Palomino *et al.*, 2008). Variation in 2C DNA content did not show significant differences, with an ANOVA test suggesting that any actual minor variations had originated during species differentiation (Biradar and Rayburn, 1993).

In the most recent study, the extent and significance of intraspecific genome size variation in 20 quinoa accessions from Ecuador, Peru, Bolivia, Argentina, Chile and the USA were analysed using flow cytometry, with propidium iodide as the DNA stain (Kolano *et al.*, 2012b). The histograms of the nuclear DNA content in young leaves contained a single distinct peak corresponding to G1 nuclei (2C DNA content) and a minor peak representing nuclei in the G2 phase (4C DNA content). Analysis of the distributions showed that more than 90% of leaf cells were in the G0/G1 phase. The mean 2C nuclear DNA content of individual accessions ranged from 2.905 pg to 3.077 pg, the mean being 2C = 2.973 ± 0.043 pg (Table 5.4). The differences between the accessions were statistically significant, but the maximum inter-accession difference between the populations with the largest and the smallest genome reached only 5.9%. The intraspecific variation revealed in *C. quinoa* (5.9%) was therefore limited and in agreement with the findings of a number of recent studies that showed genome size variation in species with rather small genomes like *Arabidopsis thaliana* (Schmuths *et al.*, 2004) and *Sesleria albicans* Kit. ex Schult. (Lysak *et al.*, 2000). The 2C values given by Kolano *et al.* (2012b) corroborate previously published values obtained with flow cytometry (Palomino *et al.*, 2008; Kolano *et al.*, 2009), and were only slightly lower than those estimated using Feulgen cytophotometry (Bhargava *et al.*, 2007), but differed greatly from the results reported by Stevens *et al.* (2006) (2C = 2.01 pg).

Table 5.4. Nuclear DNA content in various accessions of quinoa. [Reprinted from Kolano *et al.* (2012b), with kind permission from Springer Science + Business Media B.V.]

Accession	Geographical group	Country	DNA content (pg/2C) Mean	Range
P2	Southern highland	Peru	2.905	2.784–3.020
P1	Southern highland	Peru	2.909	2.791–3.022
P11	Southern highland	Peru	2.931	2.804–3.017
P6	Southern highland	Peru	2.934	2.811–3.001
P3	Southern highland	Peru	2.945	2.826–3.099
P4	Northern highland	Peru	2.949	2.858–3.094
B1	Southern highland	Bolivia	2.950	2.840–3.042
P8	Southern highland	Peru	2.955	2.881–3.022
C1	Lowland	Chile	2.962	2.897–3.022
E1	Northern highland	Ecuador	2.966	2.830–3.085
A1	Southern highland	Argentina	2.977	2.848–3.092
P7	Southern highland	Peru	2.979	2.897–3.069
P10	Southern highland	Peru	2.981	2.836–3.063
C2	Lowland	Chile	2.990	2.872–3.101
U1	Lowland	Peru	2.992	2.888–3.106
B2	Southern highland	Bolivia	2.995	2.940–3.032
C3	Lowland	Chile	3.013	2.878–3.151
P5	Northern highland	Peru	3.013	2.940–3.088
P9	Southern highland	Peru	3.046	2.985–3.108
C4	Lowland	Chile	3.077	2.997–3.154

5.6 Concluding Remarks

Chromosome counts and estimation of genome size are important requirements for progress in breeding work on any crop. It can be seen that detailed karyotypic studies can appropriately elucidate the relationship of quinoa vis-à-vis other related cultivated and wild species.

References

Belling, J. (1921) On counting chromosomes in pollen mother cells. *American Naturalist* 55, 573–574.

Bennett, M.D. (1972) Nuclear DNA content and minimum generation time in herbaceous plants. *Proceedings of the Royal Society of London B* 181, 109–135.

Bennett, M.D. and Leitch, I.J. (1995) Nuclear DNA amounts in angiosperms. *Annals of Botany* 76, 113–176.

Bennett, M.D. and Smith, J.B. (1991) Nuclear DNA amounts in angiosperms. *Philosophical Transactions of the Royal Society of London. Series B, Biological Sciences* 334, 309–345.

Bharathan, G. (1996) Reproductive development and nuclear DNA content in angiosperms. *American Journal of Botany* 83, 440–451.

Bhargava, A. (2005) Genetic divergence and cytological studies in *Chenopodium* spp. PhD. thesis, University of Lucknow, Lucknow, India.

Bhargava, A., Rana, T.S., Shukla, S. and Ohri, D. (2005) Seed protein electrophoresis of some culti-vated and wild species of *Chenopodium* (Chenopodiaceae). *Biologia Plantarum* 49, 505–511.

Bhargava, A., Shukla, S. and Ohri, D. (2006) Karyotypic studies on some cultivated and wild species of *Chenopodium* (Chenopodiaceae). *Genetic Resources and Crop Evolution* 53, 1309–1320.

Bhargava, A., Shukla, S. and Ohri, D. (2007) Genome size variation in some cultivated and wild species of *Chenopodium* (Chenopodiaceae). *Caryologia* 60, 245–250.

Biradar, D.P. and Rayburn, A.L. (1993) Heterosis and nuclear DNA content in maize. *Heredity* 71, 300–304.

Catacora, A.G. (1977) Determinacion del cariotipoen cinco lineas de quinua (*Chenopodium quinoa* Willd). Ingeniero Agronomo thesis, Universidad Nacional Technica del Altiplano, Puno, Peru.

Cavalier-Smith, T. (1985) *The Evolution of Genome Size*. Wiley, Chichester, UK.

Ceccarelli, M., Minelli, S., Maggini, F. and Cionini, P.G. (1995) Genome size variation in *Vicia faba*. *Heredity* 74, 180–187.

Darlington, C.D. (1937) *Recent Advances in Cytology*. J. & A. Churchill Ltd, London.

Darlington, C.D. and La Cour, L.F. (1960) *The Handling of Chromosomes*, 3rd edn. George Allen and Unwin, London.

Dolezel, J., Bartos, J., Voglmayr, H. and Greilhuber, J. (2003) Nuclear DNA content and genome size of trout and human. *Cytometry* 51A, 127–128.

Fedak, G. and Kim, N.S. (2008) Tools and methodologies for cytogenetic studies of plant chro-mosomes. *Цитология и генетика* 3, 64–80.

Fukui, K. and Mukai, Y. (1988) Condensation pattern as a new image parameter for the identifi-cation of small chromosomes in plants. *Japanese Journal of Genetics* 63, 359–366.

Fukui, K., Nakayama, S., Ohmido, N., Yoshiaki, H. and Yamabe, M. (1998) Quantitative karyo-typing of three diploid *Brassica* species by imaging methods and localization of 45S rDNA loci on the identified chromosomes. *Theoretical and Applied Genetics* 96, 325–330.

Gandarillas, H. (1979) Botanica. Quinua y kaniwa. Cultivos Andinos. In: Tapia, M.E. (ed.) *Serie Libros y Materiales Educativos*. Instituto Interamericano de Ciencias Agricolas, Bogota, Colombia, pp. 20–44.

Garnatje, T., Garcia, S., Vilatersana, R. and Valles, J. (2006) Genome size variation in the genus *Carthamus* (Asteraceae, Cardueae): systematic implications and additive changes during allopolyploidization. *Annals of Botany* 97, 461–467.

Gill, N., Hans, C.S. and Jackson, S. (2008) An overview of plant chromosome structure. *Cytogenetics and Genome Research* 120, 194–201.

Giusti, L. (1970) El genero *Chenopodium* en Argentina 1: numeros de cromosomas. *Darwiniana* 16, 98–105.

Greilhuber, J. (1988) 'Self-tanning': a new and important source of stoichiometric error in cytophoto-metric determination of nuclear DNA in plants. *Plant Systematics and Evolution* 158, 87–96.

Greilhuber, J. and Obermayer, R. (1997) Genome size and maturity group in *Glycine max* (soybean). *Heredity* 78, 547–551.

Greilhuber, J., Borsch, T., Müller, K., Worberg, A., Porembski, S. and Barthlott, W. (2006) Smallest angiosperm genomes found in Lentibulariaceae, with chromosomes of bacterial size. *Plant Biology* 8, 770–777.

Harbhajan, S. (1961) *Grain Amaranthus, Buckwheat and Chenopods*. Indian Council of Agricultural Research, New Delhi, India.

Haskell, G. and Wills, A.B. (1968) *Primer of Chromosome Practice: Plant and Animal Chromosomes Under the Microscope*. Oliver and Boyd, London.

Heiser, C.B. and Nelson, D.C. (1974) On the origin of cultivated chenopods (*Chenopodium*). *Genetics* 78, 503–505.

Kalendar, R., Tanskanen, J., Immonen, S., Nevo, E. and Schulman, A.H. (2000) Genome evolution of wild barley (*Hordeum spontaneum*) by BARE-1 retrotransposon dynamics in response to sharp microclimate divergence. *Proceedings of the National Academy of Sciences (USA)* 97, 6603–6607.

Kawatani, K. and Ohno, T. (1950) Chromosome numbers of genus *Chenopodium*, I. *Japanese Journal of Genetics* 25, 177–180.

Kawatani, K. and Ohno, T. (1956) Chromosome numbers of genus *Chenopodium*, II. *Japanese Journal of Genetics* 31, 15–17.

Kellogg, E.A. and Bennetzen, J.L. (2004) The evolution of nuclear genome structure in seed plants. *American Journal of Botany* 91, 1709–1725.

Knight, C.A., Molinari, N. and Petrov, D.A. (2005) The large genome constraint hypothesis: evolution, ecology and phenotype. *Annals of Botany* 95, 177–190.

Kolano, B., Pando, L.G. and Maluszynska, J. (2001) Molecular cytogenetic studies in *Chenopodium quinoa* and *Amaranthus caudatus*. *Acta Societatis Botanicorum Poloniae* 70, 85–90.

Kolano, B., Siwinska, D. and Maluszynska, J. (2009) Endopolyploidy patterns during development of *Chenopodium quinoa*. *Acta Biologica Cracoviensia/Botanica* 51, 85–92.

Kolano, B., Tomczak, H., Molewska, R., Jellen, E.N. and Maluszynska, J. (2012a) Distribution of 5S and 35S rRNA gene sites in 34 *Chenopodium* species (Amaranthaceae). *Botanical Journal of the Linnean Society* 170, 220–231.

Kolano, B., Siwinska, D., Pando, L.G., Szymanowska-Pulka, J. and Maluszynska, J. (2012b) Genome size variation in *Chenopodium quinoa* (Chenopodiaceae). *Plant Systematics and Evolution* 298, 251–255.

La Cour, L.F. (1947) Improvements in plant cytological technique. *Botanical Reviews* 13, 216–240.

Laurie, D.A. and Bennett, M.D. (1985) Nuclear DNA content in the genera *Zea* and *Sorghum*: intergeneric, interspecific and intraspecific variation. *Heredity* 55, 307–313.

Leitch, I.J., Chase, M.W. and Bennett, M.D. (1998) Phylogenetic analysis of DNA C-values evidence for a small ancestral genome size in flowering plants. *Annals of Botany* 82 (Suppl A), 85–94.

Lysak, M.A., Rostkova, A.R., Dixon, J.M., Rossi, G. and Doležel, J. (2000) Limited genome size variation in *Sesleria albicans*. *Annals of Botany* 86, 399–403.

Ohno, S. (1972) So much 'junk' DNA in our genome. In: Smith, H.H. (ed.) *Evolution of Genetic Systems. Brookhaven Symposium in Biology* 23, 366–370.

Ohri, D. (1998) Genome size variation and plant systematics. *Annals of Botany* 82, 75–83.

Ohri, D. and Khoshoo, T.N. (1986) Genome size in gymnosperms. *Plant Systematics and Evolution* 153, 119–132.

Ohri, D., Jha, S.S. and Kumar, S. (1994) Variability in nuclear DNA content within pigeon pea (*Cajanus cajan* (L.) Millsp.). *Plant Systematics and Evolution* 189, 211–216.

Orgel, L.E. and Crick, F.H.C. (1980) Selfish DNA: the ultimate parasite. *Nature* 284, 604–607.

Palmer, R.G. and Heer, H. (1973) A root tip squash technique for soybean chromosomes. *Crop Science* 13, 389–391.

Palomino, G.H., Segura, M.D., Bye, R.B. and Mercado, R.P. (1990) Cytogenetic distinction between *Teloxys* and *Chenopodium* (Chenopodiaceae). *The Southwestern Naturalist* 35, 351–353.

Palomino, G., Hernandez, L.T. and Torres, E.D. (2008) Nuclear genome size and chromosome analysis in *Chenopodium quinoa* and *C. berlandieri* subsp *nuttalliae*. *Euphytica* 164, 221–230.

Pellicer, J., Fay, M.F. and Leitch, I.J. (2010) The largest eukaryotic genome of them all? *Botanical Journal of the Linnean Society* 164, 10–15.

Price, H.J. (1988) Nuclear DNA content variation within angiosperm species. *Evolutionary Trends in Plants* 2, 53–60.

Price, H.J., Bachman, K., Chambers, K.L. and Riggs, J. (1980) Detection of intraspecific variation in nuclear DNA content in *Microseris douglassi* (Asteraceae). *Botanical Gazette* 141, 195–198.

Ricroch, A., Yockteng, R., Brown, S.C. and Nadot, S. (2005) Evolution of genome size across some cultivated *Allium* species. *Genome* 48, 511–520.

Risi, J. and Galwey, N.W. (1984) The *Chenopodium* grains of the Andes: Inca crops for modern agriculture. In: Coaker, T.H. (ed.) *Advances in Applied Biology* Vol. 10, Academic Press, London, pp. 145–216.

Ruas, P.M., Bonifacio, A., Ruas, C.F., Fairbanks, D.J. and Anderson, W.R. (1999) Genetic rela-
    tionships among 19 accessions of six species of *Chenopodium* L. by random amplified poly-
    morphic DNA fragments (RAPD). *Euphytica* 105, 25–32.
Schmuths, H., Meister, A., Horres, R. and Bachmann, K. (2004) Genome size variation among
    accessions of *Arabidopsis thaliana*. *Annals of Botany* 93, 317–321.
Sharma, A.K. and Sharma, A. (1965) *Chromosome Techniques, Theory and Practice*.
    Butterworth, London.
Simmonds, N.W. (1971) The breeding system of *Chenopodium quinoa*. I. Male sterility.
    Heredity 27, 73–82.
Singh, R.J. (2002) *Plant Cytogenetics*. CRC Press, Boca Raton, Florida.
Stebbins, G.L. (1958) Longevity, habitat and release of genetic variability in higher plants.
    Cold Spring Harbor Symposium. *Plant Biology* 23, 365–378.
Stevens, M.R., Coleman, C.E., Parkinson, S.E., Maughan, P.J., Zhang, H.B., Balzotti, M.R.,
    Kooyman, D.L., Arumuganathan, K., Bonifacio, A., Fairbanks, D.J., Jellen, E.N. and Stevens,
    J.J. (2006) Construction of a quinoa (*Chenopodium quinoa* Willd.) BAC library and its use in
    identifying genes encoding seed storage proteins. *Theoretical and Applied Genetics* 112,
    1593–1600.
Suzuka, O. (1950) Chromosome numbers in pharmaceutical plants I. *Seiken Ziho (Rept. Kihara
    Inst. Biol. Res.)* 4, 57–58.
Swaminathan, M.S., Magoon, M.L. and Mehra, K.L. (1954) A simple propiono-carmine PMC
    smear method for plant with small chromosomes. *Indian Journal of Genetics and Plant
    Breeding* 14, 87–88.
Swift, H. (1950) The constancy of deoxyribose nucleic acid in plant nuclei. *Genetics* 36, 643–654.
Tanaka, R. and Tanaka, A. (1980) Karyomorphological studies on halophytic plants. I. Some
    taxa of *Chenopodium*. *Caryologia* 45, 257–269.
Thomas, C.A. (1971) The genetic organization of chromosomes. *Annual Review of Genetics* 5,
    237–256.
Tjio, J.H. and Levan, A. (1956) The chromosome number of man. *Hereditas* 42, 1–6.
Tsuchiya, T. (1971) An improved aceto-carmine squash method, with special reference to the
    modified Rattenbury's method of making a preparation permanent. *Barley Genetics
    Newsletter* 1, 71–72.
Uotila, P. (1973) Chromosome counts on *Chenopodium* L. from SE Europe and SW Asia.
    Annales Botanici Fennici 10, 337–340.
Walker, D.J., Moñino, I., González, E., Frayssinet, N. and Correal, E. (2005) Determination of
    ploidy and nuclear DNA content in populations of *Atriplez halinus* (Chenopodiaceae).
    Botanical Journal of the Linnean Society 147, 441–448.
Wang, S., Tsuchiya, T. and Wilson, H.D. (1993) Chromosome studies in several species of
    Chenopodium from North and South America. *Journal of Genetics and Breeding* 47,
    163–170.
Wilson, E.R. (1925) *The Cell in Development and Heredity*. Macmillan, New York.
Wilson, H.D. (1988) Quinua biosystematics I. Domestic populations. *Economic Botany* 42,
    461–477.
Wilson, H.D. (1990) Quinua and relatives (*Chenopodium* sect. *Chenopodium* subsect. Cellulata).
    Economic Botany 44, 92–110.
Wilson, H.D. and Heiser, C.B. (1979) The origin and evolutionary relationships of 'Huauzontle'
    (*Chenopodium nuttalliae* Safford) domesticated chenopod of Mexico. *American Journal of
    Botany* 66, 198–206.

6 Botany

6.1 Introduction

Quinoa (Family Amaranthaceae) is an underutilized crop that has been cultivated in the Andean region for around 7000 years. It is a tetraploid plant ($2n = 4x = 36$) with a basic chromosome number $x = 9$, and has numerous wild relatives with chromosome numbers of $2n=18$, 36 and 54, indicative of its apparent tetraploid origin (Maughan *et al.*, 2007). The botanical name of quinoa is presumably derived from the word 'kinwa' or 'kinua' in the Quechua dialect and is pronounced as 'keen-wah'. Such was the importance of quinoa for the people of the Inca civilization that it was dubbed as 'mother of all grains' and 'gold of the Incas', thus denoting its spiritual significance for them. Quinoa has been declared the 'Perfect Food for Humanity' by the United Nations Educational, Scientific and Cultural Organization (UNESCO), and an excellent food for human nutritional needs by the Food and Agricultural Organization (FAO).

6.2 Vegetative Parts

Quinoa is a dicotyledonous, annual plant with broad, generally pubescent, smooth (rarely) to lobed leaves normally arranged alternately, and a stem that is variously coloured depending on the variety.

6.2.1 Root system

A well-developed, highly ramified tap root system is present in the quinoa plant which helps it to resist strong winds (Gandarillas, 1979; Risi and Galwey, 1984). After germination, the radicle forms a tap root from which secondary and tertiary roots develop. Depending on the height of the plant, the root of quinoa can remain near the surface (12.6–15 cm) or penetrate as deep as 1.5 m below the surface. The highly branched root system makes the species more resistant to drought and protects the plant in times of water scarcity.

6.2.2 Stem

The stem is cylindrical near the soil surface but becomes angular at the ramifications. The stem is green, yellow, purple or dark red coloured, or may be striped (Lescano, 1981). The reddish colour is due to the presence of betacyanins, a type of betalains (Mabry *et al.*, 1963; Gallardo *et al.*, 2000). Betalains, conferring yellow-to-red colours, are nitrogen-containing water-soluble compounds derived from tyrosine that are found only in a limited number of plant lineages and are divided into two major structural groups: red-violet betacyanins and yellow betaxanthins (Cai *et al.*, 2005; Tanaka *et al.*, 2008). The length of the stem varies from 0.5 to 2.5 m, depending on the variety and the environment (Risi and Galwey, 1984). Bhargava *et al.* (2007a) observed about 13-fold difference in the stem length (plant height) of several quinoa accessions cultivated in northern India (Table 6.1) and a considerable amount of variation in stem diameter in different quinoa germplasm lines (unpublished data). The stem of quinoa is profusely branched; branching depending mainly on the variety and plant density. The number of primary branches per plant showed a 4.5-fold variation among quinoa lines tested in northern India (Table 6.1).

6.2.3 Leaves

Quinoa exhibits great variation in leaf size (Table 6.1) (Bhargava *et al.*, 2007a). Figure 6.1 shows young quinoa leaves and an emerging inflorescence. The leaves exhibit polymorphism, the upper leaves being lanceolate while the lower leaves are rhomboidal or triangular (Hunziker, 1943; Nelson, 1968). The edges of the leaves may be smooth, toothed or serrated. The laminae of the young leaves are usually covered with a grainy vesiculate pubescence on the lower surface (Risi and Galwey, 1984). However, the pubescence may be completely absent in some varieties. The petioles of the leaves originating from the main stem are often longer than the petiole of leaves originating from primary and secondary branches (Gandarillas, 1979). Leaves on younger plants are usually green, but as the plant matures, they turn yellow, purple or red. Often red colouration is seen at the bases of the petioles and in the leaf veins, which is attributed to the presence of betalains.

6.3 Reproductive Parts

6.3.1 Inflorescence

The inflorescence is a panicle that is usually profusely branched (Fig. 6.2), 15–70 cm in length and rising from the top of the plant and in the axils of lower leaves (Bhargava *et al.*, 2007b). It has a principal axis from which secondary and tertiary branches arise. Besides the terminal inflorescence, there

Table 6.1. Mean performance for 12 morphological traits in 27 germplasm lines of *C. quinoa* and 2 lines of *C. berlandieri* subsp. *nuttalliae*. [Reprinted from Bhargava *et al.* (2007a), with permission from Elsevier.]

Germplasm lines	Origin	Days to flowering	Days to maturity	Plant height (cm)	Leaf area (cm²)	Primary branches/ plant	Inflorescence length (cm)	Inflorescence/ plant	Seed size (mm)	1000 seed weight (g)	Dry weight/ plant (g)	Harvest index	Seed yield (t/ha)
C. quinoa CHEN 58/77	–	73.55	117.67	45.41	15.71	16.56	2.93	41.19	1.58	1.81	6.31	1.07	2.11
C. quinoa CHEN 67/78	Puno, Peru	74.55	119.44	59.63	6.12	16.70	1.71	91.63	1.34	0.78	5.75	0.74	3.75
C. quinoa CHEN 71/78	Bolivia	79.33	131.67	46.33	26.94	15.44	3.39	127.73	1.97	2.85	7.21	1.43	3.27
C. quinoa CHEN 33/84	–	101.55	144.00	42.33	9.46	16.96	2.42	13.85	1.57	2.07	3.84	1.40	1.33
C. quinoa CHEN 84/79	Cuzco, Peru	86.00	121.67	86.97	17.47	22.11	1.00	117.78	2.21	3.57	10.47	1.32	3.44
C. quinoa CHEN 92/91	Colombia	81.89	123.22	77.49	24.69	14.06	2.25	64.11	2.01	3.70	10.21	0.88	2.25
C. quinoa CHEN 7/81	–	85.11	133.78	123.56	22.14	28.00	4.09	141.55	2.09	3.65	28.00	1.41	9.83
C. quinoa PI 614938	Oruro, Bolivia	71.00	109.33	11.27	5.67	10.00	1.07	11.67	1.73	1.87	1.11	1.06	0.32
C. quinoa PI 478408	La Paz, Bolivia	71.33	109.33	17.67	8.93	8.55	0.84	14.65	2.17	2.87	1.26	1.19	0.47
C. quinoa PI 478414	La Paz, Bolivia	83.66	134.11	78.98	21.53	20.55	1.60	106.48	1.81	3.03	14.00	1.25	6.07
C. quinoa PI 596498	Cuzco, Peru	83.77	129.00	65.87	20.82	17.33	2.47	90.33	2.03	3.08	19.89	0.79	3.93
C. quinoa Ames 13219	La Paz, Bolivia	81.99	129.98	53.96	11.75	19.21	2.64	114.66	2.06	3.54	15.08	0.73	2.80
C. quinoa Ames 13719	New Mexico, USA	82.21	120.28	115.52	25.03	27.74	2.67	98.00	2.15	3.65	32.03	0.99	9.33

Continued

Table 6.1. Continued.

Germplasm lines	Origin	Days to flowering	Days to maturity	Plant height (cm)	Leaf area (cm²)	Primary branches/ plant	Inflorescence length (cm)	Inflorescence/ plant	Seed size (mm)	1000 seed weight (g)	Dry weight/ plant (g)	Harvest index	Seed yield (t/ha)
C. quinoa PI 587173	Jujuy, Argentina	85.33	125.78	101.03	30.91	16.74	2.25	68.50	2.01	4.09	15.47	0.81	3.17
C. quinoa PI 510532	Peru	86.67	157.11	144.03	22.02	25.55	2.24	138.22	1.51	1.25	52.89	0.29	1.68
C. quinoa PI 614883	Jujuy, Argentina	70.78	109.89	54.89	12.33	21.89	3.61	45.89	1.73	1.77	3.03	0.97	1.00
C. quinoa PI 584524	Chile	81.33	127.00	115.89	29.64	25.00	2.51	137.55	1.58	3.02	29.86	0.90	6.60
C. quinoa Ames 22156	Chile	80.55	126.00	106.44	26.16	20.44	1.60	85.55	1.93	3.51	17.21	1.21	5.03
C. quinoa Ames 13762	New Mexico, USA	79.33	132.44	123.72	5.00	23.00	4.31	136.44	1.83	2.75	35.21	0.94	8.50
C. quinoa PI 614881	Jujuy, Argentina	87.11	127.22	113.00	25.00	24.56	3.01	114.22	2.05	2.94	24.16	1.34	8.25
C. quinoa PI 510537	Peru	84.33	124.00	100.00	14.39	25.44	1.44	136.00	1.78	2.71	13.02	1.32	4.39
C. quinoa PI 510547	Peru	82.11	131.78	66.67	16.02	14.11	2.08	68.92	1.82	3.13	12.67	1.33	4.70
C. quinoa Ames 22158	Chile	80.89	131.11	80.27	23.25	21.24	3.85	40.29	1.95	3.17	12.70	1.18	4.85
C. quinoa PI 510536	Peru	73.78	115.22	31.05	4.42	17.53	1.79	21.03	1.93	2.34	1.38	1.28	0.67
C. quinoa PI 478410	La Paz, Bolivia	82.77	126.78	101.10	17.29	22.61	0.90	118.33	1.80	2.63	29.00	0.43	3.13
C. quinoa PI 433232	Chile	81.00	130.00	108.66	23.01	20.89	4.54	74.22	1.77	2.28	13.11	1.09	3.56
C. quinoa Ames 21909	Oruro, Bolivia	82.55	152.44	82.44	25.87	21.00	2.12	132.22	1.83	3.31	15.97	1.15	9.08
C. berlandieri subsp. *nuttalliae* PI 568155	Mexico	91.33	163.33	139.44	21.44	35.74	6.47	114.78	1.58	1.28	28.94	0.26	2.01
C. berlandieri subsp. *nuttalliae* PI 568156	Mexico	85.33	152.33	135.44	13.53	29.11	4.77	103.39	1.65	1.37	15.05	0.65	2.32
Mean		81.76	129.51	83.76	18.15	20.62	2.64	88.59	1.84	2.69	16.37	1.01	4.06
+S.E.		+1.18	+2.51	+6.79	+1.44	+1.08	+0.24	+7.81	+0.03	+0.15	+2.24	+0.06	+0.52

Fig. 6.1. An emerging inflorescence and young quinoa leaves. (Courtesy: F.F. Fuentes, Universidad Arturo Prat, Chile.)

Fig. 6.2. Profusely branched inflorescence of quinoa. (Courtesy: F.F. Fuentes, Universidad Arturo Prat, Chile.)

are axillary inflorescences, which arise from the leaf axils on the lower parts of the plant and these show determinate growth, being terminated by a hermaphrodite flower (Bhargava *et al.*, 2007b). The elongation of the inflorescence is due to the intercalary growth of the axis. Short branches bearing a group of flowers originate from third-order (tertiary) axes, which are called glomeruli. The secondary and tertiary branches also bear a terminal hermaphrodite flower. An important feature of quinoa is the presence of hermaphrodite as well as unisexual female flowers (Hunziker, 1943; Simmonds, 1965; Risi and Galwey, 1984; Bhargava *et al.*, 2007b). The inflorescence in quinoa is usually of the following two types (Jacobsen, 1993; Bertero *et al.*, 1996) (Fig. 6.3): (i) amaranthiform, in which the glomeruli are inserted directly on second-order axes (secondary axis) and (ii) glomerulate, in which the glomeruli are inserted on third-order axes (tertiary axis).

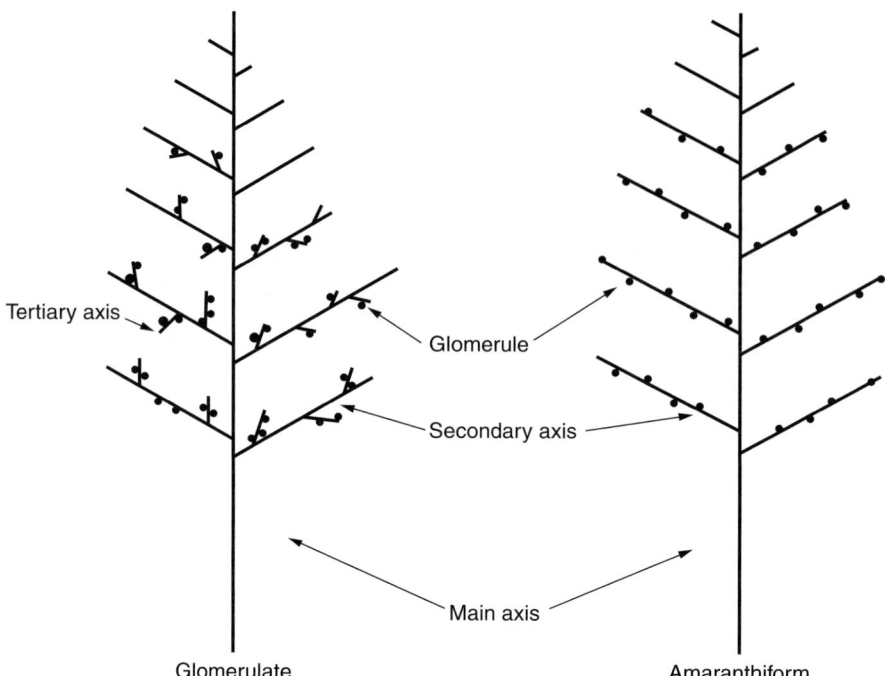

Fig. 6.3. Different types of inflorescence in quinoa. [Reprinted from Bertero *et al.* (1996), with permission from Oxford University Press.]

The colour of the inflorescences in quinoa also varies according to the genotype (Fig. 6.4). Fuentes and Bhargava (2011) reported that yellow-coloured inflorescences were the most frequent (57%), followed by red (32%), while orange-, pink- and purple-coloured inflorescences were of low relative frequency (each about 4%) in the germplasm collected from northern Chile.

6.3.2   Flower and floral types

Quinoa flowers lack petals and both pistillate and perfect flowers are known to occur. The perfect flower has five sepals, five anthers and a superior ovary from which two or three stigmatic branches emerge (Hunziker, 1943), but Bertero *et al.* (1996) observed four stigmatic branches in the gynoecial primordium. However, at anthesis only three stigmatic branches are present and one is aborted (Bertero *et al.*, 1996). Bhargava *et al.* (2007b) undertook a detailed study on the structure and the arrangement of various types of flowers on the inflorescence in quinoa. The flowers could be divided into five types, based on their being hermaphrodite or female, presence or absence of perianth and size.

1. Terminal hermaphrodite flower: This is the terminal flower, 2 mm in breadth, present on the main and axillary inflorescences, and on each cluster or group of flowers on the inflorescence.

Fig. 6.4. Variation of the inflorescence in quinoa. (Courtesy: F.F. Fuentes, Universidad Arturo Prat, Chile.)

2. Lateral hermaphrodite flower: These are dispersed among the female flowers and are present terminally on the first, second or even third branching of the dichasium. These types usually have pentamerous perianth and stamens.
3. Chlamydeous female flowers – large: These have pentamerous perianth, but lack stamens altogether and are just half the size (1 mm) of the hermaphrodite flowers.
4. Chlamydeous female flowers – small: These are present on the ultimate branches of dichasium. They are morphologically similar to type III flowers, except for their smaller size (0.5 mm).
5. Achlamydeous flowers – small: These are small naked carpels lacking perianth altogether and are present on the ultimate branches of the dichasium.

6.3.3 Flower clusters or glomeruli

In quinoa, the glomeruli are borne opposite to each other on the tertiary axes of the inflorescence and show dichasial arrangement of flowers (Bhargava *et al.*, 2007b). The dichasium is symmetrical and terminated by a hermaphrodite flower. The position of the cluster on the axis determines its size and number, and proportion of different flower types. These can be divided into 10 types, depending on the number of divisions of dichasium and the type and number of flowers on successive branches:

Type I: The initial flower in each cluster is bisexual (7.7%) and the rest are all female (Bhargava *et al.*, 2007b). The first and the second branches of the dichasium are terminated by chlamydeous large female flowers, while the third and fourth branches respectively bear chlamydeous and achlamydeous small female flowers (Fig. 6.5a).

Type II: The initial and the first branches of the dichasium are terminated by bisexual (10.6%) flowers. The second branch bears chlamydeous large female flowers, while the second and third bear female flowers (Fig. 6.5b) (Bhargava et al., 2007b).

Type III: This category shows 11.1% bisexual flowers and differs from the preceding one only in that the fourth branch bears achlamydeous small female flowers (Fig. 6.5c) (Bhargava et al., 2007b).

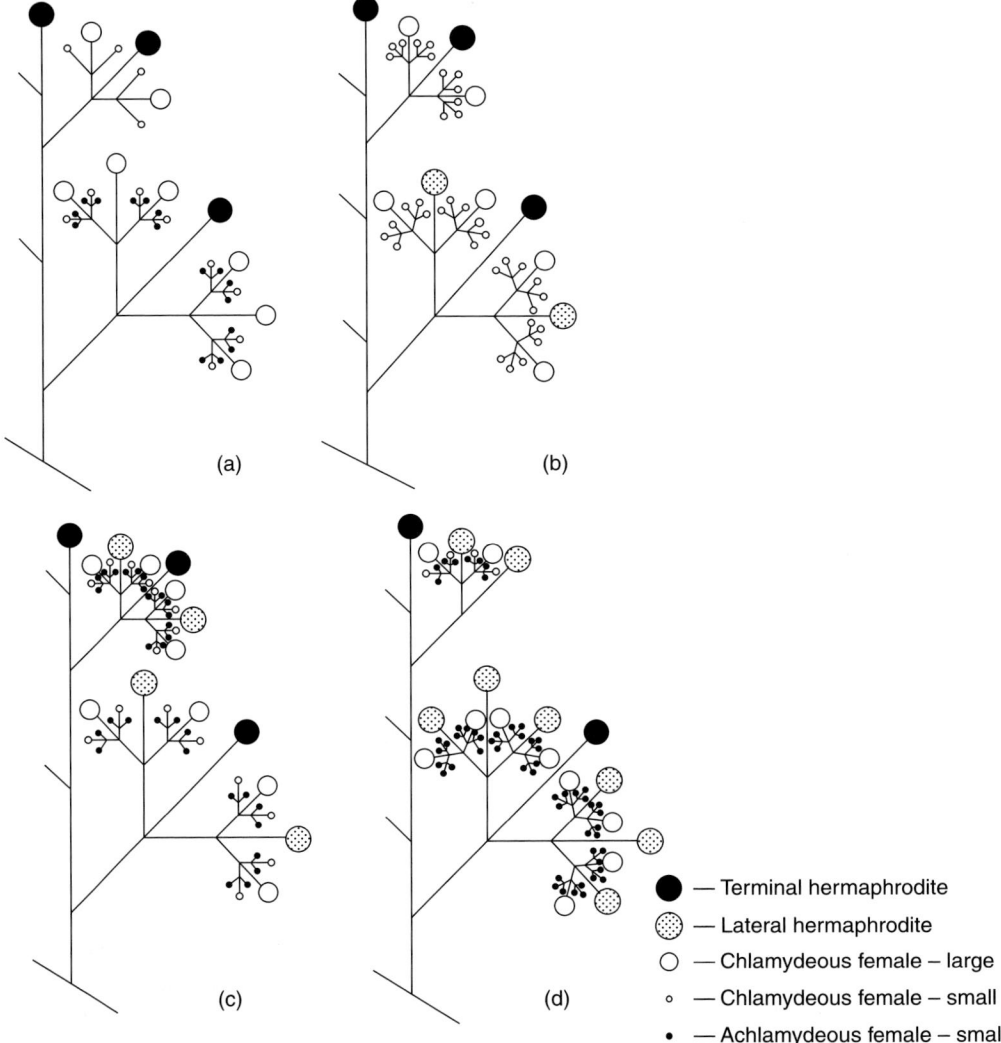

Fig. 6.5. Diagrammatic representation of floral types and flower arrangement in a glomerule in *C. quinoa* Willd. (a) Type I, (b) Type II, (c) Type III, (d) Type IV. [Reprinted from Bhargava *et al.* (2007b), with kind permission from Springer Science + Business Media B.V.]

Type IV: The dichasium is divided five times and about 12.5% of flowers are bisexual. The first and second branches bear bisexual flowers, the third branch bears chlamydeous large female flowers, while the fourth and fifth branches bear achlamydeous female flowers (Fig. 6.5d) (Bhargava *et al.*, 2007b).

Type V: This type bears 21.7% female flowers on the inflorescence. In all these accessions, the first division of the dichasium terminates in a bisexual flower, while the second and third branches bear a large and a small chlamydeous female flower respectively (Fig. 6.6a) (Bhargava *et al.*, 2007b).

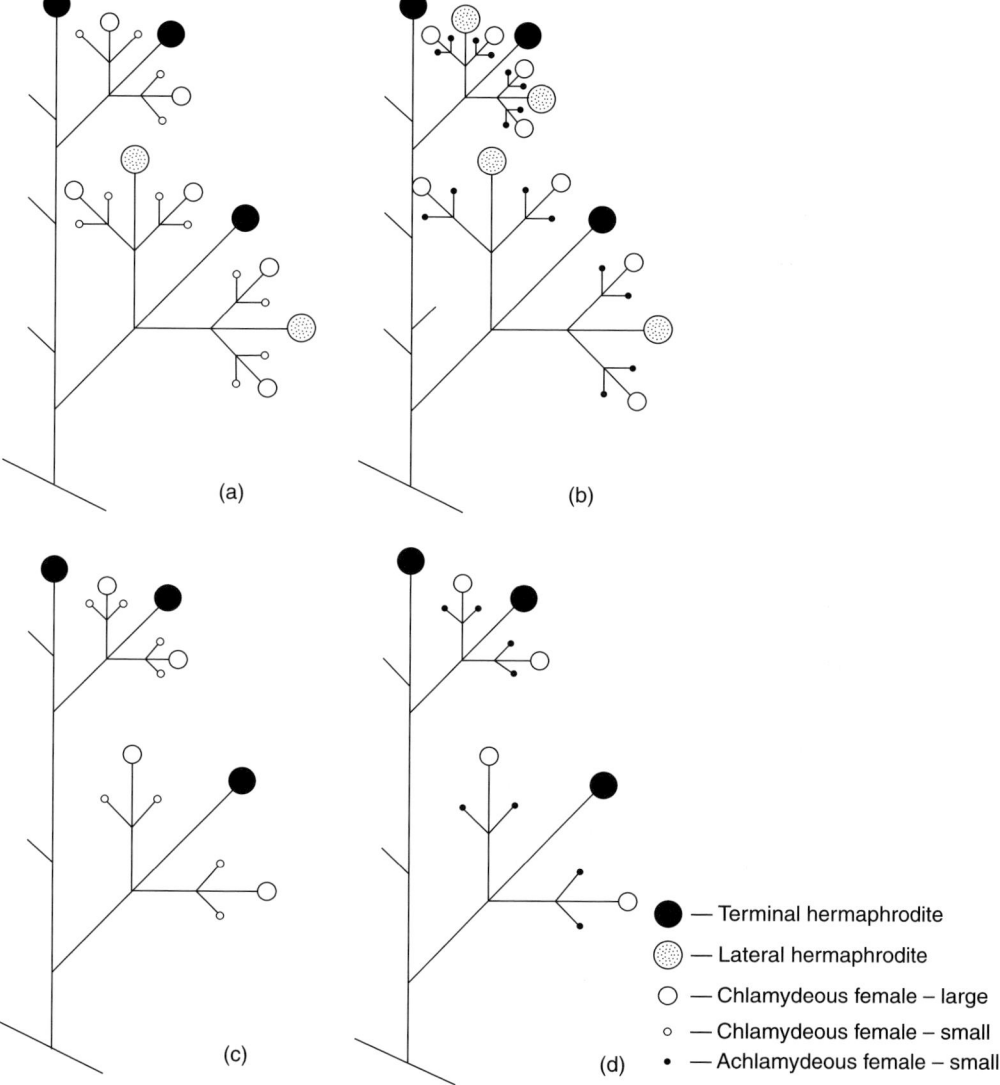

Fig. 6.6. Diagrammatic representation of floral types and flower arrangement in a glomerule in *C. quinoa* Willd. (a) Type V, (b) Type VI, (c) Type VII, (d) Type VIII. [Reprinted from Bhargava *et al.* (2007b), with kind permission from Springer Science + Business Media B.V.]

Type VI: This type differs from type V in that only the third branch bears achlamydeous small female flowers (Fig. 6.6b) (Bhargava *et al.*, 2007b).

Type VII: This type has 20.0% bisexual flowers, while the dichasium is divided only twice. The first and second branches bear chlamydeous large and small female flowers respectively (Fig. 6.6c) (Bhargava *et al.*, 2007b).

Type VIII: This type differs from type VII in that only the second branch bears achlamydeous small female flowers (Fig. 6.6d) (Bhargava *et al.*, 2007b).

Type IX: This type shows 46.6% bisexual flowers and the dichasium is divided only twice. The first branch bears a bisexual flower, while the second branch bears a large chlamydeous female flower (Fig. 6.7a) (Bhargava *et al.*, 2007b).

Type X: This type shows the most bisexual flowers (48.9%), while the dichasium is divided four times. The first three branches bear bisexual flowers, while the fourth one bears a small chlamydeous female flower (Fig. 6.7b) (Bhargava *et al.*, 2007b).

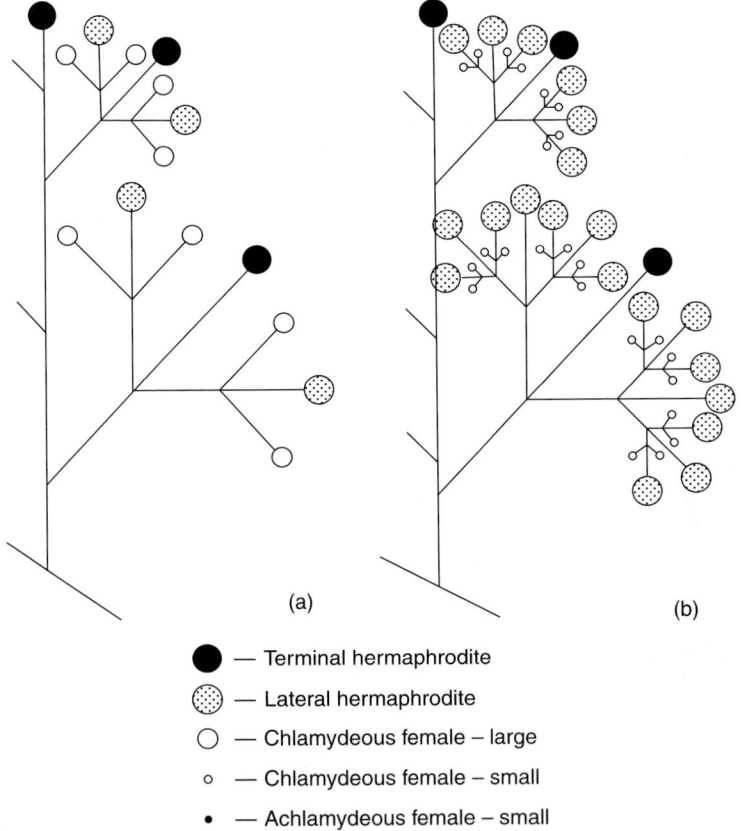

(a)   (b)

● — Terminal hermaphrodite

⊙ — Lateral hermaphrodite

○ — Chlamydeous female – large

∘ — Chlamydeous female – small

• — Achlamydeous female – small

Fig. 6.7. Diagrammatic representation of floral types and flower arrangement in a glomerule in *C. quinoa* Willd. (a) Type IX, (b) Type X [Reprinted from Bhargava *et al.* (2007b), with kind permission from Springer Science + Business Media B.V.]

6.3.4 Fruit

The fruit is an achene, comprising several layers, viz. perigonium, pericarp and episperm (Risi and Galwey, 1984), from outwards to inside. The fruit size ranges from 1.8 to 2.6 mm in diameter and its shape may be cylindrical, conical or ellipsoidal (Ignacio *et al.*, 1976). Quinoa seeds can germinate very quickly, i.e. in a few hours after having been exposed to moisture (Vega-Gálvez *et al.*, 2010). Seeds vary greatly in shape, ranging from conical, cylindrical to ellipsoidal. Seed size varies from small (<1.8 mm), medium (1.8–2.1 mm) to large (2.2–2.6 mm). A comparatively small seed size has been reported from the Indian subcontinent (Bhargava *et al.*, 2007a). Seed weight ranges from 2 to 6 mg. Seed colour in quinoa is variable, black being dominant over red and yellow, which in turn are dominant over white seed colour (Risi and Galwey, 1984; Mujica, 1994). The median longitudinal section of the seed shows that the embryo is peripheral and a basal body is present in the seed as a storage tissue or perisperm (Fig. 6.8) (Prego *et al.*, 1998). In the mature seed, the endosperm is present only in the micropylar region of the seed and consists of one to two cell-layered tissue surrounding the hypocotyl–radicle axis of the embryo. The pericarp of quinoa seed is two-layered. The cells of the outer layer are large and papillose in shape, while the cells of the inner layer are tangentially stretched (Prego *et al.*, 1998). The saponins are concentrated in the pericarp and have a bitter taste, so they need to be removed before consumption by washing or rubbing. If bitterness remains after processing the grain, it is probably due to remnants of the pericarp because the saponins are located in this layer (Villacorta and Talavera, 1976). Localization of stored reserves inside the seeds of quinoa show a marked compartmentalization similar to that seen in *Amaranthus hypocondriacus* seeds (Coimbra and Salema, 1994).

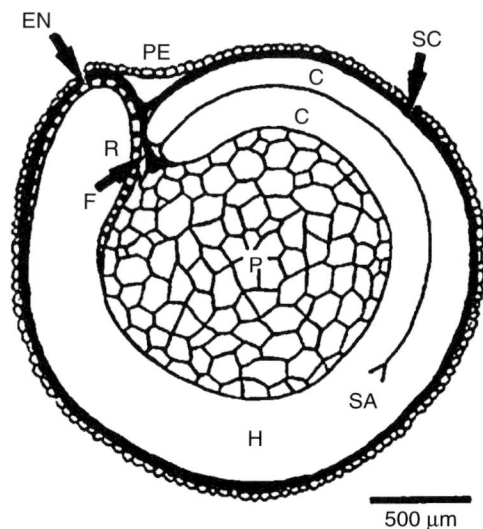

500 μm

Fig. 6.8. Median longitudinal section of the grain of quinoa. PE: Pericarp; H: Hypocotyl radicle axis; C: Cotyledons; EN: Endosperm; F: Funicle; P: Perisperm; R: Radicle; SA: Shoot apex. [Reprinted from Prego *et al.* (1998).]

Carbohydrate reserves are found principally in the perisperm, while proteins, mineral nutrients and lipid reserves are located mostly in the endosperm and embryo (Prego *et al.*, 1998).

6.4 Concluding Remarks

The botanical characteristics of quinoa can be immensely useful in selecting plants for appropriate breeding programmes with specific objectives. The self-incompatible or male sterile lines and the ones with low frequency of hermaphrodite flowers, identified by studying the flower clusters, can be selected for breeding programmes because the quinoa flowers that are small are not amenable to emasculation. However, detailed studies are needed to examine the effect of different proportions of floral types in different accessions on inbreeding and outcrossing rates and seed production.

References

Bertero, D., Medan, D. and Hall, A.J. (1996) Changes in apical morphology during floral initiation and reproductive development in quinoa (*Chenopodium quinoa* Willd.). *Annals of Botany* 78, 317–324.

Bhargava, A., Shukla, S. and Ohri, D. (2007a) Genetic variability and interrelationship among various morphological and quality traits in quinoa (*Chenopodium quinoa* Willd.). *Field Crops Research* 101, 104–116.

Bhargava, A., Shukla, S. and Ohri, D. (2007b) Gynomonoecy in *Chenopodium quinoa* Willd. (Chenopodiaceae): variation in inflorescence and floral types in some accessions. *Biologia* 62, 19–23.

Cai, Y., Sun, M. and Corke, H. (2005) HPLC characterization of betalains from plants in the Amaranthaceae. *Journal of Chromatographic Science* 43, 454–460.

Coimbra, S. and Salema, R. (1994) *Amaranthus hypochondriacus*: seed structure and localization of seed reserves. *Annals of Botany* 74, 373–379.

Fuentes, F.F. and Bhargava, A. (2011) Morphological analysis of quinoa germplasm grown under lowland desert conditions. *Journal of Agronomy and Crop Science* 197, 124–134.

Gallardo, M., González, J.A. and Prado, F.E. (2000) Presencia de betalaínas en plántulas de *Chenopodium quinoa* Willd. *Lilloa* 40, 109–113.

Gandarillas, H. (1979) Botanica. Quinua y kaniwa. Cultivos Andinos. In: Tapia, M.E. (ed.) *Serie Libros y Materiales Educativos*. Instituto Interamericano de Ciencias Agricolas, Bogota, Colombia, pp. 20–44.

Hunziker, A.T. (1943) Los species alimenticias de Amaranthus y Chenopodium cultivadas por los Indios de America. *Revista Argentina de Agronomia* 30, 297–353.

Ignacio, Q.J., Fernanadez, C.A. and Cortes, G.J. (1976) Contribucion al studio morfologico del grano de quinua. In: *Segunda Convencion Internacional de Quenopodiaceas*. Universidad Boliviana Tomas Frias, Comite Departamental de Obras Publicas de Potosi, Instituto Interamericanode Ciencias Agricolas, Potosi, Bolivia.

Jacobsen, S.-E. (1993) Quinoa: a novel crop for European agriculture. PhD. thesis. The Royal Veterinary and Agricultural University, Denmark.

Lescano, R.J.L. (1981) *Cultivo de la Quinua*. Universidad Nacional Tecnica del Altiplano, Centro de Investigaciones en Cultivos Andinos, Puno, Peru.

Mabry, T.J., Taylor, A. and Turner, B.L. (1963) The betacyanins and their distribution. *Phytochemistry* 2, 61–64.

Maughan, P.J., Bonifacio, A., Coleman, C.E., Jellen, E.N., Stevens, M.R. and Fairbanks, D.J. (2007) Quinoa (*Chenopodium quinoa*). In: Kole, C. (ed.) *Genome Mapping and Molecular Breeding in Plants, Volume 3 Pulses, Sugar and Tuber Crops.* Springer, Berlin, Germany, pp. 147–158.

Mujica, A. (1994) Andean grains and legumes. In: Hernando Bermujo, J.E. and Leon, J. (eds) *Neglected Crops: 1492 From a Different Perspective.* Plant Production and Protection Series No. 26. FAO, Rome, Italy, pp. 131–148.

Nelson, D.C. (1968) Taxonomy and origins of *Chenopodium quinoa* and *Chenopodium nuttaliae.* PhD. thesis, University of Indiana, Bloomington, Indiana.

Prego, I., Maldonado, S. and Otegui, M. (1998) Seed structure and localization of reserves in *Chenopodium quinoa. Annals of Botany* 82, 481–488.

Risi, J. and Galwey, N.W. (1984) The Chenopodium grains of the Andes: Inca crops for modern agriculture. In: Coaker, T.H. (ed.) *Advances in Applied Biology,* Vol. 10. Academic Press, London, pp. 145–216.

Simmonds, N.W. (1965) The grain chenopods of the tropical American highlands. *Economic Botany* 19, 223–235.

Tanaka, Y., Sasaki, N. and Ohmiya, A. (2008) Biosynthesis of plant pigments: anthocyanins, betalains and carotenoids. *The Plant Journal* 54, 733–749.

Vega-Gálvez, A., Miranda, M., Vergara, J., Uribe, E., Puente, L. and Martínez, E.A. (2010) Nutrition facts and functional potential of quinoa (*Chenopodium quinoa* Willd.), an ancient Andean grain: a review. *Journal of Sciences of Food and Agriculture* 90, 2541–2547.

Villacorta, C.L. and Talavera, R.V. (1976) Anatomia del grano de quinua (*Chenopodium quinoa* Willd.). *Anales Científicos* 14, 39–45.

7

Crop Production and Management

7.1 Introduction

The Andean Altiplano, lying mostly between 3500 and 4300 m above sea level, is a high plains region encompassing Lake Titicaca and extending approximately 800 km from north to south. It is one of the most difficult ecologies for modern agriculture. In this region, especially the Altiplano shared by Bolivia and Peru, agronomic and edaphic conditions for effective crop production are severely constrained by persistent, harsh environmental factors like frost, hail, torrential rains, flooding and drought (Kolata, 2009). Moreover, widespread nutrient-poor soils, lacking in nitrogen, potassium, phosphorus and organic matter, combined in some areas with excessive soil salinity, present enormous technical and economic challenges to farmers at all scales of land holding (Carney et al., 1993; Binford and Kolata, 1996, Beisboer et al., 1999; Kolata, 2009). The agroecosystems in this region are largely managed by smallholder farmers in rural communities growing a mix of old and new world crops across a range of socioeconomic and environmental settings. The three major pre-colonial domesticated food crops of the Andean region include maize, potatoes and quinoa (Hunziker, 1952). The ecological range of maize is generally limited to relatively temperate environments below 3200 m. Potatoes and quinoa are well adapted to virtually the entire altitudinal spectrum of the Andean region, including, in the case of quinoa and the related crop plants kañawa and kiwicha, elevations over 4000 m (Kolata, 2009). Quinoa has a remarkable adaptability to different agroecological zones. It adapts to different climates, growing at relative humidities from 40% to 88% and withstanding temperatures from −4°C to 38°C (FAO, 2011); temperatures ranging between 15°C and 20°C are ideal for cultivation. It is a water-efficient plant, tolerant and resistant to lack of soil moisture, and produces acceptable yields with rainfall of 100–200 mm.

© Bhargava and Srivastava 2013. *Quinoa: Botany, Production and Uses*
(A. Bhargava and S. Srivastava)

In the Andean region, various quinoa production systems are prevalent in different regions. The cultivation of quinoa in Bolivia dates back to pre-colonial times. The Bolivian Altiplano has been divided into three clearly differentiated areas: Northern Altiplano, Central Altiplano and Southern Altiplano, based on the characteristics of climate, soil, and agriculture and livestock possibilities (Rojas *et al.*, 2004). Quinoa is cultivated under the conventional production system on the Central and Northern Altiplano in rotation with other crops. The crop is generally cultivated after potatoes and in turn is followed by broad beans or a species of fodder (barley or oats). On the Northern Altiplano, quinoa is cultivated mainly for local consumption and is sown on small areas of land, with potatoes and barley being the more important crops. On the Southern Altiplano, it is the only crop available for farmers, and is cultivated under adverse conditions with little rain and frequent frosts (Rojas *et al.*, 2004). In Peru, various agroecological zones are known as 'aynokas' (community fields), 'canchas' (fields surrounded by stone walls), waru warus (high beds), 'kochas' (fields around small lakes) and 'andenes' (terraces) (Mujica *et al.*, 2003). Quinoa is sown by transplanting in the valleys, in association with beans, potatoes, barley and Andean roots, to avoid the risk of adverse environmental conditions. Quinoa in Chile had been cultivated primarily by the indigenous Aymara Indians in the northern Chilean Altiplano and by the Mapuche in the south-central area (Fuentes *et al.*, 2009). Nevertheless, the cultivation of quinoa extends to the south-central zone of Chile in a fragmented pattern. The crop in Chile is grown mainly in two ecological regions and geographically distinct areas. These are the central-south region near sea level, where there is a long photoperiod, and in the Altiplano of the Tarapacá Region, at an altitude between 3000 and 4200 m, where there is a short photoperiod.

7.2 Land Preparation

Quinoa grows well on sandy-loam to loamy-sand soils. It can also grow on marginal soils that have poor or excessive drainage, low natural fertility and extremes in pH (4.8–8.5) (Narea, 1976; Tapia, 1979). In fact, quinoa has been cultivated on acid soils with a pH of 4.5 (in Cajamarca, Peru) and alkaline soils with a pH of up to 9.5 (in Uyuni, Bolivia) (Mujica, 1994). A level, well-drained seedbed is most suited for quinoa cultivation. Land preparation is primarily a mechanical process designed to condition the land to provide a good seed bed. In the traditional system in Bolivia, the land is prepared using a tool called a 'Taquiza', 'Liukána' or 'Tank'ana' that partially or wholly removes the surface soil, depending on the production area (FAO, 2011). In the Southern Altiplano, conical mounds of 25–30 cm diameter and a height of 15–20 cm are made, although this system is generally performed on the slopes (Cossio, 2005).

7.3 Sowing

Sowing is an important initial step for the establishment of quinoa since the emergence of seedlings which impacts plant density and final yields depends on this stage. Therefore, this task must be carried out correctly and at the appropriate time. The amount of soil moisture at the time of sowing is a decisive factor for good germination and to obtain an adequate population of plants per plot (Rojas et al., 2004).

Sowing of quinoa is done in a variety of ways. The crop can be sown in rows, groups, by broadcasting or by transplanting. Seeds should be sown 1–2 cm deep in a fine-structured, moist seed bed (Jacobsen, 2003). The sowing depth is critical at emergence and the technical recommendation is to cover the seeds with less than 2 cm soil (Spehar and Rocha, 2009). Otherwise, emergence may be reduced, resulting in crop failure.

- Broadcast sowing: This is the most common method of sowing in the Andes. However, this method makes control of weeds and any mechanized processes difficult (Risi and Galwey, 1984), and requires about 10–20% more seed.
- Row sowing: In this method, the seed is placed in rows, the row width being 40–80 cm. The seeds are placed at the bottom of the furrow in dry soils or at the top of the ridge in areas with abundant rain (Risi and Galwey, 1984). The mechanized system makes the furrows, carries out the distribution and tops off the seed simultaneously. Sowing in furrows is frequently practised in Bolivia, where the furrows are opened up with the yoke or mechanical traction plough and after which the seed is distributed in a continuous stream (Rojas et al., 2004). In the Southern Altiplano, mechanical sowing is performed with the Satiri drill, which has two furrowers with chutes through which the seed is fed. The opening of the furrowers can be adjusted to a distance between 0.8 and 1 m, and the seed is deposited in the furrows, also at a distance of 0.8 to 1 m (FAO, 2011).
- Transplanting: The crop is sown in plots and seedlings are later transplanted in the main plots. This technique has long been used in the Urubamba Valley, Cuzco, Peru (Risi and Galwey 1984) and has also been successfully applied in India (unpublished results).
- Group sowing: This technique is used in areas of low rainfall like the Salares in Bolivia. In this method 4–5 seeds are planted in deep holes 70–120 cm apart and about 40 cm in diameter (Rea et al., 1979).
- Mixed sowing: This method involves cultivation with a ridger, followed by broadcast sowing. This enables the seeds to be deposited both on the ridges and in furrows, and further assures establishment under deficient or excess moisture (Risi and Galwey, 1984).

The plant-to-plant and row-to-row spacings have not been recommended and are different for different locations. In the Central and Northern Altiplano the distance between rows is 50 cm, while in the Southern Altiplano the distance between the rows varies from 1 to 1.20 m. The different spacings successfully used in different regions of the world are provided in Table 7.1.

Table 7.1. Row spacing and plant density in field experiments performed in different regions of the world with diverse agroclimatic conditions.

Area	Row spacing	Seed rate/plant density/ plant spacing	Reference
The Netherlands	25 cm	0.7 g seeds per m^2	Mastebroek *et al.* (2002)
Poland	40 cm	17 kg seeds/hectare	Gęsiński (2008)
England	40 cm	15 kg seeds/hectare	Risi and Galwey (1989)
England	20, 40, 80 cm	15, 20, 30 kg seeds/ hectare	Risi and Galwey (1991)
Brazil	50 cm	NA	Spehar and Rocha (2010)
Denmark	25, 50 cm	200 plants/m^2	Jacobsen (1997)
Chile	150 cm	25 cm plant-to-plant spacing	Fuentes and Bhargava (2010)
India	30 cm	30 plants/m^2	Bhargava *et al.* (2007)

7.4 Seeding Rates and Plant Density

Sowing density depends on various factors like genotype growth habit, sowing date, climatic conditions and soil fertility (Carbone-Risi, 1986; Santos, 1996). Different seeding rates have been recommended in South America and the USA. Current recommendations for seeding rates in the USA are 1–1.5 million/ha (Johnson and Croissant, 1985). South American recommendations are 8 million/ha for row cropping and 20 million/ha for broadcast cultural practices (Johnson and Ward, 1993). Another parameter for seeding rate is given in kilogram (kg) seed per hectare. One kilogram contains around 300,000 seeds. Optimum sowing density varies from 4–6 kg/ha in the Bolivian Altiplano to 15–23 kg/ha in Peru (Blanco, 1970; Canahua, 1977). Highest-density sowing is used in varieties with larger seeds (diameter >2 mm), while lower density sowing is used in varieties with small seeds (Aguilar and Jacobsen, 2003). High density is known to produce small, weak plants having lower yield per plant and greater lodging (Risi and Galwey, 1984; Aguilar and Jacobsen, 2003). Too few plants may result in ramified plants that may not mature in time, as well as provide more space for the growth of weeds (Aguilar and Jacobsen, 2003).

7.5 Fertilizer Application

A good response in the crop has been observed with respect to the application of nitrogenous fertilizers, but high levels of available nitrogen are reported to decrease yield due to slow maturity and intense lodging (Oelke *et al.*, 1992). However, recent studies (Berti *et al.*, 2000; Schulte auf'm Erley *et al.*, 2005) suggest that quinoa responds strongly to nitrogen fertilization and grain yield did not show a decrease with increasing N rates. Jacobsen *et al.* (1994) observed that quinoa grain yield increased with increasing nitrogen fertilization rate from 40 to 160 kg N/ha. Schulte auf'm Erley *et al.* (2005) assessed the response of quinoa to nitrogen fertilizer rates of 0, 80 and 120 kg N/ha and found that quinoa cultivars responded strongly to nitrogen application and produced up to 3.5 mg/ha yield when fertilized with 120 kg N/ha in southern Germany. Nitrogen application is known to increase seed yield as well as the protein content of the seeds, with a 0.1% increase in protein for 1 kg of ammonium nitrate applied (Johnson and Ward, 1993). A dose of 80 kg N/ha, applied in split form, 50% during the topping of the ears and 50% during the pre-flowering stage, has been recommended with respect to the fertilization of quinoa on the Central and Northern Altiplanos (Rojas *et al.*, 2004). For South America, 120 kg/ha of urea has been recommended (Johnson and Ward, 1993).

In Egypt, field trials were carried out on the Green Desert Egypt Association farm in the Wadi El-Natroon region (longitude 30.35°, latitude 30.4° and altitude 20 m) during the 2008/2009 and 2009/2010 winter crop seasons to study the best nitrogen fertilizer rate out of 0, 90, 180, 270 and 360 kg N/ha on growing quinoa under sandy soil and irrigated by 4400 m^3/ha of underground water with salinity of EC = 3.6 dS/m (Shams, 2012). Results revealed that in first and second season, respectively, fertilizing quinoa with 360 kg N/ha resulted in maximum plant height of 52.73 and 51.78 cm, grain yield of 10.070 and 8.177 g/plant, grain yield/ha of 1203 and 1088 kg/ha, biological yield of 2787 and 2322 kg/ha and field water use efficiency of 2.733 and 2.472 kg/mm. The maximum nitrogen use efficiency values of 5.367 and 3.417 kg/kg N applied were obtained when quinoa received only 90 kg N/ha, in first and second season, respectively.

Heavy doses of phosphorus and potash are known to increase vegetative growth without any increase in seed yield (Etchevers and Avila, 1979). But this could be due to excess of soil potassium in the tropical Andean soils. Gandarillas (1982) also stated that quinoa showed no response for either potassium or phosphorus. No clear response has been observed on the addition of phosphorus (Rojas *et al.*, 2004).

Organic fertilizers are being used in the Bolivian Altiplano to cater to the increasing demand for organic quinoa. This is principally done by two methods (Rojas *et al.*, 2004):

1. Introducing manure into holes at a rate of 300–500 g per hole after rotavation, leaving a mound of earth to identify the place at the time of sowing. By using this method, between 3 and 5 mt of manure per hectare are used.
2. Introducing 10 mt of manure per hectare in January and February before the soil is rotavated, spread uniformly over the entire surface area.

An increase in yield has been noted after the application of sheep and llama manure. A favourable response in yield has also been observed after the application of organic and inorganic fertilizers together, compared with either fertilizer on their own (Rojas *et al.*, 2004).

7.6   Water Requirements and Irrigation

Quinoa is a drought-tolerant crop that has a low requirement for water, although the yield is significantly affected by irrigation (Oelke *et al.*, 1992). Excessive irrigation at the seedling stage causes stunting and damping off of seedlings, while after stand establishment it produces tall plants with no improvement in yield (Oelke *et al.*, 1992). The low water requirement shows the drought-hardy nature of the crop and makes it suitable for cultivation on large tracts across the world in countries where reliable irrigation is non-existent and farmers have to depend on seasonal rains (Bhargava *et al.*, 2006). In Colorado, experiments on sandy loam soils using 128, 208, 307 and 375 mm of water produced the best yield of 1439 kg/ha with 208 mm of water (rainfall and irrigation) (Flynn, 1990). However, the study cannot be called conclusive since it was limited to a single location and soil type.

Irrigating quinoa is not a normal practice in the farming systems of the Andean region. It is also very difficult to find studies on irrigation and its impact on yield because quinoa grows under marginal soil, climate and water conditions (Geerts *et al.*, 2008). However, recent studies have demonstrated that quinoa yield can be increased by applying only a small amount of irrigation water during an appropriate growing stage (García, 2003; Geerts *et al.*, 2008). It has been demonstrated that with irrigation applications during the critical stages of growth (mainly flowering), seed production can be stabilized between 1.2 and 2 mt/ha, much larger than the current average (500 kg/ha). Thus, the application of just a little water is likely to substantially raise the seed yield. Detailed studies on farmers' willingness to accept irrigation recommendations have demonstrated that deficit irrigation technology is more likely to be picked up in areas where farmers own larger fields where they can obtain surplus production for sale and where there is already a certain type of irrigation (Taboada *et al.*, 2011).

7.7 Effect of Temperature and Photoperiod on Quinoa

Bertero *et al.* (1999a) examined the effects of photoperiod on phasic development, leaf appearance and seed growth in two quinoa cultivars, and of photoperiod × temperature interactions on seed growth in one cultivar. Photoperiods were given as 10 hours of natural daylight followed by extensions with low intensity artificial light giving either a short (SD, 10.25 hours) or long (LD, 14 or 16 hours) photoperiod. A quantitative SD response was observed for time to anthesis and total number of leaves, and more than 50% of the leaf primordia were formed after floral initiation. Plants grown in SD until anthesis produced seed four-fold larger in diameter than seed on plants grown in LD. Seed diameter was also reduced by 24% by LD applied after anthesis, and by 14% by high temperature, but the combination of high temperature with LD gave the greatest inhibition of seed growth. It was concluded that photoperiod had strong effects on all stages of plant reproduction and often acted indirectly, as shown by delayed responses expressed in later phases of development. Bertero *et al.* (1999b) showed that photoperiodic sensitivity was negatively associated with the latitude of origin of nine quinoa lines and positively associated with duration of the basic vegetative phase when temperature and photoperiod responses were taken as independent (non-interactive). However, photoperiod and temperature parameters when taken as independent (interactive) were not significantly related with latitudes of origin. Another study (Bertero *et al.*, 2000) has shown that photoperiod sensitivity of phyllochron decreased with increase in the latitude of origin of the cultivar. Temperature sensitivity was highest in cultivars originating in cold or dry climates, and lowest for cultivars from humid and warmer climates.

It was suggested that during the domestication of quinoa, photoperiod sensitivity has been selected for as a homeostatic mechanism to counteract the potentially reduced leaf area associated with early flowering under short days and high temperatures in the tropics (Bertero *et al.*, 2000).

It has been proposed that mean incident radiation affects the phyllochron in quinoa (Bertero, 2001). Radiation sensitivity was reported to be highest in cultivars from Peru, Bolivia and Southern Chile, and lowest in those from Ecuador, which had the high sensitivity to photoperiod and longest phyllochron. The responses of developmental processes to temperature and photoperiod in quinoa were ascertained by Bertero (2003) under controlled environments using nine cultivars. Different phases of development like emergence to floral initiation, floral initiation to first anthesis and first anthesis to physiological maturity were found to be sensitive to photoperiod. Leaf appearance rate was also affected by photoperiod and temperature. Sensitivity of leaf appearance rates to temperature increased, while sensitivity to photoperiod decreased, as latitude of origin of the cultivars increased.

7.8   Weed Control

Quinoa faces competition from rapidly growing weeds during the initial phenological stages in the first two weeks after emergence because it grows very slowly (Bhargava *et al.*, 2006). Therefore, utmost care should be taken in the regulation of sowing dates for the crop. An early sowing would enable quinoa to have a head start over weeds so that the plant can attain good growth during this period (Bhargava *et al.*, 2006). Pigweed, kochia, lambsquarters and sunflower are the common weeds in North America, while *Parthenium, C. album* and *Sysmbrium* are some frequently encountered weeds in the north Indian Plains. However, in some areas weeds are not regarded as a problem but rather as a source of forage that is important in the months of fodder scarcity (Aguilar and Jacobsen, 2003).

Weed control has a major impact on grain yield in quinoa. In Colorado, control of grasses increased yields from 640 kg/ha to 1822 kg/ha (Johnson, 1990). Weed control using herbicides has been effective and has yielded good results. Preliminary studies of pre-emerge herbicides such as Dual, Furloe, Sutan and Antor in Colorado not only showed control of grasses and broadleaf weeds, but also resulted in good crop safety (Westra, 1988). In England, application of Metamazide, Propachlor, Linuron, Propyzamide and aloxium sodium did not significantly reduce plant stands of two quinoa cultivars (Galwey and Risi, 1984). Post-emergent control was best for Poast, Tough and Probe, and with Tough and Probe at low application rates (Westra, 1988).

7.9 Harvesting

The plants reach physiological maturity after 5–8 months, depending on the variety and environmental conditions (Mujica, 1993; Valdivia *et al.*, 1997). The drying of the plant and shedding of leaves denotes the mature stage of the plant. The leaves turn orange, yellow or red, depending on the variety, and the grains can be seen in the panicle through the opening of the perigonium, which is characteristic at the stage of physiological maturity (Aroni, 2005). The postharvest process involves harvesting that plants and cleaning the seeds. This comprises collection, pre-drying, storage and threshing of plants, and cleaning, drying and storage of seeds (Dominguez, 2003).

There are three ways to harvest the plants, viz. by using traditional, semi-mechanized and mechanized technology. The traditional method involved pulling out the plants by hand, selecting the mature panicles from each hole or row, and proceeding to shake or hit the root on the knees in order to remove earth and stones (FAO, 2011). However, several disadvantages, such as accelerated soil erosion, lower soil fertility and the presence of impurities like mixing soil with the grain, made this method unpopular as quinoa began to be commercialized. Hand harvesting was slowly replaced by manual harvesting with a sickle. This modification involves cutting the plant around 10–15 cm above the ground and leaving the stubble in the soil, thus helping soil conservation (FAO, 2011). Semi-mechanized technology involves cutting the plants with a mechanical mower, while mechanized technology makes use of mechanized devices like harvesters and winnowers. A comparison of the three methods is provided in Table 7.2.

Table 7.2. A comparison of different systems using traditional, semi-mechanized and mechanized technology. [Reprinted from Dominguez (2003), with permission from Taylor & Francis.]

Postharvest operations	Traditional technology	Semi-mechanized technology	Mechanized technology
Collection	Manual, using a sickle	Manual and mechanical	Combine harvester
Predrying	In piles (7–15 days)	In piles	–
Storage of plants	In barns	In silos	–
Threshing	Manual (rubbing the panicle)	Motorized vehicles, stationary threshers	Combine harvester
Cleaning	Manual (air currents)	Winnower (manual)	Winnower
Drying	Natural (3 days)	Artificial or mixed	Artificial
Storage of seed	In dry areas In barns	Bulk, in warehouses In sacks	In silos In sacks
Saponin removal	Manual cleaning (rubbing with stones)	Mechanical (dry or humid)	Mechanical (dry or humid)

7.10 Postharvest

Postharvest technology involves drying or stacking, threshing, venting and storage.

7.10.1 Predrying and drying

Predrying causes a reduction in the water content of the plant and protects the plants against rain or frost (Dominguez, 2003). Immediately after harvesting, the plants are piled with the panicles inwards, and the centre elevated. The crop is left like this for 7–15 days. There are three methods of stacking or drying across the Andes, called Arcos, Taucas and Chucus (Aroni, 2005). Arcos is a time-consuming method involving stacking the quinoa plants in the form of a cross resting on a base of other native species with the panicles leaning upwards. Drying is facilitated because there is good air circulation and the ears are sufficiently exposed to the sun for drying. Taucas consists of building mounds or stacks of plants with the panicles ordered towards the same side and on a piece of canvas or nylon (FAO, 2011). This method may take a little longer to dry, but facilitates threshing because of its concentration in one place. However, non-uniform drying and exposure of the harvest to rain and wind are the major disadvantages of this method. Chucus are cone shaped mounds of quinoa plants, which are scattered throughout the field. The mounds are made by standing the plants up in a circle with the panicles toward the top to give more stability to the chucu (FAO, 2011). This method permits faster drying.

7.10.2 Threshing

Threshing involves the separation of grains from the panicle. Threshing is performed in a number of ways:

- Threshing by hand: In this method panicles are hit against a hard surface or rubbed with the hand to separate the seeds from the plant (Dominguez, 2003). The plants are beaten with a stick on a hard platform. The resulting plant material is then coarsely sifted and then winnowed to obtain the grain.
- Threshing with animals: Quinoa panicles are placed in heaps, and the animals (horses, donkeys) are allowed to pass over the material (Dominguez, 2003). A major disadvantage of this method is that the product is mixed with animal wastes, which reduces product quality.
- Semi-mechanical threshing: In this method, practised in the Southern Bolivian Altiplano, a tarpaulin is extended on the ground and the dried plants are piled on top in two parallel lines over which the wheels of a vehicle pass (FAO, 2011). The panicles are placed towards the inside of the two rows so that in multiple passes the vehicle is able to separate the grains.
- Mechanical threshing: This method, practised in the Southern Bolivian Altiplano, used threshers such as the 'Vencedora' and 'Alvan Blash' that achieve grain separation of around 95%.

7.10.3 Cleaning seeds

This method, known as winnowing, involves separation soil, stones, excreta and broken seeds from the seed material. There are three main methods of winnowing, viz. traditional, improved manual and mechanical winnowing. In traditional winnowing, the seed material falls to the ground from the hands, or on to a cloth or plastic sheet, taking advantage of the afternoon wind currents (Dominguez, 2003). Improved manual winnowing is performed with a machine, which has an air-flow regulator and separates the quinoa grain from the unwanted material from the perigonium (FAO, 2011). Mechanical winnowing has recently been adopted and has varying yield, depending mainly on the amount of wastes that the material to be winnowed contains (FAO, 2011).

7.10.4 Drying

Drying is very important in quinoa processing because it prevents fermentation. Even slight humidity can discolour the grains in a short time, thereby reducing its commercial value. Clean seeds are put on a cloth and exposed to the sun for 3 days, bringing them in at night, to eliminate the remaining excess moisture in the grains (Dominguez, 2003).

7.10.5 Storage

Appropriate storage is essential to avoid product loss by rodents and moths. Storage should be in a clean, dry and ventilated environment. The grain is normally stored in sacks made of woven llama wool, or new or good condition polypropylene bags under dry, ventilated conditions (FAO, 2011).

7.11 Concluding Remarks

The remarkable adaptability of quinoa to thrive and flourish in different agro-ecological zones makes it an ideal crop for diversification in newer areas. The crop requires few inputs and can withstand a variety of stresses without a major adverse effect on the seed yield. Harvesting in its native region is done by both traditional and modern techniques, according to the resources available. Efforts should be made to develop efficient postharvest technologies because inefficient postharvest handling results in inferior quality product and lower prices.

References

Aguilar, P.C. and Jacobsen, S.-E. (2003) Cultivation of quinoa on the Peruvian Altiplano. *Food Reviews International* 19, 31–41.
Aroni, J.C. (2005) Fascículo 5: Cosecha y poscosecha. In: PROINPA y FAUTAPO (eds) *Serie de Módulos Publicados en Sistemas de Producción Sostenible en el Cultivo de la*

Quinua: Módulo 2. Manejo Agronómico de la Quinua Orgánica. Fundación PROINPA, Fundación AUTAPO, Embajada Real de los Países Bajos, La Paz, Bolivia, pp. 87–102.

Beisboer, D.D., Binford, M.W. and Kolata, A.L. (1999) The natural and human setting. In: Kolata, A.L. (ed.) *Tiwanaku and its Hinterland: Archaeological and Paleoecological Investigations in the Lake Titicaca Basin of Bolivia.* Smithsonian Institution Press, Washington, DC.

Bertero, H.D. (2001) Effects of photoperiod, temperature and radiation on the rate of leaf appearance in quinoa (*Chenopodium quinoa* Willd.) under field conditions. *Annals of Botany* 87, 495–502.

Bertero, H.D. (2003) Response of developmental processes to temperature and photoperiod in quinoa (*Chenopodium quinoa* Willd.). *Food Reviews International* 19, 87–97.

Bertero, H.D., King, R.W. and Hall, A.J. (1999a) Photoperiod-sensitive developmental phases in quinoa (*Chenopodium quinoa* Willd.). *Field Crops Research* 60, 231–243.

Bertero, H.D., King, R.W. and Hall, A.J. (1999b) Modeling photoperiod and temperature responses of flowering in quinoa (*Chenopodium quinoa* Willd.). *Field Crops Research* 63, 19–34.

Bertero, H.D., King, R.W. and Hall, A.J. (2000) Photoperiod and temperature effects on the rate of leaf appearance in quinoa (*Chenopodium quinoa*). *Australian Journal of Plant Physiology* 27, 349–356.

Berti, M., Wilckens, R., Hevia, F., Serri, H., Vidal, I. and Mendez, C. (2000) Fertilizacion nitro-genada en quinoa (*Chenopodium quinoa* Willd.). *Ciencia Investigacion Agraria* 27, 81–90.

Bhargava, A., Shukla, S. and Ohri, D. (2006) *Chenopodium quinoa*: an Indian perspective. *Industrial Crops and Products* 23, 73–87.

Bhargava, A., Shukla, S. and Ohri, D. (2007) Genetic variability and interrelationship among various morphological and quality traits in quinoa (*Chenopodium quinoa* Willd.). *Field Crops Research* 101, 104–116.

Binford, M.W. and Kolata, A. (1996) The natural and cultural setting. In: Kolata, A. (ed.) *Tiwanaku and its Hinterland: Archaeological and Paleoecological Investigations of an Andean Civilization.* Smithsonian Institution Press, Washington, DC.

Blanco, T.C. (1970) *La Quinua. Como se Debe Cultivar.* Universidad Tecnica de Oruro, Oruro, Argentina.

Canahua, M.A. (1977) Observaciones del comportamiento de quinoa a la sequia. In: *Primer Congreso Internacional sobre cultivos Andinos.* Universidad Nacional San Cristobal de Huamanga, Instituto Interamericano de Ciencias Agricolas, Ayacucho, Peru, pp. 390–392.

Carbone-Risi, J.J.M. (1986) Adaptation of the Andean grain crop quinoa for cultivation in Britain. PhD. thesis. University of Cambridge, Cambridge.

Carney, H.J., Binford, M.W., Kolata, A.L., Marin, R.R. and Goldman, C.R. (1993) Nutrient and sediment retention in Andean raised-field agriculture. *Nature* 364, 131–133.

Cossio, J. (2005) Fascículo 1 – Preparación de suelo. In: PROINPA y FAUTAPO (eds) *Serie de Módulos Publicados en Sistemas de Producción Sostenible en el Cultivo de la Quinua: Módulo 2. Manejo Agronómico de la Quinua Orgánica.* Fundación PROINPA, Fundación AUTAPO, Embajada Real de los Países Bajos, La Paz, Bolivia, pp. 5–28.

Dominguez, S.S. (2003) Quinoa: postharvest and commercialization. *Food Reviews International* 19, 191–201.

Etchevers, B.J. and Avila, T.P. (1979) Factores que afectan el crecimiento de quinua (*Chenopodium quinoa*) en al centro-sur de Chile. *Papers of the 10th Latin American Meeting of Agricultural Sciences.*

FAO (2011) *Quinoa: An Ancient Crop to Contribute to World Food Security.* Food and Agriculture Organization, Rome, Italy.

Flynn, R.O. (1990) Growth characteristics of quinoa and yield response to increase soil water deficit. MSc. thesis, Colorado State University, Fort Collins, Colorado.

Fuentes, F.F. and Bhargava, A. (2010) Morphological analysis of quinoa germplasm grown under lowland desert conditions. *Journal of Agronomy and Crop Science* 197, 124–134.

Fuentes, F.F., Martinez, E.A., Hinrichsen, P.V., Jellen, E.N. and Maughan, P.J. (2009) Assessment of genetic diversity patterns in Chilean quinoa (*Chenopodium quinoa* Willd.) germplasm using multiplex fluorescent microsatellite markers. *Conservation Genetics* 10, 369–377.

Galwey, N.W. and Risi, J. (1984) *Development of the Andean Grain Crop Quinoa for Production in Britain*. University of Cambridge Annual Report, Cambridge.

Gandarillas, H. (1982) *Quinoa Production*. IBTA-CIID (Translated by Sierra-Blanca Association, Denver, Colorado, 1985).

García, M. (2003) Agroclimatic study and drought resistance analysis of quinoa for a deficit irrigation strategy in the Bolivian Altiplano. Doctoral dissertation, K.U. Leuven University, Leuven, Belgium.

Geerts, S., Raes, D., García, M., Vacher, J., Mamani, R., Mendoza, J., Huanca, R., Morales, B., Miranda, R., Cusicanqui, J. and Taboada, C. (2008) Introducing deficit irrigation to stabilize yields of quinoa (*Chenopodium quinoa* Willd.). *European Journal of Agronomy* 28, 427–436.

Gęsiński, K. (2008) Evaluation of the development and yielding potential of *Chenopodium quinoa* Willd. Under the climatic conditions of Europe. Part II: Yielding potential of *Chenopodium quinoa* under different conditions. *Acta Agrobotanica* 61, 185–189.

Hunziker, A.T. (1952) *Los Pseudocereales de la Agricultura Indigena de America*. ACMR Agency, Buenos Aires, Argentina.

Jacobsen, S.-E. (1997) Adaptation of quinoa (*Chenopodium quinoa*) to Northern European agriculture: studies on developmental pattern. *Euphytica* 96, 41–48.

Jacobsen, S.-E. (2003) The worldwide potential of quinoa (*Chenopodium quinoa* Willd.). *Food Reviews International* 19, 167–177.

Jacobsen, S.-E., Jørgensen, I. and Stølen, O. (1994) Cultivation of quinoa (*Chenopodium quinoa*) under temperate climatic conditions in Denmark. *The Journal of Agricultural Science* 122, 47–52.

Johnson, D.L. (1990) New grains and pseudograins. *Advances in New Crops*. Timber Press, Portland, Oregon, pp. 122–127.

Johnson, D.L. and Croissant, R.L. (1985) *Quinoa Production in Colorado*. Colorado State University, Cooperative Extension, Fort Collins, Colorado.

Johnson, D.L. and Ward, S.M. (1993) Quinoa. In: Janick, J. and Simon, J. E. (eds) *New Crops*. Wiley, New York, pp. 219–221.

Kolata, A.L. (2009) *Quinoa: Production, Consumption and Social Value in Historical Context*. Department of Anthropology, University of Chicago, Illinois.

Mastebroek, H.D., van Loo, E.N. and Dolstra, O. (2002) Combining ability for seed yield traits of *Chenopodium quinoa* breeding lines. *Euphytica* 125, 427–432.

Mujica, A. (1993) *Cultivo de Quinua*. Instituto Nacional de Investigación Agraria, Lima, Peru.

Mujica, A. (1994) Andean grains and legumes. In: Hernando Bermujo, J.E. and Leon, J. (eds) *Neglected Crops: 1492 From a Different Perspective*. Plant Production and Protection Series, FAO, Rome, Italy, pp. 131–148.

Mujica, A., Marca, S. and Jacobsen, S.-E. (2003) Current production and potential of quinoa (*Chenopodium quinoa* Willd.) in Peru. *Food Reviews International* 19, 149–156.

Narea, A. (1976) La producción de quinua en el Perú. In: *Segunda Convención, International de Quenopodiaceas*. Universidad Boliviana, Tomas Frias, Bolivia, pp. 170–176.

Oelke, E.A., Putnam, D.H., Teynor, T.M. and Oplinger, E.S. (1992) *Alternative Field Crops Manual*. University of Wisconsin Cooperative Extension Service, Madison, Wisconsin/Center for Alternative Plant and Animal Products, University of Minnesota Extension Service, St Paul, Minnesota.

Rea, J., Tapia, M. and Mujica, A. (1979) Practicas agronomicas. In: Tapia, M., Gandarillas, H., Alandia, S., Cardozo, A. and Mujica, A. (eds) *Quinoa y Kaniwa, Cultivos Andinos*. FAO, Rome, Italy, pp. 83–120.

Risi, J. and Galwey, N.W. (1984) The *Chenopodium* grains of the Andes: Inca crops for modern agriculture. *Advances in Applied Biology* 10, 145–216.

Risi, J. and Galwey, N.W. (1989) The pattern of genetic diversity in the Andean grain crop quinoa (*Chenopodium quinoa* Willd.). I. Association between characteristics. *Euphytica* 41, 147–162.

Risi, J. and Galwey, N.W. (1991) Effects of sowing date and sowing rate on plant development and grain yield of quinoa (*Chenopodium quinoa*) in a temperate environment. *The Journal of Agricultural Science* 117, 325–332.

Rojas, W., Soto, J.-L. and Carrasco, E. (2004) *Study on the Social, Environmental and Economic Impacts of Quinoa Promotion in Bolivia*. PROINPA Foundation, La-Paz, Bolivia.

Santos, R.L.B. (1996) Estudos iniciais para o cultivo de quinoa (*Chenopodium quinoa* Willd) no Cerrado. Dissertation (Masters degree). Universidade de Brasília, Faculdade de Agronomia e Medicina Veterinária, Brasília, Brazil.

Schulte-auf'm-Erley, G., Kaul, H.P., Kruse, M. and Aufhammer, W. (2005) Yield and nitrogen utilization efficiency of the pseudocereals amaranth, quinoa and buckwheat under differing nitrogen fertilization. *European Journal of Agronomy* 22, 95–100.

Shams, A.S. (2012) Response of quinoa to nitrogen fertilizer rates under sandy soil conditions. In: *Proceedings of the 13th International Agronomy Conference*, Faculty of Agriculture, Benha University, Egypt, 9–10 September 2012, pp. 195–205.

Spehar, C.R. and Rocha, J.E.S. (2009) Effect of sowing density on plant growth and development of quinoa, genotype 4.5, in the Brazilian Savannah Highlands. *Bioscience Journal* 25, 53–58.

Spehar, C.R. and Rocha, J.E.S. (2010) Exploiting genotypic variability from low-altitude Brazilian Savannah-adapted *Chenopodium quinoa*. *Euphytica* 175, 13–21.

Taboada, C., Mamani, A., Raes, D., Mathijs, E., Garcia, M., Geerts, S. and Gilles, J. (2011) Farmer's willingness to adopt irrigation for quinoa in communities of the central Altiplano of Bolivia. *Revista Latinoamericana de Desarrollo Economico* 16, 7–28.

Tapia, M.E. (1979) Historia y distribucion geographica. Quinua y kaniwa. Cultivos Andinos. In: Tapia, M.E. (ed.) *Serie Libros y Materiales Educativos*. Instituto Interamericano de Ciencias Agricolas, Bogota, Colombia, pp. 11–15.

Valdivia, R., Paredes, S., Zegarra, A., Choquehuanca, V. and Reinoso, R. (1997) *Manual del Productor de Quinua. Centro de Investigación de Recursos Naturales y Medio Ambiente*. Editorial Altiplano, Puno, Peru.

Westra, P. (1988) *Weed Control in Quinoa*. Report to Sierra Blanca Association.

8 Stress Tolerance

8.1 Introduction

The gains in agricultural output in many parts of the world have reached their ceiling, whereas the world population continues to rise. Agricultural land and water supplies are being degraded at a rapid pace as a result of intensive agricultural practices employed in developed and developing countries. The challenge of meeting future food needs is likely to be more daunting than was experienced under the green revolution. The two primary factors that stimulated growth in the green revolution were inputs in the form of irrigation water and chemical fertilizers, which cannot be expected to produce similar yield increases in the future (Brady and Weil, 1999). The rate of increase in irrigated land is expected to be minor and the worldwide use of fertilizers has levelled off; in many areas, the optimum level for application has been reached (Jacobsen *et al.*, 2003). Ideally, present agronomic practices should be changed to a more rational use of land and water resources, but this change is not expected to occur rapidly in the foreseeable future (Yamaguchi and Blumwald, 2005). Therefore, increasing the yield of crop plants in normal soils and in less productive lands is an absolute requirement for feeding the world (Yamaguchi and Blumwald, 2005). Apart from this, the development and use of crops that can tolerate high levels of stress would be a feasible option for addressing the problem.

8.2 Abiotic Stresses

Agricultural productivity in several regions is severely affected by many abiotic and biotic stresses that are harmful to plant growth and affect large areas of potential agricultural land. Abiotic stress is defined as any environmental condition, apart from the action of other organisms, that reduces the growth, survival and/or fecundity of plants (Boscaiu *et al.*, 2008). Thus, an environmental factor that limits crop productivity or causes a reduction in biomass is referred to as a stress (Grime, 1979; Robert-Seilaniantz *et al.*, 2010). Abiotic environmental factors include temperature, humidity, light intensity, water supply, mineral availability and CO_2, all of which determine the growth of a plant. The plants, being sessile, cannot escape from abiotic stress factors and are continuously exposed to different factors without any protection. Abiotic stresses adversely affect growth and productivity and trigger a series of morphological, physiological, biochemical and molecular changes in plants (Bhatnagar-Mathur *et al.*, 2008). The stress caused by abiotic factors alters plant metabolism leading to negative effects on growth, development and productivity (Rao *et al.*, 2006). If the stress becomes severe or continues for a long period, it may lead to unbearable metabolic burden on cells, reduced growth and ultimately plant death. Plant stress may vary from zero to severe through mild and moderate levels. In nature, plants may not be totally free from stresses and are expected to experience some degree of stress from any factor or factors (Rao *et al.*, 2006).

All the major crop plants, as well as most wild species, are quite sensitive to abiotic stress conditions. The domestication process and further development of crops by selective breeding for characters such as rapid growth, biomass accumulation and fruit and seed production, have not improved stress tolerance. It has been generally observed that domesticated species are probably more sensitive to stress that their wild ancestors, since inhibition of vegetative growth and reproductive development is the first and most general response to stress of plants, which invest all their resources (energy, metabolic precursors) to survive the adverse environmental conditions (Serrano and Gaxiola, 1994; Zhu, 2001). However, there are plants naturally adapted to stress conditions. Despite the unfavourable conditions to which they are subjected, these specialized plants are able to survive and complete their life cycle in their respective environments (Boscaiu *et al.*, 2008).

8.3   Quinoa as a Stress-Tolerant Crop

The Andean region covers 2 million square kilometres and extends from southern Venezuela to northern Argentina and Chile, and includes Colombia, Ecuador, Peru and Bolivia. The Andean region is characterized by a harsh climate with frequent periods of drought, night frost and cold, and by nutrient-deficient soils. Quinoa is considered to be a crop that is resistant to several of the adverse abiotic factors that constrain crop production in the Andes. However, research on the physiological mechanisms for these types of resistance, and their response to actual stress levels conferred by the environment, has only recently been initiated (Vacher *et al.*, 1994; Andersen *et al.*, 1996; Vacher, 1998; Jacobsen *et al.*, 1999; Mujica and Jacobsen, 1999; Mujica *et al.*, 2000).

8.3.1   Salt resistance in quinoa

Salinity is a major abiotic stress that adversely affects crop productivity and quality. Salinity occurs through natural or human-induced processes that result in the accumulation of dissolved salts in the soil water to an extent that inhibits plant growth (Munns, 2009). Sodicity is a secondary result of salinity in clay soils where leaching due to rainfall or irrigation water has washed soluble salts into the subsoil and left sodium bound to the negative charges of the clay. Table 8.1 provides information on salinity, sodicity and alkalinity prevalent in the soils and their effects on plant development.

Although the USDA Salinity Laboratory defines a saline soil as having an EC_e of 4 dS/m or more, many crops are affected by soil with an EC_e less than 4 dS/m, particularly in regions of high evapotranspiration leading to low soil moisture. The actual salinity of a field whose soil has an EC_e of 4 dS/m could be over 8 dS/m most of the time. According to the FAO Land and Plant Nutrition Management Service, salinity affects over 6% of the world's land. Of the current 230 million hectares of irrigated land, 45 million hectares are salt-affected (19.5%) and of the 1500 million hectares under dryland agriculture, 32 million are salt-affected to varying degrees (2.1%) (Munns, 2009). Although salinity and sodicity are common phenomena in arid and semiarid regions of the world, salt-affected soils have been recorded in almost all the climatic regions, and in a wide range of altitudes. The problem of soil salinity is increasing as a result of improper drainage in canal-irrigated wetland agroecosystems, use of poor quality water for irrigation, entry of seawater during cyclones in coastal areas, and salt accumulation in the root zone in arid and semi-arid regions due to high evaporative demand and insufficient leaching of ions because rainfall is inadequate (Chinnusamy and Zhu, 2003). Salinity is detrimental to plant growth in a variety of ways. The presence of salt in the soil solution reduces the ability of the plant to take up water, which leads to reduction in the growth rate. This is commonly referred to as the 'osmotic or water-deficit effect' of salinity. The salt-specific or ion-excess effect of salinity occurs when excessive amounts of salt enter the plant in the transpiration stream, leading to cell injury in the transpiring leaves further reducing growth (Greenway and Munns, 1980). In saline environments, ions like Na^+ and Cl^- penetrate the hydration shells of proteins and interfere with the non-covalent interactions between amino acids leading to conformational changes and loss of function of proteins (Chinnusamy

Table 8.1. Classification of saline, sodic and alkaline soils according to USDA Salinity Laboratory (USSL, 2005).

Soil type	Details	Effect on plant growth
Saline	High concentration of soluble salts $EC_e \geq 4$ dS/m	Inhibition of shoot and root growth due to osmotic and salt-specific components
Alkaline	Low concentration of soluble salts, but a high exchangeable Na^+ percentage (ESP) $EC_e \geq 15$ dS/m	Inhibition of root growth due to poor soil structure
Alkalinity	Type of sodic soil with a high pH $EC_e > 15$ dS/m, pH 8.5–10	Nutrient uptake is affected due to high pH

and Zhu, 2003). Ionic toxicity, osmotic stress and nutritional defects under salinity may lead to metabolic imbalances, which result in oxidative stress (Zhu, 2001; Chinnusamy and Zhu, 2003).

Management practices to reduce soil salt concentrations are often expensive and cannot always be applied in the world's underdeveloped and developing countries (Djanaguiraman et al., 2006). A promising alternative is the introduction of species (halophytes) capable of tolerating higher soil salinities with a good adaptability in terms of growth and yield (Ruffino et al., 2010). High concentration of salts induces dormancy in the seeds of many halophytic species (Gul and Weber, 1998; Debez et al., 2004), while seeds of glycophyte species lose their viability under these conditions. Thus, germination in both halophyte and glyophyte plants is sensitive to salinity (Pujol et al., 2000). The decrease in seed germination caused by salinity is the result of both osmotic dehydration and toxicity as a consequence of the excessive intake of ions such as Cl^- and Na^+ in tissues (Munns et al., 1995; Zhu, 2003). During germination, the joint action of the two stresses is manifested as a decrease in the percentage and velocity of germination (Huang and Redmann, 1995; Ungar, 1996; Kerepesi and Galiba, 2000; Delatorre-Herrera and Pinto, 2009). In halophytes the decrease in germination is mainly a consequence of lower osmotic potential in the soil solution, while in glycophytes, the excessive accumulation of ions in tissue causes nutritional imbalance, toxicity and death (Munns et al., 1995; Katambe et al., 1998; Dodd and Donovan, 1999). Thus, the differences in germination observed between the seeds of tolerant and sensitive plants are considered to be a result of the activation of different mechanisms of cells from the different parts of the embryonic axis in order to tolerate both osmotic and ionic stress during germination (Dodd and Donovan, 1999; Kerepesi and Galiba, 2000; Almansouri et al., 2001). During establishment of the seedling, salt modifies many biological processes such as growth, protein synthesis, osmotic homeostasis, carbon partitioning, carbohydrate metabolism, photosynthesis, lipid metabolism and gene expression (Prado et al., 2000; Munns, 2002).

Quinoa as a halophyte

Quinoa is a halophytic species that is regarded as having an unusually high tolerance to salinity. Some varieties of the crop show remarkable resistance to salt during germination. For example, Kancolla, an altiplano ecotype, reportedly has a germination rate of 75% at a concentration of 57 mS/cm (Christiansen et al., 1999; Jacobsen et al., 1999). Many varieties of this crop can grow in salt concentrations as high as those found in seawater (40 mS/cm) (Wilson et al., 2002; Jacobsen et al., 2003; Jacobsen, 2007; Adolf et al., 2010). These characteristics make it an attractive crop for regions where salinity has been recognized as a major agricultural problem (Prado et al., 2000). Quinoa has demonstrated the ability to accumulate salt ions in its tissues in order to control and adjust leaf water potential (Jacobsen et al., 2003). This enables the plants to maintain cell turgor and limit transpiration under saline conditions, avoiding physiological damage from drought and thus potential death.

Salinity also influences a number of biological processes during establishment of the seedling, such as ionic and osmotic homeostasis, pigment synthesis, carbon partitioning, photosynthesis, lipid and protein synthesis, and overall plant metabolism and growth (Prado *et al.*, 2000; Ruffino *et al.*, 2010; Adolf *et al.*, 2012a).

Salinity and its impact on agromorphological traits in quinoa

A number of studies have indicated that quinoa displays a high degree of genetic distancing and variable tolerance to salinity.

Jacobsen *et al.* (2001) evaluated the level of salt tolerance in quinoa on application of saline water throughout the entire growth period and estimated the effect of the salt on various characteristics. Plants were treated with nine levels of salinity: electrical conductivity (EC) at 25°C of <1, 5, 10, 15, 20, 25, 30, 35 and 40 mS/cm. Highly significant differences were observed between cultivars, and between cultivar and salinity levels. Stomatal conductance was very sensitive to salinity, as were leaf area, plant height, root and stem weight, leaf water potential and seed yield. Stem diameter and harvest index were the variables least affected by salinity. The highest seed yield was obtained at an electrical conductivity of 15 mS/cm. Some traits showed better responses under moderate saline conditions (10–20 mS/cm) than under lower electrical conductivity, indicating that quinoa is a facultative halophyte.

Bhargava *et al.* (2003) evaluated the promising genotypes of quinoa on normal and sodic soil to compare the yield potential, variability, genetic parameters and genetic association among the different component traits and their direct and indirect effects on yield. Four genotypes were sown on sodic soil that belonged to the family of Aeric Halaquepts, having a silty loam texture at the surface with pH range from 8.6 to 10.0 and electrical conductivity seldom above 2 dS/m. Data were recorded for seven traits: plant height, stem diameter, primary branches/plant, inflorescence/plant, inflorescence length, dry weight of plant and grain yield/plant. High heritability and moderate genetic advance was observed for inflorescence length and grain yield on sodic soil and for stem diameter, primary branches, inflorescence/plant, dry weight of plant and inflorescence length on normal soil. Stem diameter and inflorescence/plant exhibited high direct path (0.837 and 0.761, respectively) and significant positive association (0.979 and 0.967, respectively) with grain yield on sodic soil.

Adolf *et al.* (2012b) assessed varietal differences of quinoa's tolerance to salinity and investigated the physiological mechanisms conferring these differences. Responses to salinity differed greatly between the varieties. Despite the scarcity of information on intra-species differences in salt tolerance in quinoa, the great genetic variability within the species suggests that differences exist with respect to this trait (Adolf *et al.*, 2012b). The variability present in the species may help elucidate the key mechanisms involved in salt tolerance by comparing cultivars of different resistance under saline conditions: this knowledge could be immensely useful for further understanding mechanisms of salt tolerance and facilitating the selection of varieties with improved tolerance.

Gómez-Pando *et al.* (2010) studied the effect of NaCl on the germination of 182 quinoa accessions. All accessions showed less than 60% germination when seeds were irrigated with saline water at 30 dS/m. In contrast, irrigation with 25 dS/m saline water allowed over 60% germination in 15 accessions. The overall coefficients of variation indicated that quinoa genotype and salt treatment dramatically influenced root dry mass per plant, but did not noticeably affect the length of the plant's life cycle. Unexpectedly, salt treatment resulted in increased plant height, leaf dry mass and grain yield. Using Euclidean distance for the simultaneous selection of these five agronomic traits, four accessions proved to be the best performing genotypes under salt stress.

The influence of NaCl on the germination rate, growth and soluble sugar content in quinoa seeds, cotyledons and embryonic axes during the first phases of germination was investigated in detail by Prado *et al.* (2000). The germination of seeds decreased markedly under saline conditions. In the presence of NaCl, maximum germination occurred later than in the control, the extent of delay depending on the salt concentration used. The percentage of germination was only 14% after 14 hours in the presence of 0.4 mM NaCl, which was in sharp contrast to 87% germination for the control over the same duration. The percentage of aborted seeds after 14 hours in NaCl was lower than in distilled water (7% and 16%, respectively). A high percentage (67%) of the ungerminated seeds from saline treatment germinated after washing with distilled water (Prado *et al.*, 2000). Total soluble carbohydrate content changed under salinity stress in embryonic axes and cotyledons during germination and seedling growth. Under saline conditions the sugar content in both embryonic axes and cotyledons decreased remarkably in the first 6 hours, and then increased between 6 and 14 hours, reaching maxima at 10 hours and 14 hours, respectively. The more substantial differences were observed in glucose and fructose contents. Glucose and fructose content in two tissues in distilled water exhibited a marked increase from the initial germination stage up to 10 hours, with 4.3- and 4.9-fold increases for glucose and fructose, respectively. In cotyledons, the glucose and fructose contents showed only slight differences between the two tissues. The total soluble sugar/dry weight ratio in embryonic axes and cotyledons during the first 6 hours of seed germination increased dramatically in controls and then decreased rapidly. Under salt stress, both tissues showed an initial decrease in total soluble sugar/dry weight ratios, followed by recovery in later stages (Prado *et al.*, 2000).

Salinity sensitivity studies of quinoa were conducted in the greenhouse on the cultivar 'Andean Hybrid' to determine if quinoa had useful mechanisms for salt-tolerance studies (Wilson *et al.*, 2002). Salinity treatments ranging from 3, 7, 11 to 19 dS/m were achieved by adding $MgSO_4$, Na_2SO_4, NaCl and $CaCl_2$ to the base nutrient solution. These salts were added incrementally over a 4-day period to avoid osmotic shock to the seedlings. Solution pH was uncontrolled and ranged from 7.7 to 8.0. With respect to salinity effects on growth in quinoa, no significant reduction was observed in plant height, leaf area or fresh weight until the electrical conductivity exceeded 11 dS/m. In fact the plants exhibited an increase in leaf area at 11 dS/m as compared with 3 dS/m controls, which confirmed that the growth was characteristic of a halophyte (Wilson *et al.*, 2002).

Koyro and Eisa (2008) investigated the influence of salinity on seed composition. Ion accumulation in seeds was determined at various salinity levels (grown at 0, 100, 200, 300, 400 and 500 mol/m^3) to ascertain if quinoa regulated accumulation of growth-inhibiting concentrations of sodium and chloride ions. The results showed that quinoa was able to complete its life cycle and produce seeds even at about 500 mol/m^3 NaCl, which is comparable to sea water. However, growth was significantly reduced in the presence of high salt concentrations. An interested finding was that although salinity affected seed size and seed structure, the seeds of salt-treated quinoa plants were better adjusted to germination in saline media than seeds from plants that had not previously experienced salinity. Yield (total fresh weight of seeds in g/plant), number of seeds (n), fresh weight of seeds and dry weight of seeds decreased at high NaCl salinity. Salinity also induced significant reduction in seed size. High salinity affected the embryo mainly by a reduction of the amplitude of the leaf lamina and of the leaf surface (higher leaf mass to area ratio), suggesting a reduced capacity for the storage of abundant proteins and lipids in the leaf. Salinity did not affect the integrity of the seed cover (seed coat and perisperm) or lead to a reduction of the thickness of the pericarp and the seed coating. The reduction in dry matter was compensated by an increase of the water and ash content. The concentration of proteins and total nitrogen increased significantly in the seeds, whereas the concentration of total carbohydrates decreased remarkably, leading to the overall decreased C/N ratio (Koyro and Eisa, 2008). There was a stable accumulation of K and other essential nutrients such as P, Mg and S even at high salinity. There were high concentrations of the cations K, Ca and Mg in the pericarp and the anions S and P in the seed interior (perisperm, cotyledons and hypocotyl). At high salinity the seed coat hindered the passage of NaCl to the seed interior. Na and Cl were located primarily in the pericarp. K was present in all tissues, including the pericarp at high salinity and higher than Na (except seed coat), leading to a higher K/Na ratio in the seed interior. The results indicate a highly protected seed interior that leads to high salinity resistance in quinoa seeds (Koyro and Eisa, 2008).

Compatible solutes or osmoprotectants are compounds involved in osmoregulation during salt stress and have been shown to be involved in salt stress in stress-tolerant plants (McNeil *et al.*, 1999; Trinchant *et al.*, 2004; Chen *et al.*, 2007). The compatible solutes comprise a wide range of organic compounds, such as simple sugars (fructose and glucose), sugar alcohols (glycerol and methylated inositols), complex sugars (trehalose, raffinose and fructans), polyols, quaternary ammonium compounds (proline, glycine betaine, â-alanine betaine, proline betaine) and tertiary sulfonium compounds (Rhodes and Hanson, 1993; Nuccio *et al.*, 1999). Osmoprotectants buffer the effects of salt in a variety of ways. When accumulated in high amounts, osmoprotectants can offset the osmotic imbalance caused by a high accumulation of salt in the intercellular space. As salt is excluded from the cell, it builds up in the supernatant, which causes water to move out of the cell. Compatible solutes are accumulated in the cytoplasm in response to high salt concentrations outside the cell and prevent cellular water loss by balancing the osmotic potential (Yancey, 1994). Osmoprotectants can also provide enzyme protection and maintain membrane integrity under salt stress (Sakamoto and Murata, 2002). One type of osmoprotectant is glycine betaine, an amphoteric compound, electrically neutral over a wide range of pH and extremely soluble in water, allowing it to interact with both hydrophilic and hydrophobic regions of macro-molecules (Sakamoto and Murata, 2002). Glycine betaine is accumulated in many halophytic species in order to balance the osmotic potential difference between vacuole and the cytoplasm (Flowers *et al.*, 1977). The accumulation of glycine betaine in halophytes is induced by salt stress and increases with a raise of salinity (Khan *et al.*, 1998, 2000) in the halophyte. One of glycine betaine's derivatives, trigonelline, has also been identified, though not quanti-fied, in quinoa seeds (Dini *et al.*, 2006) and may act as an important compo-nent of salt tolerance in the plant.

Ruffino *et al.* (2010) investigated the effect of salinity on the establish-ment of quinoa seedlings. Growth, water status, photosynthetic pigments, soluble sugars, proline, glycine, betaine, protein and ion content were ana-lysed in cotyledons of growing seedlings to investigate the hypothesis that quinoa seedlings are well adapted to grow under salinity because of their abil-ity to adjust the metabolic functionality of their cotyledons. Seedlings were grown for 21 days at 250 mM NaCl from the start of germination. The final germination percentage of quinoa seeds was not affected by salt treatment, however, physiological and metabolic parameters of cotyledons in young seed-lings were altered. Relative water content, chlorophyll, carotenoids, lipids and proteins were significantly lower under salinity. Total soluble sugars, sucrose and glucose concentrations were higher in salt-treated than in control cotyle-dons. The content of Na^+ and Cl^- in cotyledons of quinoa significantly increased in the presence of 250 mM NaCl, whereas K^+ content showed a slower decrease. The cotyledons of salt-treated quinoa seedlings accumulate more proline than glycine betaine. However, the reported levels for both proline and glycine betaine (7–13 mM and 3–6 mM, respectively) appear to be far too low to contribute significantly to cell osmotic adjustment. It was concluded that high adaptability to soil salinity of quinoa seedlings is a consequence of better metabolic control, as compared to non-halophytic species, based on the coty-ledon's functionality, ion absorption, osmolyte accumulation and osmotic adjustment (Ruffino *et al.*, 2010).

In a more recent study, ionic and osmotic relations in quinoa were studied by exposing plants to six salinity levels (0–500 mM NaCl) for 70 days (Hariadi *et al.*, 2011). Quinoa plants responded to salinity treatment in a manner typi-cal of halophytes (Fig. 8.1). Optimal plant growth was achieved at NaCl con-centrations of 100 mM, regardless of whether salt stress was administered abruptly or by a gradual increase in NaCl levels. The results showed that quinoa possesses a very efficient system to adjust osmotically for abrupt increases in NaCl stress. Up to 95% of osmotic adjustment in old leaves and 80–85% of osmotic adjustment in young leaves was achieved by means of accumulation of inorganic ions (Na^+, K^+, and Cl^-) at these NaCl levels, while the contribution of organic osmolytes was very limited. Consistently higher K^+ and lower Na^+ levels were found in young compared with old leaves, for all salinity treatments. The shoot sap K^+ progressively increased with increased salinity in old leaves;

Control   100 mM   200 mM   300 mM   400 mM   500 mM

Fig. 8.1. Quinoa plants (genotype 5206) grown at various NaCl levels for 70 days. [Reprinted from Hariadi *et al.* (2011), with permission from Oxford University Press.]

this was interpreted as evidence for the important role of free K^+ in leaf osmotic adjustment under saline conditions. A 5-fold increase in salinity level (from 100 mM to 500 mM) resulted in only a 50% increase in the sap Na^+ content, suggesting either very strict control of xylem Na^+ loading or efficient Na^+ removal from leaves. Both these mechanisms have often been discussed as key determinants of salinity tolerance in various plant species (Munns and Tester, 2008). A very strong correlation between NaCl-induced K^+ and H^+ fluxes was observed in quinoa root, suggesting that a rapid NaCl-induced activation of H^+-ATPase was needed to restore otherwise depolarized membrane potential and prevent further K^+ leak from the cytosol. This work emphasized the role of inorganic ions for osmotic adjustment in halophytes and called for more in-depth studies of the mechanisms of vacuolar Na^+ sequestration, control of Na^+ and K^+ xylem loading, and their transport to the shoot. It was concluded that osmotic adjustments of quinoa are primarily the result of balanced levels of inorganic ions across cellular membranes.

The physiological responses to salt stress and growth response of quinoa were compared to the model halophyte, *Thellungiella halophila* (salt cress) under varying salt treatments (Morales *et al.*, 2011). In a greenhouse experiment, NaCl was applied to quinoa varieties, Chipaya and KU-2, and to *T. halophile*, to assess their relative responses to salt stress. Height and weight data from a 7-week period demonstrated that both quinoa cultivars exhibited greater tolerance to salt stress than the model plant *T. halophila* in these specific conditions (Fig. 8.2). In a separate growth chamber experiment, selected quinoa cultivars were grown hydroponically and evaluated for physiological responses to different salt stress treatments (Morales *et al.*, 2011).

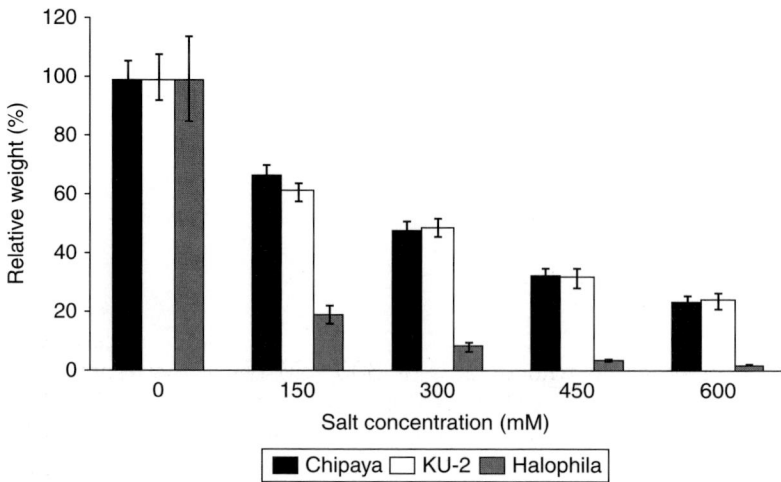

Fig. 8.2. Quinoa and *T. halophila* height and weight relative to their respective control. Plants' height and weight were combined over all time points and the average weight relative to the control was calculated. Bars represent standard error for each treatment/plant combination; n=35. (Source: Morales *et al.*, 2011.)

Tissues collected from the growth chamber experiments were used to obtain leaf water content, tissue ion concentrations, compatible solute concentrations and RNA for real-time PCR. Quinoa was able to accumulate salt using water retention mechanisms such as decreased stomatal conductance, sequestration and compatible solutes. Betaine, trehalose and trigonelline were shown to increase in response to salt as a coping mechanism in the species, with trigonelline appearing to have the greatest correlation. Trigonelline production in quinoa was very high (5000–7000 µmol/g dry weight), which likely had a significant impact on the soil osmotic gradient across the membrane. The accumulation of high levels of trigonelline, a known osmoprotectant, in the high salt treatment suggested a key role in salt tolerance of quinoa. The expression profiles of genes involved in salt stress showed constitutive expression in leaf tissue and up-regulation in root tissue in response to salt stress. These data suggested that quinoa tolerates salt through a combination of salt exclusion and accumulation mechanisms.

The physiological responses to abiotic stress in quinoa were analysed to verify whether they participated in the salt stress response of the plant (Orsini *et al.*, 2011). It was observed that *in vitro* seed germination was initially delayed by a 150 mM NaCl treatment but eventually reached the same level as the control (0 mM NaCl), whereas seedling root growth was enhanced; both parameters were moderately inhibited (~35–50%) by 300 mM NaCl. In pot experiments, plant size was reduced on increase in salinity (0–750 mM NaCl). Transpiration and stomatal conductance were decreased at the highest salinity levels tested in conformation with reduced stomatal density and size. Tissue contents of Na⁺ and Cl⁻ increased with salt treatment, but resulted in only a

50% increase in Na$^+$ from 150 to 750 mM NaCl. Internal K$^+$ was unaffected up to 450 mM NaCl, but increased at the highest salinity levels tested. Proline accumulation and a concomitant reduction in total polyamines and putrescine efflux occurred in NaCl-treated plants. The results were significant since they confirm the importance of inorganic ions for osmotic adjustment, the plant's ability to maintain K$^+$ levels and the involvement of putrescine efflux in maintaining ionic balance under conditions of high salinity. However, proline and ion excretion appeared to play a minor role.

Four genotypes from coastal central and southern regions of Chile were compared for their growth, physiological and molecular responses to NaCl at seedling stage (Ruiz-Carrasco *et al.*, 2011). Seeds were sown on agar plates supplemented with 0, 150 or 300 mM NaCl. Germination was significantly reduced by NaCl only in one accession, BO78. Shoot length was reduced by 150 mM NaCl in three out of four genotypes, and by over 60% at 300 mM. Root length was hardly affected or was even enhanced at 150 mM in all four genotypes, but inhibited, especially in BO78, by 300 mM NaCl. Thus, the root/shoot ratio was differentially affected by salt, with the highest values in PRJ, and the lowest in BO78. Biomass was also less affected in PRJ than in the other accessions, the genotype with the highest increment in proline concentration on salt treatment. Free putrescine declined dramatically in all genotypes under 300 mM NaCl; however (spermidine + spermine)/putrescine ratios were higher in PRJ than BO78. Quantitative RT-PCR analyses of two sodium transporter genes, *CqSOS1* and *CqNHX*, revealed that their expression was differentially induced at the shoot and root level, and between genotypes, by 300 mM NaCl.

Salinity stress and dehydrins in quinoa

Dehydrins are part of a large group of highly hydrophilic proteins known as LEA (Late Embryogenesis Abundant). Dehydrins are also known as group II LEA or RAB (responsive to ABA) proteins (Rorat, 2006; Battaglia *et al.*, 2008). The accumulation of dehydrins is one of the prominent components of plant adaptation to severe environmental conditions. LEAs have been found in many plants in nearly all vegetative tissues under both normal and stressed growth conditions, such as dehydration, cold and high salinity (Allagulova *et al.*, 2003; Rorat, 2006). Some recent reports have confirmed the presence of dehydrins in quinoa seeds. Carjuzáa *et al.* (2008) characterized the dehydrin content in mature embryos of two quinoa cultivars adapted to two contrasting environments, viz. high altitude vs sea level. The results showed the presence of several dehydrin bands at molecular masses of approximately 30, 32, 50 and 55 kDa in both the cultivars. Burrieza *et al.* (2012) studied the effect of salt on the dehydrin content of mature embryos of *C. quinoa* cv. Hualhuas plants grown at salinities ranging from 0 to 500 mM NaCl, with salt concentration being the only variable parameter. Dehydrins were analysed in quinoa embryos by both western blot and *in situ* immunolocalization. Western blot analysis detected at least four dehydrins (55, 50, 34 and 30 kDa) in seeds harvested from quinoa salt-stressed plants treated under a wide range of salinities. The 30 kDa dehydrin increased its accumulation in high salt stress (300 and 500 mM NaCl) as revealed by densitometric analyses. Dehydrin subcellular localization was mostly nuclear at 500 mM of NaCl. A phosphatase treatment of protein extracts caused a mobility shift of the 34 and 30 kDa dehydrin bands, which opens the possibility for a modulation of the dehydrin function/localization based on the phosphorylation of these inducible isoforms. *In situ* immunolocalization of dehydrin-like proteins in meristematic tissues, as well as in cotyledons and axis tissues, showed that the accumulation of dehydrins was elevated in all tissues under increasing salinity, but dehydrin accumulation was less in meristematic tissues compared with cotyledons or axis tissues. In all tissues, dehydrins were detected in both the nucleus and cytoplasm.

8.3.2   Drought resistance in quinoa

Drought is one of the greatest worldwide environmental constraints for agriculture and will be a major abiotic factor affecting global crop yields (Sharma and Lavanya, 2002; Umezawa *et al.*, 2006; Manavalan *et al.*, 2009). One-third of the world's population resides in water-stressed regions, and with elevated CO_2 levels in the atmosphere and climatic changes predicted in the future, drought could become more frequent and severe.

Plants use various mechanisms to cope with drought stress and are classified into three groups: drought escape, drought avoidance and drought tolerance (Turner *et al.*, 2001). Drought escape allows the plant to complete its life cycle during the period of sufficient water supply, i.e. before the onset of drought (Manavalan *et al.*, 2009). In these plants the life cycle is short and plants set some seeds instead of complete crop failure. Drought avoidance involves strategies that help the plant maintain high water status during periods of water stress, either by efficient water absorption from roots or by reducing evapotranspiration from aerial parts (Manavalan *et al.*, 2009). Drought tolerance allows the plant to maintain turgor and continue metabolism even at low water potential (Chaves *et al.*, 2003). This can be accomplished by protoplasmic tolerance or synthesis of osmoprotectants, osmolytes or compatible solutes (Nguyen *et al.*, 1997).

In large parts of the Andean region, water shortage arises from a combined effect of low rainfall, a relatively high evapotranspiration rate and poor soils with low water retaining capacity (Jacobsen *et al.*, 2003; Geerts, 2008). Drought occurs both as intermittent drought, which is highly unpredictable from year to year, and as terminal drought (Jacobsen *et al.*, 2009). Early drought after emergence may lead to a re-sowing and cause an increased risk of experiencing drought conditions during seed maturation, a delayed harvest and crop loss (Garcia *et al.*, 2007). The agroecosystem of quinoa presents an ancestral technology for management of the crop under stressful conditions. This technique makes it possible to store water in the soil, breaking the surface capillarity. In addition, the Andean farmers have quinoa species with great genetic variability that are adapted to these harsh conditions (Jacobsen *et al.*, 2003). Several mechanisms related to drought resistance are present in quinoa, including drought escape, tolerance and avoidance. Drought escape may appear as an increase in the growth cycle as a response to droughts in early vegetative stages or as early maturity in response to drought stress in later growth stages (Jacobsen and Mujica, 2001; Jacobsen *et al.*, 2003). Thus, maturity is an important characteristic in areas with risk of drought at either the beginning or end of the growth season (Jacobsen *et al.*, 2003). Quinoa tolerates drought mainly through tissue elasticity and low osmotic potential. The plant also avoids negative effects associated with drought through its deep dense root system, reduction of its leaf area by leaf dropping, presence of vesicles containing calcium oxalate that are hygroscopic in nature and reduce transpiration, small and thick walled cells preserving turgor even at severe water losses and dynamic stomatal behaviour (Canahua, 1977; Jensen *et al.*, 2000). The vesicular glands of calcium oxalate in the crop needs special mention. Certain quinoa varieties produce glands filled with calcium oxalate on their leaf surface during the vegetative phase as a drought-tolerant mechanism (Jacobsen and Mujica, 1999; Mujica *et al.*, 2001; Garcia, 2003). The possible hypotheses for their role in drought tolerance are:

- Due to hygroscopicity of calcium oxalate, the glands create an artificial humid boundary layer close to the guard cells of stomata on the leaf surface, which maintains photosynthesis at a high rate (high stomatal conductance) under drought conditions (Garcia, 2003; Geerts, 2008). The humid boundary layer keeps transpiration losses at low levels due to smaller potential difference between the leaf and boundary layer (Geerts, 2008).
- The glands make the leaf surface white in colour, substantially increasing the albedo and reducing crop evapotranspiration under non-stressed conditions and the effects of high direct irradiation in high altitude areas (Geerts, 2008).

Numerous studies have shed light on the stomatal behaviour and leaf water relations in drought stressed quinoa plants. Garcia *et al.* (1991) showed that under irrigation predawn water potential (ψ_1) ranged from -0.5 to -1.0 and in stressed conditions came down to -1.5. The effects of drought on leaf water potential, stomatal conductance, transpiration, photosynthesis rate and crop yield were investigated in quinoa under climatic conditions of an agricultural dry season (Vacher, 1998). It was found that stomatal conductance remained relatively stable with low but on-going gas exchange under very dry conditions and low leaf water potentials. The rapid stomatal closure with an associated reduction in transpiration to limit the water loss that occurred in quinoa is a typical mechanism of drought avoidance. Quinoa maintained high leaf water use efficiency to compensate for the decrease in stomatal conductance and thus optimized carbon gain with a minimization of water losses. Thus, quinoa was drought tolerant under intense water stress (Vacher, 1998).

Jensen *et al.* (2000) studied the effects of soil drying on leaf conductance, net photosynthesis and leaf water relations in quinoa. High net photosynthesis and specific leaf area values during early vegetative growth probably resulted in early vigour of the plant supporting early water uptake and thus tolerance to a following drought. The leaf water relations were characterized by low osmotic potentials and low turgid weight/dry weight ratios during later growth stages sustaining a potential gradient for water uptake and turgor maintenance under high evaporation demands.

Garcia *et al.* (2003) calculated the crop coefficient (Kc) and seasonal yield response factor (Ky) for quinoa based on water use efficiency. The Kc value for quinoa was lower than that reported for cereals, which meant that the evapotranspiration of quinoa is less than that of cereals. The Ky value for quinoa (0.67) was reportedly lower than any other crop. This low Ky value for quinoa indicated that a minor drought stress does not result in a large yield decrease.

Jacobsen *et al.* (2009) undertook a study to ascertain how chemical and hydraulic signalling from the root system controlled gas exchange in a drying soil in quinoa. It was observed that during soil drying, relative stomatal conductance (g_s) and photosynthesis A_{max} (drought stressed/fully watered plants) equalled 1, until the fraction of transpirable soil water (FTSW) decreased to 0.82 ± 0.152 and 0.33 ± 0.061, respectively, at bud formation, indicating that photosynthesis was maintained after stomata closure. The relationship between relative g_s and relative A_{max} at bud formation was represented by a logarithmic function ($r^2 = 0.79$), which resulted in a photosynthetic water use efficiency ($WUEA_{max}/g_s$) of 1 when FTSW > 0.8, and increased by 50% with soil drying to FTSW 0.7–0.4. Mild soil drying slightly increased abscisic acid (ABA) in the xylem. It was concluded that during soil drying, quinoa plants had a sensitive stomatal closure, by which the plants were able to maintain leaf water potential (ψ_l) and A_{max}, resulting in an increase of water use efficiency. Root originated ABA played a role in stomata performance during soil drying. ABA regulation seemed to be one of the mechanisms utilized by quinoa when facing drought-inducing decrease of turgor of stomata guard cells.

Gonzalez *et al.* (2009) undertook a detailed investigation to analyse the response of quinoa to drought and its effect on dry matter partitioning. Plants under drought stress showed higher values of total soluble sugars and glucose than controls, while fructose, sucrose and starch contents showed no significant differences. Leaf area and specific leaf area were also higher in drought-stressed plants. Proline, the osmolyte that accumulates in response to drought, was also reported to accumulate more in drought-stressed quinoa plants as compared to controls, a fact earlier reported by Aguilar *et al.* (2003). They reported that proline content was the highest in varieties of eco-geographic locations with very unfavourable conditions of drought. Proline is thought to oxidate quickly in turgid tissues, whereas its oxidation is inhibited under water-deficit conditions.

Two quinoa landraces (Don Javi and Palmilla) were re-introduced to arid Chile and evaluated for seed saponin content and grain yields under low irrigation (Martinez *et al.*, 2009). Treatments included low (40–75 mm) and high (150–250 mm) irrigation and were distributed across the five cultivation months. Grain yields were not different between landraces under the same treatments. Yields were instead affected by microclimate, irrigation and fertilization. Although higher yields corresponded with higher irrigation, higher yields were also found under low irrigation at the more humid site. It was concluded that re-introduction of quinoa in arid Chile was feasible even under the prevailing conditions of low rainfall and deficient soils, but better yields will need some irrigation and addition of organic matter.

Deficit irrigation

Deficit irrigation is an optimization strategy in which irrigation is applied during drought-sensitive growth stages of a crop (Geerts and Raes, 2009). Outside these periods, irrigation is limited or even unnecessary if rainfall provides a minimum supply of water. Water restriction is limited to drought-tolerant phenological stages, often the vegetative stages and the late ripening period. By restricting the water applications to the drought-sensitive growth stages, yields can be significantly increased and stabilized in years of severe precipitation deficits (Garcia *et al.*, 2003). The practice of deficit irrigation requires a thorough and precise knowledge of crop yield response to water (Fereres and Soriano, 2007). It has been demonstrated that deficit irrigation is a valuable option for yield stabilization of quinoa in large tracts of South America, especially the Bolivian Altiplano, where intra-seasonal dry spells are of large importance (Garcia, 2003; Geerts *et al.*, 2006). The Bolivian Altiplano is a semi-arid to arid region at an average altitude of 4000 m above sea level, with agricultural yield instability due to the harsh cropping conditions (Garreaud *et al.*, 2003; Garcia *et al.*, 2004; Geerts *et al.*, 2006). Nevertheless, quinoa has been cultivated in this region for more than 7000 years (Pearsall, 1992) because it is well adapted to grow under unfavourable soil and climatic conditions (Garcia *et al.*, 2003; Bhargava *et al.*, 2006).

The effects on total quinoa production of concentrated drought stress at various phenological stages were assessed using above-ground mini-lysimeters under controlled conditions in the semi-arid Central Bolivian Altiplano (Geerts *et al.*, 2008a). The deficit irrigation strategy was tested in the field by following local agricultural practices and compared with full irrigation and rainfed treatments. The performance of quinoa was assessed by measuring total seed yield, seed size, harvest index and water use efficiency (WUE). The milky grain phase was observed as being most sensitive to drought stress, followed by the flowering stage. Drought stress after emergence until the 6-leaf stage and from the 6-leaf stage until the 12-leaf stage did not cause lower yields and resulted in equal or higher WUE than full irrigation. A significant, negative linear relation was demonstrated between pre- and post-anthesis actual evapotranspiration (ET$_a$) and WUE, indicating the extra harmful character of droughts during flowering and grain maturation if the crop did not suffer drought stress before flowering. It was concluded that quinoa yields can be stabilized at 1.2 up to 2 Mg/ha with the help of deficit irrigation by applying only half of the irrigation water required for full irrigation. Thus, Geerts *et al.* (2008a) proved that deficit irrigation can be very beneficial for quinoa in the semi-arid Central Bolivian Altiplano.

The effects of deficit irrigation on quinoa yield and WUE, as well as the contribution of water from a shallow water table to seasonal crop evapotranspiration, were assessed in experimental and farmers' fields in the Southern Bolivian Altiplano where water resources are limited and often saline (Geerts *et al.*, 2008b). Rainfed quinoa and quinoa with irrigation restricted to the flowering and early grain maturation were studied during the growing seasons of 2005–2006 and 2006–2007 in a location with (Irpani) and without (Mejillones) water contribution from a shallow water table. It was demonstrated that additional irrigation during flowering and early grain maturation was only beneficial if a basic crop water demand is fulfilled, around 55% of the seasonal crop water requirements. Below this threshold, yields, total water use efficiency and marginal irrigation water use efficiency of quinoa with deficit irrigation were low. Capillary rise from groundwater was assessed using the one-dimensional UPFLOW model of Raes and Deproost (2003). The contribution of water from capillary rise in the region of Irpani ranged from 8 to 25% of seasonal crop evapotranspiration of quinoa, and depended mostly on the depth of the groundwater table and the amount of rainfall during the rainy season. Deficit irrigation with poor quality water and cultivation of crops in fields with a shallow saline groundwater table pose a serious threat for sustainable quinoa farming. When the groundwater table and the irrigation water are saline, deficit irrigation can quickly result in soil salinization if there is insufficient rainfall. Impact assessments based on modelling indicated that in an arid environment, 1 year of fallow was insufficient to leach a significant amount of salts out of the root zone. Farming systems with 2 or more years of fallow are more likely to keep soil salinity under control. On the other hand, irrigation before sowing will not only leach salts out of the root zone, but will also create good initial soil water conditions, resulting in the good establishment of the seedling and early root development. Therefore, although potentially beneficial, deficit irrigation of quinoa in arid regions should be considered with caution.

8.3.3 Cold stress

The majority (>80%) of the Earth's biosphere is cold and exposed to temperatures below 5°C throughout the year (Margesin *et al.*, 2007). Freezing stress occurs in many parts of the world in many forms, such as in the winters of North America, Northern Europe and Asia, and in high altitudes of South America, Asia, Africa and Europe (Jacobsen *et al.*, 2003). Low temperatures, especially freezing temperatures, can dramatically affect plants from cellular to ecosystem scales (Loik *et al.*, 2004). The impact of cold and frost on plant physiology and development is crucial in many aspects of seed germination, leaf appearance, plant–water relations, biochemical changes, biomass accumulation and partitioning (Bois *et al.*, 2006). Environmental conditions such as cold intensity and duration, or canopy structure, and plant physiological and phenological status, which partly rely on previous cold acclimation, all have an effect (Bois *et al.*, 2006). Low temperatures and frost set limits for agricultural crop species and in marginal areas can cause severe yield losses (Margesin *et al.*, 2007).

The Andean region is no exception to cold stress, with many areas being exposed to extreme cold. The harsh freezing conditions in the Andean highlands remain the main hazard for quinoa (Jacobsen *et al.*, 2005). Temperatures below freezing occur over large tracts of the Andean highlands for long durations, and frost at night up to 200 days a year is quite common. For example, the mean night temperature on the Bolivian Altiplano at altitudes between 3600 and 4100 m during the cropping season is around 0°C, with frequent freezing spells (Le Tacon *et al.*, 1992). Also the probability of frost with temperatures below −2°C during the vegetative growth season is greater than 30% (Le Tacon *et al.*, 1992). Frost in the Andean region has two origins:

- White/radiative frost: caused by a local cooling of the air above the ground to below 0°C. It occurs under high relative humidity and high dew-point temperature.
- Black/convective frost: caused by the effect of the penetration of cold air masses from the south, consisting of air at below freezing temperatures. It occurs when the air is dry and the temperature does not reach the dew point temperature.

The periods of major frost risk are at the beginning of the growth season and towards the end when winter begins. Quinoa is one of the few crop species that can tolerate frost to some extent. The plant possesses a supercooling capacity of 5°C, a mechanism that prevents immediate damage by low temperatures (Jacobsen *et al.*, 2003).

The main mechanism for frost resistance in quinoa seems to be its ability to tolerate ice formation in cell walls and subsequent dehydration of the cells, without suffering irreversible damage (Jacobsen *et al.*, 2003). A high content of soluble sugars implies a high level of frost tolerance and causes a reduction in the freezing temperature and the mean lethal temperature (TL50). Soluble sugars, such as fructans, sucrose and dehydrins, may be good indicators of frost tolerance in breeding material of quinoa (Jacobsen *et al.*, 2003). Apart from their role in osmotic adjustment, compatible solutes like sugars also have osmoprotective functions. Due to their specific hydrophilic structure, compatible solutes are capable of replacing water on the surfaces of proteins, protein complexes or membranes, thus preserving their biological functions (Rosa *et al.*, 2009). Most compatible solutes also seem to play an important role in hydroxyl radical scavenging, thus defending seedlings against oxidative damage, which is a common consequence of many abiotic stresses (Morsy *et al.*, 2007). The accumulation of sugars in quinoa was amply demonstrated by Jacobsen *et al.* (2005), who undertook a detailed study to determine the damage caused to quinoa by various intensities and durations of frost under different humidity conditions. The results showed that severity of frost was influenced by relative humidity, with a stronger effect under dry conditions. Under frost, quinoa accumulated soluble sugars and proteins, the high sugar content being associated with greater hardiness.

Phenological stage is important in determining the extent of frost damage in quinoa plants. Minimum temperatures that quinoa can resist in various phenological stages usually define the adaptation of different cultivars to specific agroecological zones (Jacobsen *et al.*, 2005). It was suggested by Canahua and Rea (1979) that quinoa is more susceptible to frost damage at flower bud and anthesis compared with the 5-leaved vegetative stage. Quinoa exhibited frost resistance in the 2- and 5-leaf stage, with no damage from low temperatures. Tapia *et al.* (1979) also stated that even a mild frost of −2°C at anthesis was sufficient to cause serious damage to the crop. Later studies also confirmed that quinoa resists frost without major damage before the flower bud formation stage, but is susceptible to frost during and after anthesis (Limache, 1992).

A comparison of the effect of frost damage on seed yield at different phenological stages has shown that plants at the 2-leaf stage kept at −4°C for 4 h were only slightly affected, with a seed yield 9.2% less than the control grown continuously at 19°C. At the 12-leaf stage and at anthesis, yield decrease at −4°C was as much as 50.7 and 65.7%, respectively (Jacobsen *et al.*, 2003).

Bois *et al.* (2006) examined the effects of temperature on the germination, phenology and growth of several quinoa cultivars originating from the Andean Altiplano of Bolivia. Germination was evaluated in 10 cultivars at temperatures between 2 and 20°C. Plant growth and development were examined under minimum temperatures between 8 and 13°C and maximum temperatures between 20 and 28°C. The thermal time concept was used to compare the various treatments and estimate the phyllochron, as well as the base temperature and optimum temperature for leaf appearance, time to flowering and leaf width growth. Two cultivars at the vegetative stage were compared for night freezing tolerance down to −6°C, recording leaf exotherms and plant survival rate (Bois *et al.*, 2006). The results showed that quinoa seeds germinated rapidly even at low temperatures, with the base temperature for germination lower than 0°C for 9 cultivars out of 10. This is an adaptive feature in the Andean Altiplano where radiative frosts are frequent at the beginning of the cropping season. Following seed germination, successive phenological processes showed a base temperature of 1°C and an optimal temperature of 22°C. At 10°C, which is close to the base temperature of many tropical plants, leaf growth rate in quinoa was still substantial at around 30% of its maximum value. This trait sustains early seedling growth and contributes to the better cold and frost tolerance of quinoa in comparison with other Andean crops. Freezing experiments showed that no plant could survive after 4 h at −6°C, while no serious effect was noted down to −3°C. Leaf exotherms confirmed that ice nucleation occurred between −5 and −6°C in most of the plants, the traditional landrace showing a lower freezing tolerance than the selected line (Bois *et al.*, 2006).

8.3.4   Other stresses

Hail and snow are sporadic and localized and sometimes cause irreversible damage, especially when the crop is near maturity (Jacobsen *et al.*, 2003). When hail occurs in the vegetative phase, leaf and stem may be damaged. However, there are cultivars with a greater tolerance to hail, mainly because of a minor leaf angle and greater thickness and resistance of leaves and stem. Flooding occasionally occurs in rainy years on flat areas. Flooding produces root rot, greatly reducing yields (Jacobsen *et al.*, 2003). Wind affects crop productivity by causing plants to fall, especially in the arid region of the Altiplano and in some inter-Andean valleys. Wind is also responsible for erosion and drying both soil and plants. When quinoa is cultivated in deserts and hot areas, high temperatures can cause flowers to abort and the death of pollen (Jacobsen *et al.*, 2003). Fortunately, the genetic variability of quinoa makes it possible to select cultivars with greater tolerance to each of these environmental factors.

8.4 Concluding Remarks

With the steady increase in human population, especially in developing countries, and the declining availability of new agricultural land, there is an urgent need to confront and minimize the effects of various environmental stresses on plant growth and crop yield. The capability of quinoa to endure various biotic and abiotic stresses makes the crop a potential alternative crop for many stress-affected areas of the world, where these problems have become of increasing significance due to intensified agricultural production (Bhargava *et al.*, 2006). The great potential of this crop has not yet been fully exploited, mainly because of the lack of research on sustainable cropping systems and on management of biotic and abiotic constraints to production (Jacobsen *et al.*, 2003). The improved knowledge of the mechanisms of resistance of quinoa to adverse abiotic factors will help develop techniques for overcoming the constraints imposed by harsh environments existing not only in the Andes but in other regions worldwide (Jacobsen *et al.*, 2003; Bhargava *et al.*, 2006).

References

Adolf, V.I., Jacobsen, S.-E., Liu, F., Jensen, C.R. and Andersen, M.N. (2010) Effects of salinity in two contrasting quinoa cultivars. In: *Memorias III Congreso Mundial de la Quinoa*, 16–19 March 2010, Oruro, Bolivia, p. 110.

Adolf, V.I., Jacobsen, S.-E. and Shabala, S. (2012a) Salt tolerance mechanisms in quinoa (*Chenopodium quinoa* Willd.). *Environmental and Experimental Botany* DOI: http://dx.doi.org/10.1016/j.envexpbot.2012.07.004

Adolf, V.I., Shabala, S., Andersen, M.N., Razzaghi, F. and Jacobsen, S.-E. (2012b) Varietal differences of quinoa's tolerance to saline conditions. *Plant and Soil* 357, 117–129.

Aguilar, P.C., Cutipa, Z., Machaca, E., Lopez, M. and Jacobsen, S.-E. (2003) Variation of proline content of quinoa (*Chenopodium quinoa* Willd.) in high beds (Waru Waru). *Food Reviews International* 19, 121–127.

Allagulova, Ch.R., Gilamov, F.R., Shakirova, F.M. and Vakhitov, V.A. (2003) The plant dehydrins: structure and functions. *Biochemistry (Moscow)* 68, 945–951.

Almansouri, M., Kinet, J.M. and Lutts, S. (2001) Effect of salt and osmotic stresses on germination in durum wheat (*Triticum durum* Desf.). *Plant and Soil* 231, 243–254.

Andersen, S.D., Rasmussen, L., Jensen, C.R., Mogensen, V.O., Andersen, M.N. and Jacobsen, S.-E. (1996) Leaf water relations and gas exchange of field grown *Chenopodium quinoa* Willd. during drought. In: Stolen, O., Pithan, K. and Hill, J. (eds) *Small Grain Cereals and Pseudocereals*. Workshop at KVL, Copenhagen, Denmark.

Battaglia, M., Olvera-Carrillo, Y., Garciarrubio, A., Campos, F. and Covarrubias, A.A. (2008) The enigmatic LEA proteins and other hydrophylins. *Plant Physiology* 148, 6–24.

Bhargava, A., Shukla, S., Katiyar, R.S. and Ohri, D. (2003) Selection parameters for genetic improvement in *Chenopodium* grain on sodic soil. *Journal of Applied Horticulture* 5, 45–48.

Bhargava, A., Shukla, S. and Ohri, D. (2006) *Chenopodium quinoa*: an Indian perspective. *Industrial Crops and Products* 23, 73–87.

Bhatnagar-Mathur, P., Vadez, V. and Sharma, K.K. (2008) Transgenic approaches for abiotic stress tolerance in plants: retrospect and prospects. *Plant Cell Reports* 27, 411–424.

Bois, J.F., Winkel, T., Lhomme, J.P., Raffaillac, J.P. and Rocheteau, A. (2006) Response of some Andean cultivars of quinoa (*Chenopodium quinoa* Willd.) to temperature: effects on germination, phenology, growth and freezing. *European Journal of Agronomy* 25, 299–308.

Boscaiu, M., Lull, C., Lidon, A., Bautista, I., Donat, P., Mayoral, O. and Vicente, O. (2008) Plant responses to abiotic stress in their natural habitats. *Bulletin UASVM, Horticulture* 65, 53–58.

Brady, N.C. and Weil, R.P. (1999) Soil organic matter. In: Brady, N.C. and Weil R.P. (eds) *The Nature and Properties of Soils.* Prentice Hall, Upper Saddle River, New Jersey, pp. 446–490.

Burrieza, H.P., Koyro, H.-W., Tosar, L.M., Kobayashi, K. and Maldonado, S. (2012) High salinity induces dehydrin accumulation in *Chenopodium quinoa* Willd. cv. Hualhuas embryos. *Plant and Soil* 354, 69–79.

Canahua, A. (1977) Observaciones del comportamiento de la quinua a la sequía. In: *Congreso Internacional de Cultivos Andinos*, Ayacucho, Peru, pp. 390–392.

Canahua, A. and Rea, J. (1979) Quinuas resistentes a heladas. In: *Actas del II Congreso Internacional de Cultivos Andinos*, 4–8 June 1979. IICA, Riobamba, Ecuador, pp. 14–35.

Carjuzáa, P., Castellión, M., Distéfano, A.J., del Vas, M. and Maldonado, S. (2008) Detection and subcellular localization of dehydrin-like proteins in quinoa (*Chenopodium quinoa* Willd.) embryos. *Protoplasma* 233, 149–156.

Chaves, M.M., Maroco, J.P. and Pereira, J. (2003) Understanding plant responses to drought: from genes to the whole plant. *Functional Plant Biology* 30, 239–264.

Chen, Z., Cuin, T.A., Zhou, M., Twomey, A., Naidu, B.P. and Shabala, S. (2007) Compatible solute accumulation and stress-mitigating effects in barley genotypes contrasting in their salt tolerance. *Journal of Experimental Botany* 58, 4245–4255.

Chinnusamy, V. and. Zhu, J.-K. (2003) Plant salt tolerance. *Topics in Current Genetics* 4, 241–270.

Christiansen, J.L., Ruiz-Tapia, E.N., Jornsgard, B. and Jacobsen, S.-E. (1999) Fast seed germination of quinoa (*Chenopodium quinoa*) at low temperature. In: *COST 814-Workshop: Alternative Crops for Sustainable Agriculture*, Turku, Finland, pp. 220–225.

Debez, A., Hamed, K.B., Grignon, C. and Abdelly, C.H. (2004) Salinity effects on germination, growth and seed production of the halophyte *Cakile maritime*. *Plant and Soil* 262, 179–189.

Delatorre-Herrera, J. and Pinto, M. (2009) Importance of ionic and osmotic components of salt stress on the germination of four quinua (*Chenopodium quinoa* Willd.) selections. *Chilean Journal of Agricultural Research* 69, 477–485.

Dini, I., Tenore, G.C., Trimarco, E. and Dini, A. (2006) Two novel betaine derivatives from Kancolla seeds (Chenopodiaceae). *Food Chemistry* 98, 209–213.

Djanaguiraman, M., Sheeba, J.A., Shanker, A.K., Devi, D.D. and Bangarusamy, U. (2006) Rice can acclimate to lethal level of salinity by pre-treatment with sublethal level of salinity through osmotic adjustment. *Plant and Soil* 284, 363–373.

Dodd, G. and Donovan, L. (1999) Water potential and ionic effects on germination and seedling growth of two cold desert shrubs. *American Journal of Botany* 86, 1146–1153.

Fereres, E. and Soriano, M.A. (2007) Deficit irrigation for reducing agricultural water use. Special issue on integrated approaches to sustain and improve plant production under drought stress. *Journal of Experimental Botany* 58, 147–159.

Flowers, T.J., Troke, P.F. and Yeo, A.R. (1977) The mechanism of salt tolerance in halophytes. *Annual Review of Plant Physiology* 28, 89–121.

Garcia, M. (2003) Agroclimatic study and drought resistance analysis of quinoa for an irrigation strategy in the Bolivian Altiplano. Dissertationes de Agricultura. Faculty of Applied Biological Sciences, K.U. Leuven, Belgium.

Garcia, M., Vacher, J.J. and Hidalgo, J. (1991) Estudio comparative del comportamiento hidrico de dos variedades de quinua en al altiplano central. In: *Actas del VII Congreso Internacional Sobre Cultivos Andinos*, La Paz, Bolivia, pp. 57–62.

Garcia, M., Raes, D. and Jacobsen, S.-E. (2003) Evapotranspiration analysis and irrigation requirements of quinoa (*Chenopodium quinoa*) in the Bolivian highlands. *Agricultural Water Management* 60, 119–134.

Garcia, M., Raes, D., Allen, R. and Herbas, C. (2004) Dynamics of reference evapotranspiration in the Bolivian highlands (Altiplano). *Agricultural and Forest Meteorology* 125, 67–82.

Garcia, M., Raes, D., Jacobsen, S.-E. and Michel, T. (2007) Agroclimatic constraints for rainfed agriculture in the Bolivian altiplano. *Journal of Arid Environments* 71, 109–121.

Garreaud, R., Vuille, M. and Clement, A.C. (2003) The climate of the Altiplano: observed current conditions and mechanisms of past changes. *Palaeogeography, Palaeoclimatology, Palaeoecology* 194, 5–22.

Geerts, S. (2008) Deficit irrigation strategies via crop water productivity modeling: field research of quinoa in the Bolivian Altiplano. Dissertationes de Agricultura 814. Faculty of Bio-Science Engineering, K.U. Leuven, Belgium.

Geerts, S. and Raes, D. (2009) Deficit irrigation as an on-farm strategy to maximize crop water productivity in dry areas. *Agricultural Water Management* 96, 1275–1284.

Geerts, S., Raes, D., Garcia, M., Del Castillo, C. and Buytaert, W. (2006) Agro-climatic suitability mapping for crop production in the Bolivian Altiplano: a case study for quinoa. *Agricultural and Forest Meteorology* 139, 399–412.

Geerts, S., Raes, D., Garcia, M., Vacher, J., Mamani, R., Mendoza, J., Huanca, R., Morales, B., Miranda, R., Cusicanqui, J. and Taboada, C. (2008a) Introducing deficit irrigation to stabilize yields of quinoa (*Chenopodium quinoa* Willd.). *European Journal of Agronomy* 28, 427–436.

Geerts, S., Raes, D., Garcia, M., Condori, O., Mamani, J., Miranda, R., Cusicanqui, J., Taboada, C., Yucra, E. and Vacher, J. (2008b) Could deficit irrigation be a sustainable practice for quinoa (*Chenopodium quinoa* Willd.) in the Southern Bolivian Altiplano? *Agricultural Water Management* 95, 909–917.

Gómez-Pando, L.R., Álvarez-Castro, R. and Eguiluz-de la Barra, A. (2010) Effect of salt stress on Peruvian germplasm of *Chenopodium quinoa* Willd.: a promising crop. *Journal of Agronomy and Crop Science* 196, 391–396.

Gonzalez, J.A., Gallardo, M., Hilal, M., Rosa, M. and Prado, F.E. (2009) Physiological responses of quinoa (*Chenopodium quinoa* Willd.) to drought and waterlogging stresses: dry matter partitioning. *Botanical Studies* 50, 35–42.

Greenway, H. and Munns, R. (1980) Mechanisms of salt tolerance in non-halophytes. *Annual Reviews in Plant Physiology* 31, 149–190.

Grime, J.P. (1979) *Plant Strategies and Vegetation Process*. Wiley, New York.

Gul, B. and Weber, D. (1998) Effect of dormancy relieving compounds on the seed germination of non-dormant *Allenrolfea occidentalis* under salinity stress. *Annals of Botany* 82, 555–560.

Hariadi, Y., Marandon, K., Tian, Y., Jacobsen, S.-E. and Shabala, S. (2011) Ionic and osmotic relations in quinoa (*Chenopodium quinoa* Willd.) plants grown at various salinity levels. *Journal of Experimental Botany* 62, 185–193.

Huang, J. and Redmann, R.E. (1995) Salt tolerance of *Hordeum* and *Brassica* species during germination and early seedling growth. *Canadian Journal of Plant Science* 75, 815–819.

Jacobsen, S.-E. (2007) Quinoa's world potential. In: Ochatt, S. and Jain, S.M. (eds) *Breeding of Neglected and Under-utilized Crops, Spices and Herbs*. Science Publishers, Enfield, UK, pp. 109–122.

Jacobsen, S.-E. and Mujica, A. (1999) Fisiología de la Resistencia a Sequía en Quinua (*Chenopodium quinoa* Willd.). *I Curso Internacional sobre Fisiologia de la Resistencia a Sequía en la Quinua*, 1–6 December, 1997, Universidad Nacional del Altiplano, Puno, Peru.

Jacobsen, S.-E. and Mujica, A. (2001) Avances en el conocimiento de resistencia a factores abióticos adversos en la quinua (*Chenopodium quinoa* Willd.). In: Jacobsen, S.-E. and Portillo, Z. (eds) *Memorias, Primer Taller Internacional sobre Quinua: Recursos Geneticos y Sistemas de Producción*, 10–14 May 1999. Universidad Nacional Agraria La Molina, Lima, Peru.

Jacobsen, S.-E., Nuñez, N., Stølen, O. and Mujica, A. (1999) Que sabemos sobre la resistencia de la quinua a la sequfa? In: Jacobsen, S.E. and Mujica, A. (eds) *Fisiología de la Resistencia a Sequía en Quinua (Chenopodium quinoa Willd.)*. CIP, Lima, Peru, pp. 65–69.

Jacobsen, S.-E., Quispe, H. and Mujica, A. (2001) Quinoa: an alternative crop for saline soils in the Andes. In: *Scientists and Farmer-Partners in Research for the 21st Century*. CIP Program Report 1999–2000, pp. 403–408.

Jacobsen, S.-E., Mujica, A. and Jensen, C.R. (2003) The resistance of quinoa (*Chenopodium quinoa* Willd.) to adverse abiotic factors. *Food Reviews International* 19, 99–109.

Jacobsen, S.-E., Monteros, C., Christiansen, J.L., Bravo, L.A., Corcuera, L.J. and Mujica, A. (2005) Plant responses of quinoa (*Chenopodium quinoa*) to frost at various phenological stages. *European Journal of Agronomy* 22, 131–139.

Jacobsen, S.-E., Liu, F. and Jensen, C.R. (2009) Does root-sourced ABA play a role for regulation of stomata under drought in quinoa (*Chenopodium quinoa* Willd.). *Scientia Horticulturae* 122, 281–287.

Jensen, C.R., Jacobsen, S.-E., Andersen, M.N., Nunez, N., Andersen, S.D., Rasmussen, L. and Mogensen, V.O. (2000) Leaf gas exchange and water relation characteristics of field quinoa (*Chenopodium quinoa* Willd.) during soil drying. *European Journal of Agronomy* 13, 11–25.

Katambe, W.J., Ungar, L.A. and Mitchell, J.P. (1998) Effect of salinity on germination and seedling growth of two *Atriplex* species (*Chenopodiaceae*). *Annals of Botany* 82, 167–175.

Kerepesi, I. and Galiba, G. (2000) Osmotic and salt stress induced alteration in soluble carbohydrate content in wheat seedlings. *Crop Science* 40, 482–487.

Khan, M.A., Ungar, I.A., Showalter, A.M. and Dewald, H.D. (1998) NaCl induced accumulation of glycine betaine in four subtropical halophytes from Pakistan. *Physiologia Plantarum* 102, 487–492.

Khan, M.A., Ungar, I.A. and Showalter, A.M. (2000) Effects of salinity on growth, water relations and ion accumulation of the subtropical perennial halophyte, *Atriplex griffithi* var. *stocksii*. *Annals of Botany* 85, 225–232.

Koyro, H.-W. and Eisa, S.S. (2008) Effect of salinity on composition, viability and germination of seeds of *Chenopodium quinoa* Willd. *Plant and Soil* 302, 79–90.

Le Tacon, P., Vacher, J.J., Eldin, M. and Imaña, E. (1992) Los riesgos de helada en el Altiplano boliviano. In: Morales, D. and Vacher, J.J. (eds) *Actas del VII congreso internacional sobre cultivos andinos*, 4–8 February 1992. ORSTOM, La Paz, Bolivia, pp. 287–291.

Limache, J. (1992) Tolerancia a heladas de 14 ecotipos y 2 variedades de quinua (*Chenopodium quinoa* Willd.), en Waru-Waru de Caritamaya-Acora. Tesis Ingeniero Agrónomo. Universidad Nacional del Altiplano. Puno, Peru.

Loik, M.E., Still, C.J., Huxman, T.E. and Harte, J. (2004) In situ photosynthetic freezing tolerance for plants exposed to a global warming manipulation in the Rocky Mountains, Colorado, USA. *New Phytologist* 162, 331–341.

Manavalan, L.P., Guttikonda, S.K., Tran, L.S.P. and Nguyen, H.T. (2009) Physiological and molecular approaches to improve drought resistance in soybean. *Plant Cell Physiology* 50, 1260–1276.

Margesin, R., Neuner, G. and Storey, K.B. (2007) Cold-loving microbes, plants, and animals: fundamental and applied aspects. *Naturwissenschaften* 94, 77–99.

Martinez, E.A., Veas, E., Jorquera, C., San Martin, R. and Jara, P. (2009) Re-introduction of *Chenopodium quinoa* Willd. into arid Chile: cultivation of two lowland races under extremely low irrigation. *Journal of Agronomy and Crop Science* 195, 1–10.

McNeil, S.D., Nuccio, M.L. and Hanson, A.D. (1999) Betaines and related osmoprotectants: targets for metabolic engineering of stress resistance. *Plant Physiology* 120, 945–949.

Morales, A.J., Bajgain, P., Garver, Z., Maughan, P.J. and Udall, J.A. (2011) Evaluation of the physiological responses of *Chenopodium quinoa* to salt stress. *International Journal of Plant Physiology and Biochemistry* 3, 219–232.

Morsy, R.M., Jouve, L., Hausman, J.F., Hoffmann, L. and Stewart, J.M. (2007) Alteration of oxidative and carbohydrate metabolism under abiotic stress in two rice (*Oryza sativa* L.) genotypes contrasting in chilling tolerance. *Journal of Plant Physiology* 164, 157–167.

Mujica, A. and Jacobsen, S.-E. (1999) Resistencia de la quinua a la sequía y otros factores abioticos adversos, y su mejoramiento. In: Jacobsen, S.-E. and Mujica, A. (eds) *Fisiología de la Resistencia a Sequía en Quinua (Chenopodium quinoa Willd.)*. CIP, Lima, Peru, pp. 71–78.

Mujica, A., Jacobsen, S.-E. and Cardenas, G. (2000) Selección de cultivares de quinua (*Chenopodium quinoa* Willd.) por su resistencia a sequía. In: *Proc. II Congreso Internacional de Agricultura en Zonas Aridas*, Universidades Arturo Prat, Iquique, Tarapaca, Arica and Antofagasta, 16–21 October 2000.

Mujica, A., Jacobsen, S.-E. and Izquierdo, J. (2001) Resistencia a factores adversos de la quinua. In: Mujica, A., Jacobsen, S.-E., Izquierdo, J. and Marathee, J.P. (eds) *Quinua (Chenopodium quinoa Willd.) – Ancestral Cultivo Andino, Alimento del Presente y Futuro*. FAO, UNA-Puno, CIP, Santiago, Chile, pp. 162–183.

Munns, R. (2002) Comparative physiology of salt and water stress. *Plant Cell and Environment* 25, 239–250.

Munns, R. (2009) Strategies for crop improvement in saline soil. *Salinity and Water Stress* 44, 99–110.

Munns, R. and Tester, M. (2008) Mechanisms of salinity tolerance. *Annual Reviews in Plant Biology* 59, 651–681.

Munns, R., Schachtman, D.P. and Condon, A.G. (1995) The significance of two-phase growth response to salinity in wheat and barley. *Australian Journal of Plant Physiology* 22, 561–569.

Nguyen, H.T., Babu, R.C. and Blum, A. (1997) Breeding for drought resistance in rice: physiology and molecular genetics considerations. *Crop Science* 37, 1426–1434.

Nuccio, M.L., Rhodes, D., McNeil, S.D. and Hanson, A.D. (1999) Metabolic engineering of plants for osmotic stress tolerance. *Current Opinions in Plant Biology* 2, 128–134.

Orsini, F., Accorsi, M., Gianquinto, G., Dinelli, G., Antognoni, F., Carrasco, K.B.R., Martinez, E.A., Alnayef, M., Marotti, I., Bosi, S. and Biondi, S. (2011) Beyond the ionic and osmotic response to salinity in *Chenopodium quinoa*: functional elements of successful halophytism. *Functional Plant Biology* 38, 818–831.

Pearsall, D. (1992) The origins of plant cultivation in South America. In: Cowan, C.W. and Watson, P.J. (eds) *The Origins of Agriculture. An International Perspective*. Smithsonian Institution Press, Washington, DC, pp. 173–205.

Prado, F.E., Boero, C., Gallardo, M. and Gonzalez, J.A. (2000) Effect of NaCl on germination, growth, and soluble sugar content in *Chenopodium quinoa* (Willd.) seeds. *Botanical Bulletin of Academia Sinica* 41, 27–34.

Pujol, J.A., Calvo, J.F. and Ramírez-Díaz, L. (2000) Recovery of germination from different osmotic conditions by four halophytes from south eastern Spain. *Annals of Botany* 85, 279–286.

Raes, D. and Deproost, P. (2003) Model to assess water movement from a shallow water table to the root zone. *Agricultural Water Management* 62, 79–91.

Rao, K.V.M., Raghavendra, A.S. and Reddy, K.J. (2006) *Physiology and Molecular Biology of Stress Tolerance in Plants*. Springer, Dordrecht, The Netherlands.

Rhodes, D. and Hanson, A.D. (1993) Quaternary ammonium and tertiary sulfonium compounds in higher plants. *Annual Review in Plant Physiology and Plant Molecular Biology* 44, 357–384.

Robert-Seilaniantz, A., Bari, R. and Jones, J.D.G. (2010) Abiotic or biotic stresses. In: Pareek, A., Sopory, S.K., Bohnert, H.J. and Govindjee (eds) *Abiotic Stress Adaptation in Plants: Physiological, Molecular and Genomic Foundation.* Springer, New York, pp. 103–122.

Rorat, T. (2006) Plant dehydrins: tissue location, structure and function. *Cellular and Molecular Biology Letters* 11, 536–556.

Rosa, M., Hilal, M., González, J.A. and Prado, F.E. (2009) Low temperature effect on enzyme activities involved in sucrose–starch partitioning in salt-stressed and salt acclimated cotyledons of quinoa (*Chenopodium quinoa* Willd.) seedlings. *Plant Physiology and Biochemistry* 47, 300–307.

Ruffino, A.M.C., Rosa, M., Hilal, M., Gonzalez, J.A. and Prado, F.E. (2010) The role of cotyledon metabolism in the establishment of quinoa (*Chenopodium quinoa*) seedlings growing under salinity. *Plant and Soil* 326, 213–224.

Ruiz-Carrasco, K., Antognoni, F., Coulibaly, A.K., Lizardi, S., Covarrubias, A., Martínez, E.A., Molina-Montenegro, M.A., Biondi, S. and Zurita-Silva, A. (2011) Variation in salinity tolerance of four lowland genotypes of quinoa (*Chenopodium quinoa* Willd.) as assessed by growth, physiological traits, and sodium transporter gene expression. *Plant Physiology and Biochemistry* 49, 1333–1341.

Sakamoto, A. and Murata, N. (2002) The role of glycine betaine in the protection of plants from stress: clues from transgenic plants. *Plant Cell and Environment* 25, 163–171.

Serrano, R. and Gaxiola, R. (1994) Microbial models and salt stress tolerance in plants. *Critical Reviews in Plant Sciences* 13, 121–138.

Sharma, K.K. and Lavanya, M. (2002) Recent developments in transgenics for abiotic stress in legumes of the semi-arid tropics. JIRCAS (Japan International Research Centre for Agricultural Sciences) Working Report, JIRCAS, Japan.

Tapia, M., Gandarillas, H., Alandia, S., Cardozo, A., Mujica, R., Ortiz, R., Otazu, J., Rea, J., Salas, B. and Zanabria, E. (1979) *Quinoa y kañiwa: Cultivos andinos.* CIID-IICA, Bogota, Colombia.

Trinchant, J.C., Boscari, A., Spennato, G., Van De Sype, G. and Le Rudulier, D. (2004) Proline betaine accumulation and metabolism in alfalfa plants under sodium chloride stress. Exploring its compartmentalization in nodules. *Plant Physiology* 135, 1583–1594.

Turner, N.C., Wright, G.C. and Siddique, K.H.M. (2001) Adaptation of grain legumes (pulses) to water-limited environments. *Advances in Agronomy* 71, 193–231.

Umezawa, T., Fujita, M., Fujita, Y., Yamaguchi-Shinozaki, K. and Shinozaki, K. (2006) Engineering drought tolerance in plants: discovering and tailoring genes unlock the future. *Current Opinions in Biotechnology* 17, 113–122.

Ungar, I.A. (1996) Effect of salinity on seed germination, growth, and ion accumulation of *Atriplex patula* (Chenopodiaceae). *American Journal of Botany* 83, 604–607.

Vacher, J.J. (1998) Responses of two main Andean crops, quinoa (*Chenopodium quinoa* Willd.) and papa amarga (*Solanum juzepczukii* Buk.) to drought on the Bolivian Altiplano: significance of local adaptation. *Agriculture, Ecosystems and Environment* 68, 99–108.

Vacher, J., Imana, E. and Canqui, E. (1994). Las características radiativas y la evapotranspiración potencial en el Altiplano Boliviano. *Revista de Agricultura*, Facultad de Ciencias Agrícolas, Pecuarias, Forestales y Veterinarias, Universidad Mayor de San Simón, La Paz, Bolivia.

Wilson, C., Read, J.J. and Abo-Kassem, E. (2002) Effect of mixed-salt salinity on growth and ion relations of a quinoa and a wheat variety. *Journal of Plant Nutrition* 25, 2689–2704.

Yamaguchi, T. and Blumwald, E. (2005) Developing salt-tolerant crop plants: challenges and opportunities. *Trends in Plant Science* 10, 615–620.

Yancey, P.H. (1994) Compatible and counteracting solutes. In: Strange, K. (ed.) *Cellular and Molecular Physiology of Cell Volume Regulation.* CRC Press, Boca Raton, Florida, pp. 81–109.

Zhu, J.K. (2001) Plant salt tolerance. *Trends in Plant Science* 6, 66–71.

Zhu, J.K. (2003) Regulation of ion homeostasis under salt stress. *Current Opinions in Plant Biology* 6, 441–445.

9 Diseases and Pests

9.1 Introduction

Pests and diseases continue to affect the productivity and quality of crops around the world despite many years of research and development on improved methods for their control (Lucas, 2011). It has been estimated that approximately 0.2–0.3% of the crop yield is lost annually from the field (Oerke, 2006), even in crops where pesticides and cultivars with improved genetic resistance to pests and diseases are used. The losses are usually greater in subsistence agriculture, where crop protection measures are often not applied. In subsistence agriculture, either the appropriate measures are not available, or the expertise and infrastructure to diagnose and control pest and disease problems are not in place. In commercial cultivation, the biotic agents of disease are moving targets that evolve in response to agricultural practices and environmental change (Lucas, 2011). The emergence and spread of new pests and diseases, or more aggressive or pesticide-resistant biotypes, are examples of such evolution.

9.2 Diseases

Quinoa is exposed to attack from a range of microbes, with variable intensity, depending on the environment. Viruses are known to infect the plant, but reports of significant damage are absent.

9.2.1 Bacterial diseases

Quinoa is seldom infected by bacteria and instances of bacterial attack have rarely been reported. In the region around Puno, bacterial blight causes significant damage during the grain maturing period. The disease is seed borne and is thought to be caused by *Pseudomonas* spp. (Alandia *et al.*, 1979). The spread of disease in the field is aided by wounds caused by hail, rain and damp soil (Salas *et al.*, 1977; Alandia *et al.*, 1979).

9.2.2 Fungal diseases

Downy mildew
Downy mildew is the most severe pathogen on quinoa worldwide and is known to cause yield reduction of 33–58%, even in the most resistant cultivars (Danielsen *et al.*, 2000). Under conditions favourable for the pathogen, crop losses can be as high as 100%. Mildew in quinoa is caused by *Peronospora farinosa* f.sp. *chenopodii* Byford, an oomycete (egg fungus). Oomycetes are a diverse group of eukaryotic organisms that have colonized many ecological niches (Thines and Kamoun, 2010). More than 60% of the known oomycete species are parasites of plants, and some, such as the potato late blight agent *Phytophthora infestans*, are reported to cause diseases of great importance to mankind. Recently, oomycetes have been excluded from the fungi kingdom due to differences in the composition of the cellular wall and its ploidy level. Oomycetes have most likely evolved in a marine environment, as the most basal lineages of the Oomycota are predominantly marine parasites of seaweeds, crustaceans, diatoms and nematodes (Thines and Kamoun, 2010). Despite their great genetic divergence, these basal lineages account for only about 5% of oomycete species, whereas most species are parasitic on plants mostly members of the Asteraceae, Brassicaceae, Cucurbitaceae, Rosaceae, Solanaceae, Chenopodiaceae and hundreds more dicotyledonous families. Parasitism of plants has evolved at least three times independently in different lineages of the Oomycota (Fig. 9.1).

P. farinosa is a highly polyphyletic, unresolved group, which is undergoing major taxonomic revision (Danielsen and Lübeck, 2010). The fungus belongs to the family Peronosporaceae, order Peronosporales whose members are highly specialized obligate parasites (biotrophs) that parasitize vascular plants causing mildew in a limited range of species (Danielsen and Ames, 2004). The pathogen is subdivided into three groups based on cross-inoculation experiments (Byford, 1967): (i) *P. farinosa* f.sp. *betae* on *Beta* spp.; (ii) *P. farinosa* f.sp. *spinaciae* on *Spinacia* spp.; and (iii) *P. farinosa* f.sp. *chenopodii* on *Chenopodium* spp.

This classification is also supported by other workers because sugar beet mildew does not infect quinoa under greenhouse conditions (Johanson, 1983) and quinoa mildew does not infect canihua, spinach or beetroot (Alandia *et al.*, 1979).

SYMPTOMS OF DOWNY MILDEW. Downy mildew attacks the foliage of the plant causing yellowing or reddening of the leaves, depending on the genotype, and in extreme cases eventually results in 100% defoliation (Danielsen and Ames, 2000; Kitz, 2008). Soft, grey patches of sporangia usually emerge on the underside of the leaves acting as the primary source of inoculum, which is

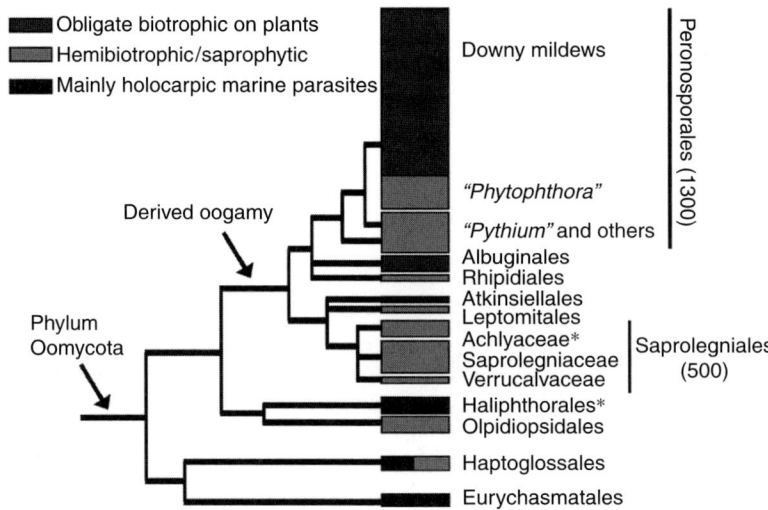

Fig. 9.1. Schematic overview of the current knowledge on oomycete evolutionary history. Number of species in two largest orders are indicated in parentheses. Names of genera in inverted commas denote taxa in need of taxonomic revision. Taxa that are not yet formally introduced are marked with an asterisk. [Reprinted from Thines and Kamoun (2012), with permission from Elsevier.]

spread by wind and rain. The disease spreads through sporangia found in lesions on the underside of leaves, and through oospores. These structures can remain dormant for long periods of time in soil and quinoa seeds. Reduced photosynthetic ability weakens the plant and halts seed production. Defoliation is the major cause of yield loss, which in some cultivars the loss can be as much as 99% (Danielsen *et al.*, 2000). Downy mildew of quinoa requires specific conditions for germination and infection. It proliferates in humidity above 80% and in cool temperatures between 15 and 20°C (Danielsen and Ames, 2000). During dry and warm conditions, the infection lies dormant as oospores in quinoa seeds and surrounding soil. It can also survive in host tissues until environmental conditions are favourable for continued growth and spore production.

Downy mildew on quinoa was first reported in 1947 by Garcia in Peru. The pathogen is endemic in Bolivia, Chile, Colombia, Ecuador and Peru (Alandia *et al.*, 1979; Aragón and Gutiérrez, 1992), although it has been reported in North America, Europe and Asia (Tewari and Botyetchko, 1990; Danielsen *et al.*, 2002; Pańka *et al.*, 2004; Kumar *et al.*, 2006; Danielsen and Lübeck, 2010). In most of the Andean region the climate is ideal for the growth of mildew during the rainy months (October to April) (Danielsen and Ames, 2000). The highlands in the south of Bolivia, near the salted lakes, are an exception where the annual precipitation is so low that the mildew is rarely present. High humidity and moderate temperatures also exist in Denmark, where downy mildew is frequently found on *C. album*, a closely related and very common weed species (Danielsen and Lübeck, 2010).

The downy mildew infection in quinoa plants was first reported in Canada by Tewari and Botyetchko (1990). The diseased leaves showed chlorotic lesions (Fig. 9.2). The dichotomously branched conidiophores culminated in pointed, slightly curved branches that produced conidia. The conidiophores ranged from 175 to 425 µm in height and were 8–14 µm wide at the bases. The conidia were pyriform and 28.6–36.7 × 16.3–26.5 µm in size (Fig. 9.2).

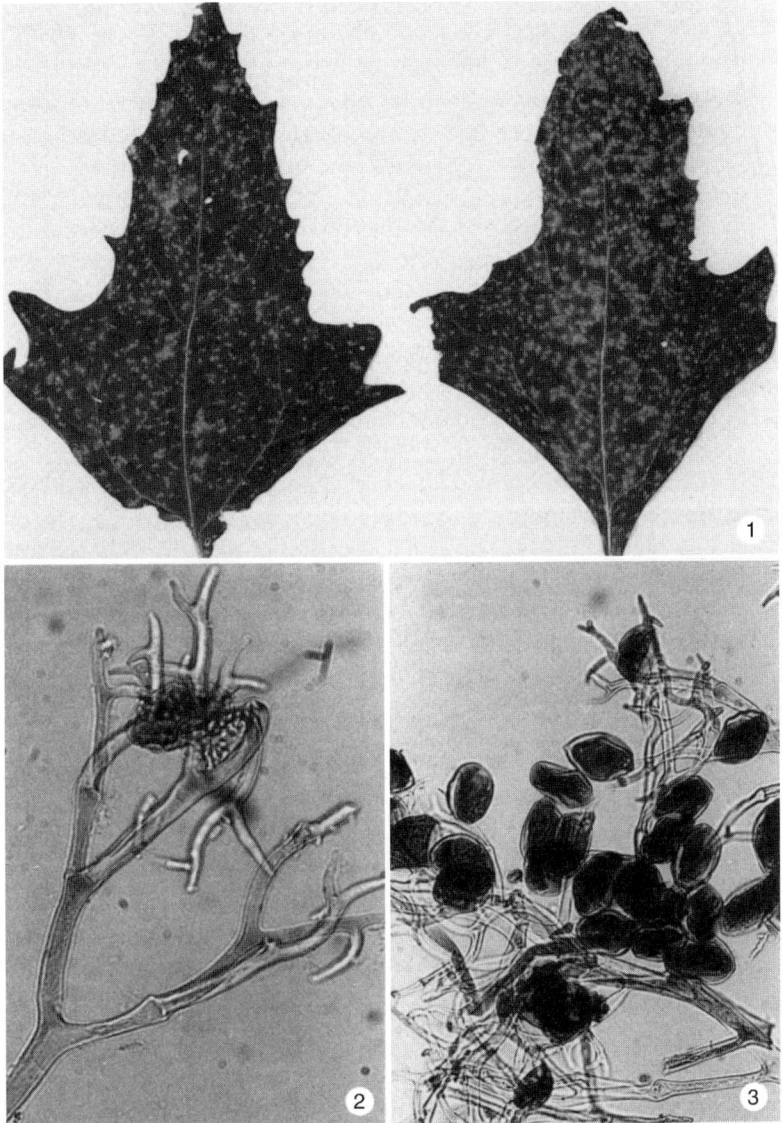

Fig. 9.2. Downy mildew on quinoa. (1) Chlorotic lesions on the upper leaf surfaces; (2) dichotomously branched conidiophore of *P. farinosa*; (3) branches of conidiophore and pyriform conidia of *P. farinose*. [Reprinted from Tewari and Boyetchko (1990), with permission from Canadian Plant Disease Survey.]

The nature and types of resistance to downy mildew in quinoa were studied by Ochoa et al. (1999) in Ecuador using 20 downy mildew isolates. A scale for scoring quinoa downy mildew infections was developed in which six different reaction types could be distinguished, ranking from fully resistant (0) to completely susceptible (5). Of the 60 lines studied, 22 showed susceptibility to all isolates and 29 showed mixed reactions, i.e. the reaction types of the leaves of the three seedlings of one line differed markedly from one another. In Ecuador, the distribution of virulence groups appeared to be restricted to the different quinoa growing regions. In the south, where quinoa was not extensively cultivated, the isolates collected belonged only to virulence group 2, which was less virulent (Ochoa et al., 1999). In the central region, where both landraces and newly released cultivars are grown, only isolates of virulence group 4 were collected. In northern Ecuador, which is the traditional quinoa growing region, isolates of all four virulence groups were collected. The virulence groups appeared to be restricted to those areas in which suitable plant genotypes were present. The range of virulence present in the northern region suggested that the fungus was able to adapt to new resistance factors. Resistance factor R3 was found most frequently in seedlings of 13 high-yielding Ecuadorian quinoa lines. No effective resistance against group 4, the most virulent group, was observed. The presence of R3 in many of the lines indicated that new resistance factors were being used in the quinoa breeding programme (Ochoa et al., 1999).

A field experiment with eight quinoa cultivars was carried out with and without application of fungicides to quantify the impact of the disease on grain yield under natural conditions (Danielsen et al., 2000). Area under the disease progress curve (AUDPC) values were calculated based on evaluations of disease severity (percentage leaf area affected). The cultivar Utusaya, originating from the Bolivian salt desert (200 mm annual rainfall), was strongly affected by downy mildew, which caused complete defoliation, premature maturation and a yield loss of 99%. Even in the most resistant cultivar, the yield was reduced by 33%, indicating the destructiveness of this disease. The correlation analysis showed a strongly significant negative correlation between AUDPC and yield. The study supported the view that late cultivars in general are more resistant to downy mildew than early cultivars.

Danielsen and Munk (2004) tested seven disease assessment methods to measure downy mildew severity in quinoa in eight cultivars in two field experiments in Peru. Two levels of downy mildew were established by means of fungicide treatments. AUDPC values were calculated for each method and correlated to the grain yield. AUDPC based on the three-leaf method (mean disease severity of three leaves per plant selected randomly from the lower, middle and upper part of the plant) showed the highest negative correlation to yield ($r = -0.736$) and was regarded as the quickest, easiest and best method to predict yield loss caused by downy mildew.

Kumar et al. (2006) first reported the presence of downy mildew on quinoa plants cultivated in India. Thirty-four accessions comprising 27 of C. quinoa, two of C. berlandieri subsp. nuttalliae and one each of C. bushianum, C. ugandae, C. strictum, C. ficifolium and C. opulifolium were evaluated and screened for downy mildew disease resistance under natural epiphytotic field conditions in the subtropical climate of mid-Eastern India. Two different methods of disease assessment were tested and compared based on AUDPC. In the first method, AUDPC (cal) was calculated utilizing all data points gathered over the entire epidemic, while in the second method AUDPC (est) was estimated using only two assessment dates. The progress curve was noted as bell shaped in both highly resistant and highly susceptible accessions, with appearance of disease symptoms 30 days after sowing (DAS) (Fig. 9.3). It was recognized as the start of the epidemic and showed a gradual increase up to assessment six (65 DAS) when it attained peak severity. Later, the curve gradually declined to assessment 12 (107 DAS). The accessions were resolved by both methods into classes ranging from highly susceptible to immune, suggesting the presence of both major gene resistance and rate-reducing resistance among them. The two-point method of disease assessment was found suitable and easy and could be employed for future screening programmes in the region. Apart from accessions of other species, four accessions of quinoa were

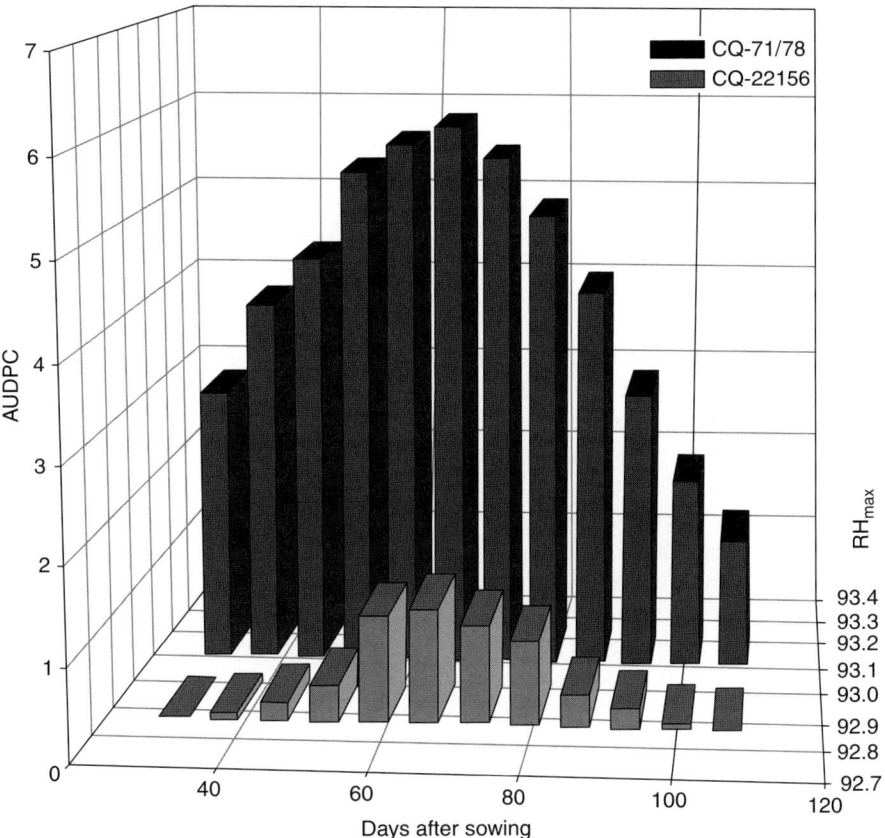

Fig. 9.3. AUDPC depicting highly susceptible and highly resistant accession *C. quinoa.* [Reprinted from Kumar *et al.* (2006), with permission from Elsevier.]

found immune/resistant to downy mildew, suggesting physiological specialization of current pathotypes for *C. quinoa* (Kumar *et al.*, 2006). The naturally
existing isolates of the pathogen showed no cross pathogenicity with wild species that appeared immune to downy mildew, a fact supported by Aragón and
Gutiérrez (1992). The identification of the critical growth period attaining peak
severity/epidemic was solved through a disease severity–temporal relationship
model at flower bud initiation stage. The study concluded that four accessions
of quinoa could be ideal sources for the introgression of resistance genes in
high-yielding but downy mildew disease-susceptible accessions through backcross breeding or molecular approaches.

The susceptibility of selected quinoa cultivars and lines to downy mildew
infection was assessed in Polish conditions by Pańka *et al.* (2004). Significant
variability was recorded in the infection of 24 quinoa cultivars and lines with
P. farinose. The infection index ranged from 12.5 to 80%. However 16 out of
24 cultivars studied showed a low insignificant infection. The group included all
the European cultivars and lines that showed their relatively high resistance to
infection with *P. farinosa*. The cultivars that originated from the Andean
region, however, appeared most susceptible to infection.

Thirty-six *Peronospora* isolates from quinoa with different geographic
origins (including Argentina, Bolivia, Denmark, Ecuador and Peru) were
compared with *Peronospora* species from other *Chenopodium* species at
morphological and molecular level (Choi *et al.*, 2010). Haustoria were
hyphal, branched and without sheaths (Fig. 9.4). Conidia were pale brown
to olivaceous, varied shaped, mostly broadly ellipsoidal to ellipsoidal, sometimes appearing as obovoid or napiform due to a distinct pedicel (Fig. 9.4).
The morphological comparison between *Peronospora* sp. from quinoa,
P. variabilis from *C. album* and *P. farinosa* s.l. from *Atriplex patula* showed
that the quinoa pathogen was clearly distinguished from *P. farinosa* on
A. patula by larger size and higher length/width ratio in conidia. However,
no morphological difference was found between *Peronospora* isolates from
C. quinoa and *P. variabilis* from *C. album*. Phylogenetic analysis based on
internal transcribed spacer (ITS) rDNA sequences placed the quinoa pathogen within the same clade as *P. variabilis* (Fig. 9.5). Within the ITS rDNA
sequences of the quinoa pathogens, two base substitutions were found,
which separated the majority of the Danish isolates from isolates from South
America, but no sequence difference was found among the isolates from
different cultivars of quinoa.

Danielsen and Lübeck (2010) compared the genotypic variation among
Andean and Danish *Peronospora* isolates from quinoa using universally primed
PCR (UP-PCR). A total of 64 markers were obtained of which 35 were polymorphic and 17 occurred at less than 5% frequency. The marker size ranged
between 200 and 2000 bp. Three markers were found exclusively in Danish
isolates and five markers exclusively in Andean isolates. The Danish isolates
were recovered from quinoa cultivars of European origin grown entirely in
Denmark for several years and were therefore considered to be of Danish origin. A combined analysis of markers separated the Danish and Andean isolates
in two distinct clusters (Fig. 9.6).

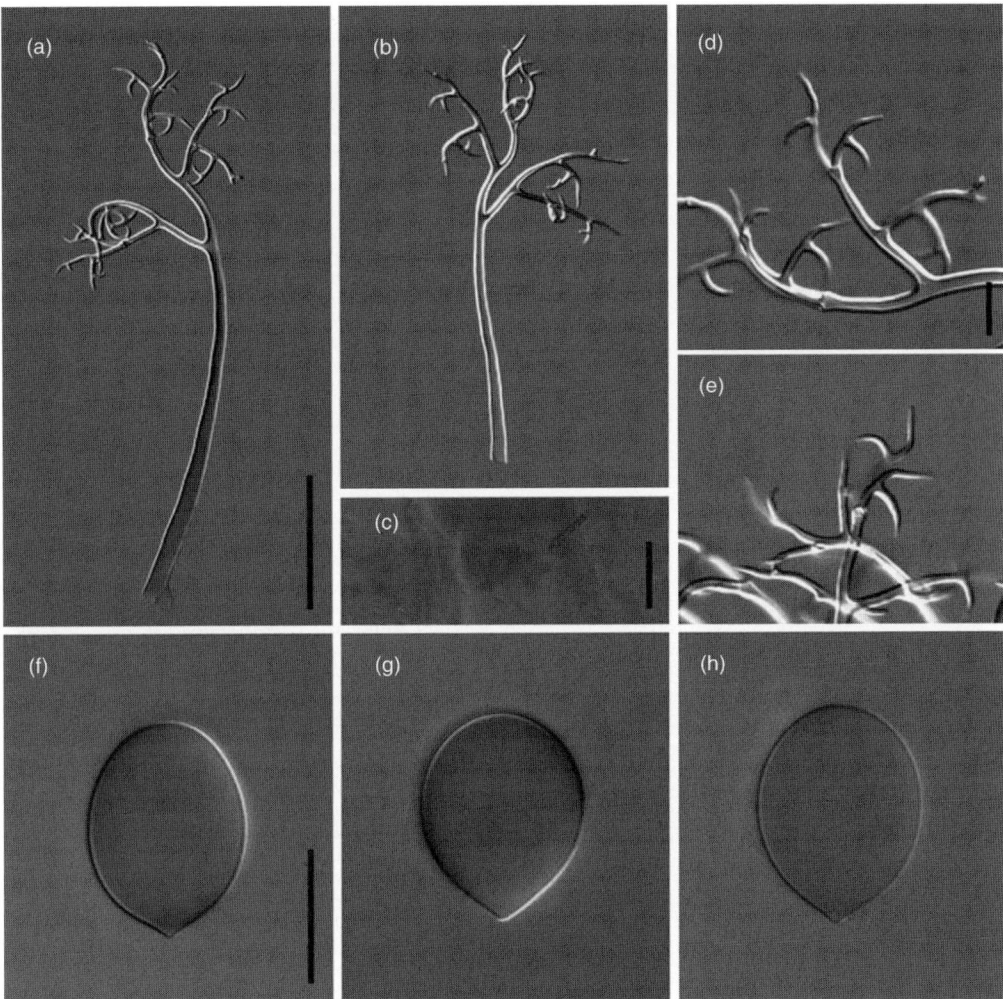

Fig. 9.4. *Peronospora variabilis* on *Chenopodium quinoa.* (a, b) Conidiophores – scale 100 µm; (c) haustorium – scale 10 µm; (d, e) branches – scale 20 µm; (f, g, h) conidia – scale 20 µm. [Reprinted from Choi *et al.* (2010), with kind permission from Springer Science + Business Media B.V.]

Damping off

This disease was first observed in quinoa plants in 1980. *Sclerotium rolfsii,* the causal organism, was isolated from the affected seedlings (Risi and Galwey, 1984). Another fungus *Fusarium* was isolated, along with *Rhizoctonia solani,* from a field of quinoa at the International Potato Center in Peru (Danielsen *et al.,* 2003). A more recent study showed that *Pythium aphanidermatum* and *Fusarium avenaceum* were the causal agents of the damping-off of quinoa seedlings (Drimalkova, 2003; Drimalkova and Veverka, 2004). Quinoa was highly susceptible to *P. apha-nidermatum,* especially during germination until the end of the cotyledon

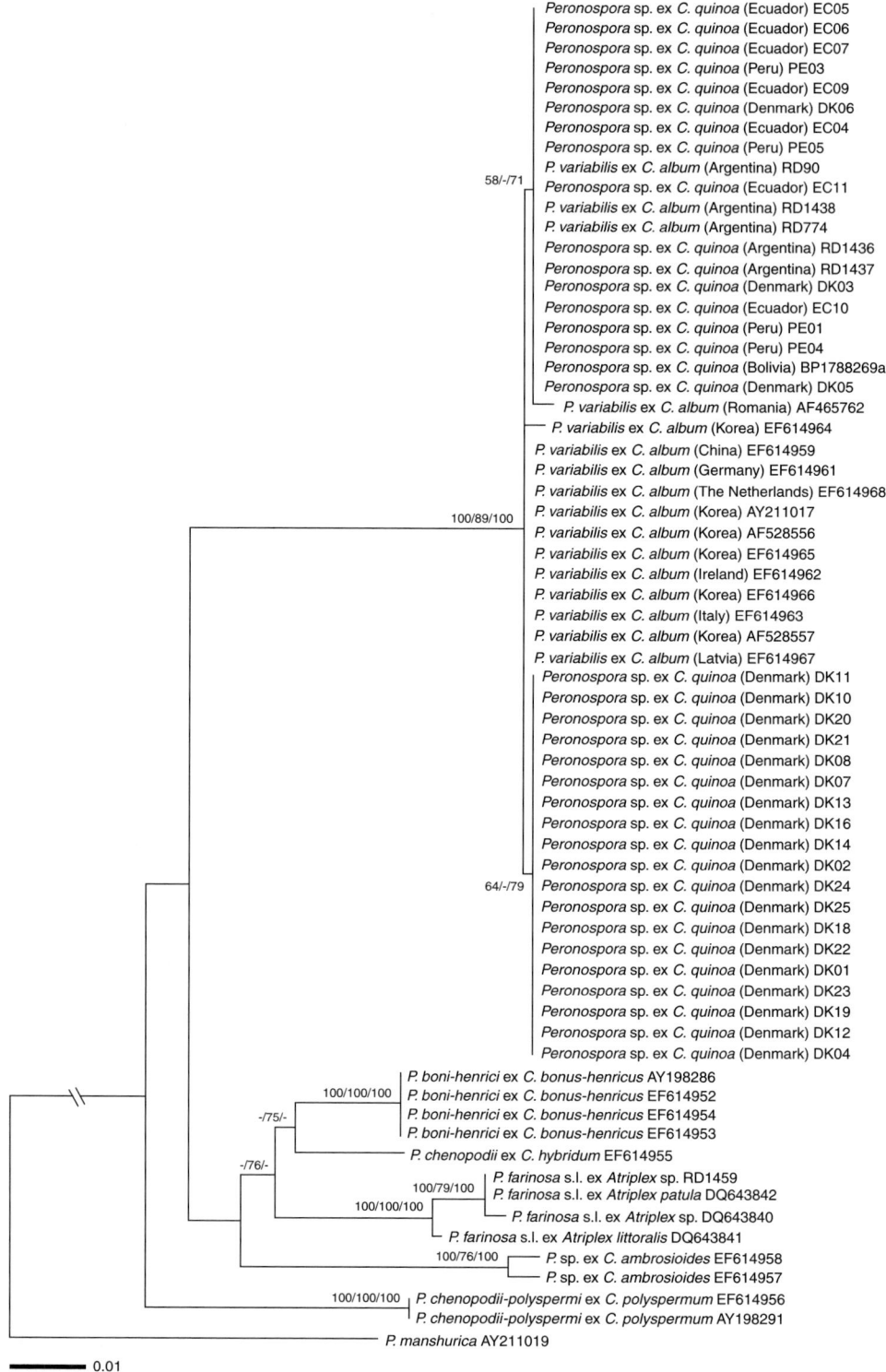

Fig. 9.5. Phylogenetic tree inferred from maximum likelihood analysis of the complete ITS region (ITS1, 5.8S rDNA, and ITS2). [Reprinted from Choi *et al.* (2010), with kind permission from Springer Science + Business Media B.V.]

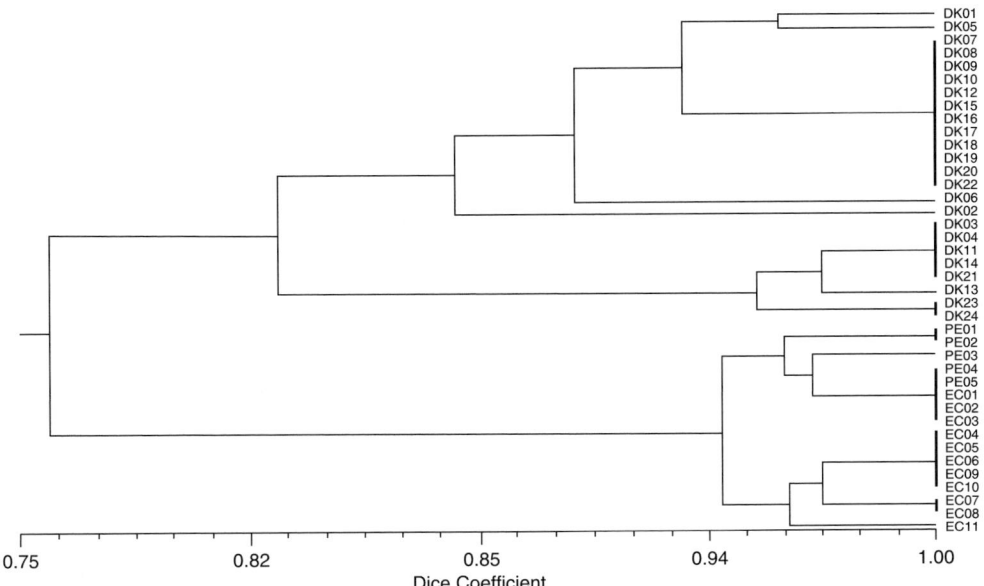

Fig. 9.6. Dendrogram showing genotypic similarity of *Peronospora farinosa* isolates from Denmark (DK) and the Andes based on UPGMA (unweighted pair-group method with arithmetic mean) using Dice's coefficient (NTSYS ver. 2.10z). (From Danielsen and Lübeck, 2010, with kind permission from John Wiley and Sons.)

leaves' formation. However, it then became more susceptible to *F. avenaceum* at the stage of cotyledon leaves and the first pair of true leaves.

Brown stalk rot

Phoma exigua var. *foveata* is the causal agent of brown stalk rot. The main symptoms are 5–15 cm long dark brown lesions with a vitreous edge on the stem and inflorescence (Risi and Galwey, 1984). Chlorosis, defoliation and shrunken stem are some of the other symptoms that may occur. This disease was first observed in Puno (Peru) in 1974–75 and has been frequently reported on the Peruvian Altiplano (Alandia *et al.*, 1979), as well as in quinoa growing areas of North America and the UK (AFRD, 2005). Low temperatures, high humidity and the presence of wounds favour the fungus (Otazu and Salas, 1977).

Stem gothic spot

The causal agent of this disease has been identified as *Phoma cava*, a coelomycetous soil fungi favoured by high relative humidity (Salas *et al.*, 1977; Alandia *et al.*, 1979). Light grey lesions with brown edges and surrounded by a vitreous halo occur on stems, inflorescence branches and floral peduncles (Risi and Galwey, 1984; Valencia-Chamorro, 2003). In severe cases, the lesions can coalesce to girdle the stem, causing leaves and inflorescence branches to collapse (Salas *et al.*, 1977).

Leaf spot

This disease is caused by *Ascochyta hyalospora*, an anamorphic fungus of Ascomycota (Boerema *et al.*, 1977; Połeć and Ruszkiewicz-Michalska, 2011). The fungus is seed borne and is favoured by high temperatures. The main symptoms of quinoa leaf spot include light spots of indefinite area on the leaves that coalesce until the spots cover the whole leaf. The fungus also causes mild to severe browning in the roots and hypocotyls of seedlings (Boerema *et al.*, 1977). Transmission occurs when pycnidia are produced in the inner tissues. Another species, *Ascochyta caulina*, has been isolated from quinoa seeds of Czech origin (Drimalkova, 2003). Germination testing showed that the fungus produced mild to severe browning in the root and hypocotyls, and some of the most severely infected seedlings died. In USA, *Passaflora dubia* has recently been reported from quinoa trials in Pennsylvania (Testen *et al.*, 2012). Foliar symptoms were round to oval, brown to grey-black lesions, less than 1 cm in diameter, with darker brown, reddish margins.

Grey mould
Botrytis cinerea, a nectrotrophic fungus, was observed on the variety 'Baer' in Cambridge, UK (Risi and Galwey, 1984). The main symptoms include grey lesions of irregular shape on the stem and inflorescence of mature plants, leaf senescence and reduced weight (Johanson, 1983; Risi and Galwey, 1984).

9.2.3 Viral diseases

Since quinoa is used as an indicator plant for viral testing, its response to a large number of viruses is known (Risi and Galwey, 1984). However, most viruses do not cause significant damage to the plant in the field.

9.3 Pests

A wide range of quinoa pests are known throughout the world. Problems in the field include miners and borers eating leaves, stems, roots and grains, chewing and sucking insects on foliage, and stem cutters. Birds and rodents feeding on mature grains also cause considerable damage to the crop (Gullan and Cranston, 1994; Zanabria and Banegas, 1997; Mujica *et al.*, 1998). Pest attacks in storage may also be significant.

9.3.1 Insects

Quinoa insect pests are classified as cutworms, biting and defoliating insects, and mining and grain-destroying insects (Zanabria and Mujica, 1977). Insect pests attacking quinoa can cause damage ranging from 8% to 40% (Ortiz and Zanabria, 1979). Quinoa is mainly attacked during vegetative growth, reproductive growth and seed storage (Jacobsen and Mujica, 2000; Rasmussen *et al.*, 2003). However, the actual loss caused by insects depends on many factors and varies over and within season and location.

The wide range of insects reported from Peru and Bolivia include members of Coleoptera (Chrysomelidae, Curculionidae, Meloidae, Melyridae,

Tenebrionidae), Diptera (Agromyzidae), Homoptera (Aphididae, Cicadellidae), Lepidoptera (Gelechiidae, Geometridae, Noctuidae, Pyralidae) and Thysanoptera (Thripidae) (Zanabria and Mujica, 1977; Mujica, 1993; Mujica *et al.*, 1998; Zanabria and Banegas, 1997; Yabar *et al.*, 2002) (Table 9.1).

However, a range of insects have also been reported from other parts of the world (Table 9.2).

Eurysacca melanocampta (Meyrick) and *E. quinoae* Povolny (Gelechiidae), found throughout Peru and Bolivia, are one of the most serious insect pests that attack quinoa in the region. These pests, recognized by their wing patterns, are abundant and destructive at all stages of plant growth, particularly during the maturation of grains. *E. melanocampta* is distributed throughout xeromontane habitats (approximately between 1900 and 4350 m above sea level) from Chile and Argentina in the south to as far as Colombia in the north (Povolny and Valencia, 1986; Povolny, 1997), while *E. quinoae* is actually widespread in Peru (Povolny, 1997; Rasmussen *et al.*, 2000). Species of

Table 9.1. Insects and pests, their infective stages causing diseases on various quinoa plant parts.

Type	Causal organism	Stage	Plant part affected	Reference
Leaf miner	*Lyriomiza brasiliensis*	Larva	Leaf	Ortiz and Zanabria (1979)
Leaf sticker	*Eurysacca* spp.	Larva	Inflorescence, stored grain	Rasmussen *et al.* (2000, 2001, 2003); Sequeiros (2001); Costa *et al.* (2009)
Cutworm	*Feltia experta* *Spodoptera* spp.	Caterpillar	Stem, leaf	Zanabria and Mujica (1977) Rasmussen *et al.* (2003)
Looper	*Perisoma sordescens*	Caterpillar	Leaf, seed, inflorescence	Zanabria and Mujica (1977)
Leaf and inflorescence caterpillar	*Hymenia recurvalis* *Pachyzancla bipunctales*	Caterpillar	Leaf, inflorescence	Ortiz and Zanabria (1979)
Defoliating insects	*Epithrix subcrinita* *Epicauta* spp.	Adult	Leaf, inflorescence	Zanabria and Mujica (1977) Ortiz and Zanabria (1979)
Piercing and cutting insects	*Macrosiphum* spp. *Myzus persicae* *Bergallia* spp. *Franklinella tuberosi*	Adult	Whole plant	Ortiz and Zanabria (1979)

Table 9.2. Insect pests reported on quinoa from various regions of the world.

Pest	Order	Country/region	Reference
Epitrix subcrinita	Coleoptera	Southern Europe	Rasmussen et al. (2003)
Aphis fabae	Hemiptera	Northern Europe	Gesinski (2000); Jacobsen
Lygus rugulipennis	Hemiptera	(Denmark, Poland)	(1993); Wrzesinska
Cnephasia sp.	Lepidoptera		et al. (2001)
Epitrix subcrinita	Coleoptera	Italy, Greece	Rasmussen et al. (2003)
Atomoscelis modestus	Coleoptera	USA	Cranshaw et al. (1990)
Scrobipalpa atriplicella	Lepidoptera	Denmark	Sigsgaard et al. (2008)
Cassida nebulosa	Coleoptera		
Plant bugs	Hemiptera	Africa	Rasmussen et al. (2003)
Termites	Isoptera		

Eurysacca are usually not present until the first quinoa inflorescences appear, and remain present in the field with increasing population numbers until harvesting. The females lay their eggs on the leaves in the pre-flowering state, when the grains have developed. The larvae of the first stages feed on the leaves and subsequently on the developing and mature grains (Melo, 2001; Rasmussen et al., 2003). The male finds a receptive adult female using search strategies such as finding the host plant of the female (by chemical or visual means), seeking sources of the interaction of odours of the female and host plants, or following sources of odour (pheromones) in the search exclusively for the female (Costa et al., 2009). Mating occurs over quinoa plants or over other plants situated nearby. The loss due to attack by *Eurysacca* spp. is severe and is estimated to range from an average yield loss of 15–18% up to 50% in dry years in Peru (Blanco, 1982; Mujica, 1993; Zanabria and Banegas, 1997). The relation between loss and infestation level is best described as an exponential growth function describing percentage loss, $F(x)$, due to the presence of x larvae:

$$F(x) = 2.1445*exp(0.0420*x)(F = 772.86, P < 0.001)$$

However, the model does not take into account plant stress, duration of infestation, physiological age of plants or the importance of source-sink areas for pest populations, as well as the influence of temperature and rainfall (Teng, 1987; Cockburn, 1991; Rasmussen et al., 2003). The economic threshold level has been estimated to be around 3–15 larvae per plant (Villanueva, 1978; Blanco, 1994). In heavily infested fields, single plants have been recorded to host up to 150–200 larvae, with an average of 46 larvae during the peak season in favourable years (Saravia and Germán, 1988; Mujica, 1993; Zanabria and Banegas, 1997).

9.3.2 Birds

Birds cause major losses in yield through foraging and damaging mature inflorescences, causing yield losses of up to 60% (Orellano and Tillmann, 1984;

Zanabria and Banegas, 1997; Rasmussen *et al.*, 2003). A wide range of species can be observed late in the season, primarily at the inflorescence stage, that feed on the soft grains. These include siskins (*Carduelis* spp.), eared dove (*Zenaida auriculata*), rufous-collared sparrow (*Zonotrichia capensis*) and white-throated sierra-finch (*Phrygilus erythronotus*) (Fjeldsa and Krabbe, 1990; Rasmussen *et al.*, 2003). But quinoa is conferred with a chemical defence with the production of bitter saponins in the pericarp that are thought to deter feeding birds and insects (Risi and Galwey, 1984; Oelke *et al.*, 1992).

9.3.3 Rodents

Rodents feed on quinoa grains but are not commonly encountered in quinoa fields (Risi and Galwey, 1984). These postharvest pests become more serious when grains are being dried and are in storage (Zanabria and Banegas, 1997).

9.3.4 Nematodes

Nacobbus aberrans and *Thecavermiculatus andinus* are nematodes that cause significant damage to quinoa in the Andean region (Franco, 2003). Both *N. aberrans* and *T. andinus* damage quinoa and also attack potato crops (Franco and Mosquera, 1992/1993a, b; Cespedez *et al.*, 1999), thereby affecting the production capacity of Andean farmers. *N. aberrans* (commonly known as False Root Knot nematode) causes host-plant root nodules (Franco, 2003). The enlarged females are enclosed in these nodules, along with a protruding gelatinous matrix containing the eggs, which remain in the soil inside small root tissues during the survival stage. The nematode is adapted to a wide range of climatic conditions and has a host range of 17 families and 65 species (Alarcon and Jatala, 1977; Franco and Main, 2008). In South America, this nematode has been detected mainly in the western countries, including Peru, Bolivia, Chile, Ecuador and Argentina. Detailed studies conducted in North and South America have indicated that *N. aberrans* populations can be separated into pathotypes with each populations of a pathotype having distinct host preferences.

T. andinus (commonly known as the 'nematode of the oca') is distributed in farms around Lake Titicaca and attacks a range of Andean crops like quinoa, potatoes, lupine, oca and olluco (Franco and Main, 2008). The host range of the nematode comprises 11 angiospermic families and 86 species. Greenhouse studies have shown a detrimental effect of the nematode on plant height, stem diameter, leaf fresh and dry weight, along with a significant reduction in yield (Franco and Main, 2008).

Another nematode *Globodera* spp., commonly known as the 'potato cyst nematode', only attacks plants of the Solanaceae family and has a limited host range. Nevertheless, various quinoa cultivars have been identified as 'trap' or 'antagonist'. Thus, while quinoa is a non-host of *Globodera* spp., it has been found that certain lines behave as 'traps' because their roots are efficient hatching stimulators and roots are invaded, but no nematode development or multiplication occur (Franco and Main, 2008).

9.4   Control

For nematodes, knowledge about the relationships (host/non-host and trap/ antagonist crops) between quinoa and other regional crops and their most important parasitic nematodes is very important. The resistant cultivars/lines could play an important role in future implementation of the integrated management of soil nematodes, especially if traditional methods become less efficient, or unsuitable, for protection. Appropriate choice of intercropping and rotation schemes with other crops may provide long-term benefits. For pest control, a number of cultural practices like recommendation sowing date, nutrient management, irrigation, planting density, crop rotation, mixed cropping, phytosanitation and tillage practices make the environment less favourable for pest invasion (Dent, 1995; Rasmussen *et al.*, 2003). The potential for utilizing natural enemies as biocontrol agents in the control of *Eurysacca* spp. and *Copitarsia turbata* is also quite promising, because high levels of parasitism have been found (Rasmussen *et al.*, 2003). Studies on the activity periods, efficiency and usefulness of the main parasitoids should be conducted in depth along with studies on how to encourage the activity of natural enemies. Potential control of pests with biopesticides involves the use of pheromones, bacteria, viruses and antifeedants. Chemical control methods include the use of the synthetic pyrethroids pesticides, use of kerosene, burning of rubber in the field and application of ashes to the soil after sowing (Rasmussen *et al.*, 2003). Strangely, integrated pest management for the control of pest attacks in quinoa is still not well implemented. An integrated pest management strategy for quinoa can be adopted, but basic research is urgently needed before we can proceed with scientifically approved management recommendations (Rasmussen *et al.*, 2003). Advances in quinoa pest research should try to adopt experiences from other crops in order to develop recommendations that are accessible to poor, small-scale Andean farmers.

9.5   Concluding Remarks

Although much work has been performed with respect to downy mildew infestation in quinoa, there is scant knowledge about other quinoa pathogens that infect seeds or seedlings and lead to decreased crops. These diseases, though less widespread than downy mildew, are considered potential production constraints, particularly when the crop is introduced in areas outside its traditional growing regions. Even though many diseases on quinoa are well known, many aspects of the disease and host–pathogen interaction are still unknown and need to be investigated. Therefore, it is necessary to apply proper methodology to handle the pathogen and study its interaction with the host, as well as the environment's effect on disease development (epidemiology), identifying specific strains, identifying the factors of pathogen resistance and the survival of oospores. This will allow us to understand the role of these factors in the pathogenesis process. Likewise future work should address the relative importance of plant architecture and phenology on the abundance of insect pests.

References

AFRD (2005) *Quinoa: The Next Cinderella Crop for Alberta?* Agriculture, Food and Rural Development, Alberta, Canada.

Alandia, S., Otazu, V. and Salas, B. (1979) Quinua y kaniwa. Cultivos Andinos. In: Tapia, M.E. (eds) *Serie Libros y Materiales Educativos*. Instituto Interamericano de Ciencias Agricolas, Bogota, Colombia, pp. 137–148.

Alarcon, C. and Jatala, P. (1977) Efecto de la temperatura en la resistencia de *Solanum andigena* a *Nacobbus aberrans*. *Nematropica* 7, 2–3.

Aragón, L. and Gutiérrez, W. (1992) El mildiu en cuatro especies de *Chenopodium*. *Fitopatología* 27, 104–109.

Blanco, A. (1994) Umbral económico de kcona kcona *Eurysacca melanocampta* (Lepidoptera, Gelechiidae) en quinua (*Chenopodium quinoa* Willd.). Thesis Ingeniero Agrónomo. UNA, Puno, Peru.

Boerema, G.H., Mathur, S.B. and Neergaard, P. (1977) *Ascochyta hyalospora* (Cooke & Ell.) comb. nov. in seeds of *Chenopodium quinoa*. *Netherland Journal of Plant Pathology* 83, 153–159.

Byford, W.J. (1967) Host specialization of *Peronospora farinose* on *Beta*, *Spinacia*, and *Chenopodium*. *Transactions of the Brazilian Mycology Society* 50, 603–607.

Cespedez, L., Franco, J. and Montalvo, R. (1999) Comportamiento de diferentes especies vegetales a la invasion y desarrollo de *Nacobbus aberrans* (Thorne, 1935), Thorne and Allen, 1944. *Nematropica* 28, 165–171.

Choi, Y.-J., Danielsen, S., Lübeck, M., Hong, S.B., Delhey, R. and Shin, H.D. (2010) Morphological and molecular characterization of the causal agent of downy mildew on quinoa (*Chenopodium quinoa*). *mycopathologia* 169, 403–412.

Cockburn, A. (1991) *An Introduction to Evolutionary Ecology*. Blackwell Science, Oxford.

Costa, J.F., Cosio, W., Cardenas, M., Yábar, E. and Gianoli, E. (2009) Preference of quinoa moth: *Eurysacca melanocampta* Meyrick (Lepidoptera: Gelechiidae) for two varieties of quinoa (*Chenopodium quinoa* Willd.) in olfactometry assays. *Chilean Journal of Agricultural Research* 69, 71–78.

Cranshaw, W.S., Kondratieff, B.C. and Qian, T. (1990) Insects associated with quinoa, *Chenopodium quinoa*, in Colorado (USA). *Journal of Kansas Entomology Society* 63, 195–199.

Danielsen, S. and Ames, T. (2000) *Mildew (Peronospora farinosa) of quinua (Chenopodium quinoa) in the Andean Region: Practical Manual for the Study of the Disease and the Pathogen*. International Potato Center, Lima, Peru.

Danielsen, S. and Ames, T. (2004) *Mildew of Quinua in the Andean Region*. Benson Agriculture and Food Institute, Brigham Young University, Provo, Utah.

Danielsen, S. and Lübeck, M. (2010) Universally Primed-PCR indicates geographical variation of *Peronospora farinosa* ex. *Chenopodium quinoa*. *Journal of Basic Microbiology* 50, 104–109.

Danielsen, S. and Munk, L. (2004) Evaluation of disease assessment methods in quinoa for their ability to predict yield loss caused by downy mildew. *Crop Protection* 23, 219–228.

Danielsen, S., Jacobsen, S.-E., Echegaray, J. and Ames, T. (2000) *Impact of Downy Mildew on the Yield of Quinoa*. CIP Program Report 1999–2000. CIP, Lima, Peru. pp. 397–401.

Danielsen, S., Jacobsen, S.-E. and Hockenhull, J. (2002) First report of downy mildew of quinoa caused by *Peronospora farinosa* f.sp. *chenopodii* in Denmark. *Plant Disease* 86, 1175.

Danielsen, S., Bonifacio, A. and Ames, T. (2003) Diseases of quinoa (*Chenopodium quinoa*). *Food Reviews International* 19, 43–59.

Dent, D. (1995) *Integrated Pest Management*. Chapman and Hall, London.

Drimalkova, M. (2003) Mycoflora of *Chenopodium quinoa* Willd. seeds. *Plant Protection Science* 39, 146–150.

Drimalkova, M. and Veverka, K. (2004) Seedlings damping-off of *Chenopodium quinoa* Willd. *Plant Protection Science* 40, 5–10.

Fjeldsa, J. and Krabbe, N. (1990) *Birds of the High Andes*. Zoological Museum and Apollo, Copenhagen and Svendborg.

Franco, J. (2003) Parasitic nematodes of quinoa in the Andean region of Bolivia. *Food Reviews International* 19, 77–85.

Franco, J. and Main, G. (2008) Management of tubers of Andean tuber and grain crops. In: Ciancio, A. and Mukerji, K.G. (eds) *Integrated Management and Biocontrol of Vegetable and Grain Crops Nematodes*. Springer, Dordrecht, The Netherlands, pp. 99–117.

Franco, J. and Mosquera, P. (1992/1993a) Ampliación de la gama de hospedantes del 'Nematodo de la Oca' (*Thecavermiculatus andinus* sp.n., Golden *et al.*, 1983) en los Andes Peruanos. *Revista Latinoamericana de la Papa* 5/6, 39–45.

Franco, J. and Mosquera, P. (1992/1993b) Patogenicidad del 'Nematodo de la Oca' (*Thecavermiculatus andinus* sp.n.) en cuatro cultivos andinos. *Revista Latinoamericana de la Papa* 5/6, 30–38.

Garcia, R.G. (1947) *Fitopatologia Agricola del Peru*. Estacion Agricola de La Molina, Ministerio de Agricultura, Lima, Peru.

Gesinski, K. (2000) *American and European Test of Quinoa (*Chenopodium quinoa *Willd.) in Poland*. Bydgoszcz, Poland.

Gullan, P.J. and Cranston, P.S. (1994) *The Insects: An Outline of Entomology*. Chapman and Hall, London.

Jacobsen, S-E. (1993) Quinoa: a novel crop for European agriculture. PhD. thesis, Royal Veterinary and Agricultural University, Copenhagen, Denmark.

Jacobsen, S.-E. and Mujica, A. (2000) Almacenamiento de la semilla de quinua. In: Jacobsen, S.-E. and Portillo, Z. (eds) *Primer Taller Internacional sobre Quinua: Recursos Genéticos y Sistemas de Producción*, 10–14 May 1999, UNALM, Peru.

Johanson, A. (1983) The resistance of quinoa (*Chenopodium quinoa*) to downy mildew (*Peronospora farinosa*), brown stalk rot (*Phoma exigua* var. *foveata*) and grey mould (*Botrytis cinerea*). BA research project, Department of Applied Biology, University of Cambridge, Cambridge.

Kitz, L. (2008) Evaluation of downy mildew (*Peronospora farinosa* f.sp. *chenopodii*) resistance among quinoa genotypes and investigation of *P. farinose* growth using scanning electron microscopy. MSc. thesis, Brigham Young University, Provo, Utah.

Kumar, A., Bhargava, A., Shukla, S., Singh, H.B. and Ohri, D. (2006) Screening of exotic quinoa accessions for downy mildew resistance under mid-eastern conditions of India. *Crop Protection* 25, 879–889.

Lucas, J.A. (2011) Advances in plant disease and pest management. *Journal of Agricultural Sciences* 149, 91–114.

Melo, L.A. (2001) Insectos asociados al cultivo de quinua en Cusco. Tesis Bachiller en Ciencias Biológicas. Universidad Nacional de San Antonio Abad del Cusco, Facultad de Ciencias Biológicas, Cusco, Peru.

Mujica, A. (1993) *Cultivo de Quinua*. INIA, Lima, Peru.

Mujica, A., Jacobsen, S.-E., Izquierdo, J. and Marathee, J.P. (1998) *Prueba Americana y Europea de Quinoa* (Chenopodium quinoa *Willd.*). FAO, Puno, Peru.

Ochoa, J., Frinking, H.D. and Jacobs, T. (1999) Postulation of virulence groups and resistance factors in the quinoa/downy mildew pathosystem using material from Ecuador. *Plant Pathology* 48, 425–430.

Oelke, E.A., Putnam, D.H., Teynor, T.M. and Oplinger, E.S. (1992) *Alternative Field Crops Manual.* University of Wisconsin Cooperative Extension Service, Madison, Wisconsin/Center for Alternative Plant and Animal Products, University of Minnesota Extension Service, St Paul, Minnesota.

Oerke, C. (2006) Crop losses to pests. *Journal of Agricultural Sciences* 144, 31–43.

Orellano, H. and Tillmann, H.J. (1984) La quinua en Yanamarca, prov. de Jauja. *Boletin de Lima* 35, 55–64.

Ortiz, R.V. and Zanabria, E. (1979) Plagas. Quinua y kaniwa. Cultivos Andinos. In: Tapia, M.E. (eds) *Serie Libros y Materiales Educativos.* Instituto Interamericano de Ciencias Agricolas, Bogota, Colombia, pp. 121–136.

Otazu, V. and Salas, B. (1977) Brown stalk rot of *Chenopodium quinoa* caused by *Phoma exigua* var. *foveata. Fitopatologia* 12, 54–58.

Pańka, D., Lenc, L. and Gęsiński, K. (2004) Preliminary observations in quinoa (*Chenopodium quinoa*) health status in Poland. *Phytopathologia Polonica* 31, 61–66.

Połeć, E. and Ruszkiewicz-Michalska, M. (2011) Some interesting species of the genus *Ascochyta. Acta Mycologia* 46, 187–200.

Povolny, D. (1997) *Eurysacca quinoae* sp.n.: a new quinoa-feeding species of the tribe Gnorimoschemini (Lepidoptera, Gelechiidae) from Bolivia. *Steenstrupia* 22, 41–43.

Povolny, D. and Valencia, L. (1986) Una palomilla de papa nueva para Colombia. *Memorias Del Curso Sobre Control Integrado de Plagas de Papa.* Bogota, Colombia.

Rasmussen, C., Jacobsen, S.-E. and Lagnaoui, A. (2000) Las polillas de la quinua. Especies en Perú de *Eurysacca* (Lepidoptera: Gelechiidae) en la quinua (*Chenopodium quinoa* Willdenow). In: *XLII Convención Nacional de Entomología*, Tarapoto, Peru 22–26 October 2000. Sociedad Entomológica del Peru, Lima, Peru.

Rasmussen, C., Jacobsen, S.-E. and Lagnaoui, A. (2001) Las polillas de la quinua (*Chenopodium quinoa* Willd.) en el Perú: *Eurysacca* (Lepidopt.: Gelechiidae). *Revista Peruana de Entomologia* 42, 57–59.

Rasmussen, C., Lagnaoui, A. and Esbjerg, P. (2003) Advances in the knowledge of quinoa pests. *Food Reviews International* 19, 61–75.

Risi, J. and Galwey, N.W. (1984) The *Chenopodium* grains of the Andes: Inca crops for modern agriculture. *Advances in Applied Biology* 10, 145–216.

Salas, B., Otazú, V. and Vilca, A. (1977) Enfermedades de la quinua. In: *Curso de Quinua.* Fondo Simón Bolivar, Ministerio de Alimentación, Instituto Interamericano de Ciencias Agricolas, Universidad Nacional Técnica del Altiplano, Puno, Peru, pp. 143–149.

Saravia, R. and Germán, M. (1988) Fluctuaciones poblacionales de larvas de insectos asociados al cultivo de la quinua en salinas de Gacia Mendoza. In: *VI Congreso Internacional Sobre Cultivos Andinos*, Quito, Ecuador, pp. 76–79.

Sequeiros, P.A. (2001) Fluctuación poblacional de plagas insectiles en quinua y sus controladores naturales. Tesis Biólogo. Universidad Nacional de San Antonio Abad del Cusco, Facultad de Ciencias Biológicas, Cusco, Peru.

Sigsgaard, L., Jacobsen, S.-E. and Christiansen, J. (2008) Quinoa, *Chenopodium quinoa*, provides a new host for native herbivores in northern Europe: case studies of the moth, *Scrobipalpa atriplicella*, and the tortoise beetle, *Cassida nebulosa. Journal of Insect Science* 8, 1–4.

Teng, P.S. (1987) *Crop Loss Assessment and Pest Management.* American Phytopathological Society, St Paul, Minnesota.

Testen, A.L., McKemy, J.M. and Backman, P. (2012) First report of *Passalora* leaf spot of quinoa caused by *Passalora dubia* in the United States. DOI: 10.1094/PDIS-05-12-0472-PDN

Tewari, J.P. and Boyetchko, S.M. (1990) Occurrence of *Peronospora farinosa* f.sp. *chenopodii* on quinoa in Canada. *Canadian Plant Disease Survey* 70, 127–128.

Thines, M. and Kamoun, S. (2010) Oomycete–plant coevolution: recent advances and future prospects. *Current Opinion in Plant Biology* 13, 1–7.

Valencia-Chamorro, S.A (2003) Quinoa. In: Caballero, B. (ed.) *Encyclopedia of Food Science and Nutrition* Vol. 8. Academic Press, Amsterdam, The Netherlands, pp. 4895–4902.

Villanueva, S. (1978) Determinación del 'Umbral económicó y 'nivel critico' de 'Kcona kcona' (*Scrobipalpula* sp.) en quinua (*Chenopodium quinoa* Willd.). Thesis for Ingeniero Agronomo, UNA, Puno, Peru.

Wrzesinska, D., Wawrzyniak, M. and Gesinski, K. (2001) Population of true bugs (Heteroptera) on the inflorescences of quinoa (*Chenopodium quinoa* Willd.). *Journal of Plant Protection Research* 41, 333–336.

Yabar, E., Gianoli, E. and Echegaray, E.R. (2002) Insect pests and natural enemies in two varieties of quinua (*Chenopodium quinoa*) at Cusco, Peru. *Journal of Applied Entomology* 126, 275–280.

Zanabria, E. and Banegas, M. (1997) *Entomología Economica Sostenible*. UNAP, Puno, Peru.

Zanabria, E. and Mujica, S.A. (1977) Plagas de la quinua. In: *Curso de Quinua*. Fondo Simon Bolivar, Ministerio de Alimentacion, Instituto Interamericano de Ciencias Agricolas, Universidad Nacional Tecnica del Altiplano, Puno, Peru, pp. 129–142.

10 Breeding

10.1 Introduction

Plant breeding is a deliberate effort by humans to nudge nature, with respect to the heredity of plants, towards an advantage (Acquaah, 2007). It is a branch of agricultural science that focuses on manipulating plant heredity to develop new and improved plant types for use by society. Plant breeding as a human endeavour has its origins in antiquity, starting off simply as discrimination among plant types to select and retain plants with the most desirable features. The early breeders relied solely on experience and intuition to select and multiply plants they thought had superior qualities. Technology and knowledge advancement enabled modern breeders to increasingly depend on science to take the guesswork out of the selection process, or at least to reduce it. Modern plant breeding is a discipline that is firmly rooted in the science of genetics. For approximately 10,000 years, human beings have modified the traits of plants to raise hundreds of domesticated breeds that today form the foundation of the world's food supply (McCouch, 2004). Techniques vary between crop species, but the scientific principles of plant breeding have remained unchanged since Mendel's experiments that selected parent plants for cross-pollination to combine desired characteristics.

The primary aim of plant breeding is to improve the quality, diversity and performance of agricultural and horticultural crops. The reasons for manipulating plant attributes or performance change according to the needs of society. Some of these are (Acquaah, 2007):

- improving food security for a growing world population;
- to adapt to environmental stresses;
- to adapt crops to specific production systems;
- developing new horticultural plant varieties;
- satisfying industrial and other end-use requirements.

A number of techniques are available to achieve this aim. The conventional tools include diversity analysis, selection, hybridization, wild crossing, chromosomal doubling and use of male sterility. The advanced tools include mutagenesis, tissue culture, plant transformation, DNA sequencing, use of molecular markers and marker-assisted selection.

10.1.1 Breeding objectives in quinoa

The basic objective of breeding in quinoa is the development of a variety with high grain yield accompanied by high protein and low saponin content (Bhargava *et al.*, 2006). In addition, the plant should be short, to facilitate mechanical harvesting, non-branching and uniformly early maturing (Jacobsen *et al.*, 1996). However, this is not an easy task because of the self-pollinating nature of the crop. The problem is compounded since the flowers are very small, which means emasculation and hybridization is very tedious and difficult. In spite of these difficulties, mass selection and hybridization has been practised in quinoa (Risi and Galwey, 1984). Various techniques of emasculation in quinoa have been suggested, but these are cumbersome. An easy approach can be the utilization of morphological markers to distinguish the hybrid from the parents.

10.2 Outcrossing in Quinoa

The plant is mainly self-pollinated, but several reports have provided information on the extent of outcrossing in the crop. Although predominantly autogamous, outcrossing occurs in quinoa over considerable distances. Outcrossing can occur up to a distance of 1 m and occasionally up to 20 m (Gandarillas, 1979). Gandarillas (1979) calculated outcrossing rates at different sowing distances in the Bolivian Altiplano. The outcrossing percentage ranged from 0.5 to 9.9%. The average rate of outcrossing worked out by Lescano (1980) in eight landraces and five varieties of quinoa was 5.8%. A recent study estimating the outcrossing (allogamy) rate (α) of *Chenopodium quinoa* cv. Sajama gave a rates as high as 17.36% (Silvestri and Gil, 2000). This is further corroborated by the absence of any inbreeding depression in characteristics like the weight and height of above-ground parts of the plant, development of inflorescence and seeds, or the homogeneity of offspring in six chenopod species studied by Dostalek (1987). However, some extreme cases of complete self-pollination through cleistogamy (Nelson, 1968) and obligate outcrossing by self-incompatibility and male sterility (Nelson, 1968; Gandarillas, 1969; Simmonds, 1971) have also been reported. The rate of outcrossing is influenced by wind speed, the proportion of different flower types and the extent of self-incompatibility prevalent in the plant (Risi and Galwey, 1984). As a result of outcrossing, the quinoa landraces have a high level of heterogeneity and heterozygosity and more advanced cultivars are difficult for farmers who propagate their own seeds to keep true to type (Lindhout and Danial, 2006).

10.3 Male Sterility

Male sterility is defined as the failure of plants to produce functional anthers, male gametes or pollen grains. The specific mechanisms causing male sterility in plants vary from species to species and are subject to influence by environment, and nuclear and cytoplasmic genes. Male sterility can be manifested in a variety of ways (Mehdi and Anwar, 2009): (i) absence or malformation of male organs in bisexual plants or absence of male flowers in dioecious plants; (ii) failure of the plant to develop anthers; (iii) abnormal microsporogenesis viz. deformed or unviable pollen; (iv) abnormal pollen maturation; inability of the pollen to germinate on compatible stigmata; (v) non-dehiscent anthers containing viable pollen; (vi) barriers other than incompatibility preventing the pollen from reaching the ovule.

Male sterility in flowering plants was first documented by Kolreuter in 1763, when he observed anther abortion within species and hybrids. Male sterility is of three types: (i) genetic/nuclear male sterility: the pollen sterility that is caused by nuclear genes is termed genetic male sterility; (ii) cytoplasmic male sterility: controlled solely by the specific cytoplasm genes whose action is not influenced by nuclear or other genes; and (iii) cytoplasmic-genetic male sterility: controlled by both nuclear and cytoplasmic genes.

10.3.1 Male sterility in quinoa

In quinoa, the clustering of numerous small-sized flowers on the inflorescence makes hybridization by manual emasculation extremely difficult. Therefore, development of male sterile quinoa lines for use as maternal parent in hybrid production has been proposed by many researchers as an alternative to the cumbersome emasculation and artificial hybridization procedure (Wilson, 1980; Risi and Galwey, 1984; Fleming and Galwey, 1995; Bhargava *et al.*, 2006). In such a system, manual removal of flowers, flower parts and manual fertilization of female flowers by breeders would no longer be necessary.

Both cytoplasmic and genic male sterility have been reported in quinoa, although there are fewer reports of nuclear male sterility. An account of a recessive nuclear gene controlling male sterility in Bolivian quinoa germplasm lines was provided by Gandarillas (1969). A male sterile line crossed and backcrossed to five different male fertile pollen parents produced all male sterile offspring. However, no detail was provided about the stage at which anther development aborted. It was concluded that the male sterile characteristic was under cytoplasmic control.

Rea (1969) described a male sterile plant having yellowish white or light brown empty anthers but did not study the inheritance of this trait.

Simmonds (1971) extensively studied the genetics of male sterility in an unnamed quinoa of Bolivian origin grown in Scotland and reported three loci: R (red plant) r (green plant); Ax (axil spot) ax (none); and Ms (hermaphrodite) ms (male sterile). The plants of genotype MsMs and Msms were fertile and the recessive msms were male sterile since they carried an erratically expressed or transmitted cytoplasmic factor for male sterility.

Risi and Galwey (1984) described the work carried out by Aguilar in Peru on male sterility in quinoa. Cytoplasmic male sterility was observed in the line 'UNTA 292'. This line produced male sterile progeny in the F1 and on two successive backcrosses to male fertile pollen parents. However, fertility was apparently restored when the Bolivian cultivar 'Sajama' was used as the pollen donor.

Ward and Johnson (1993) isolated a cultivar 'Apelawa' carrying normal and male sterile cytoplasms. The plants having male sterile cytoplasm produced flowers characterized by complete absence of anthers and prominent exsertion of stigmas. The cross between male sterile and normal male fertile donors consistently produced male sterile offspring, while interspecific crosses between male sterile quinoa plants and related wild species C. berlandieri produced offspring with partial restoration of male fertility. In spite of obtaining male sterile cytoplasm, a reliable restorer system could not be found. Further, Ward and Johnson (1994) reported that plants of Bolivian quinoa (C. quinoa Willd.) cv. 'Amachuma' carried a single nuclear recessive gene that in homozygous state produces normal anthers devoid of pollen grains. The plants heterozygous at this locus are indistinguishable from homozygous male fertile ones and further segregation for male sterility follows a normal Mendelian single gene segregation pattern. However, the authors suggested that this form of male sterility is of limited use in hybrid quinoa production due to poor stigma exsertion resulting in inefficient pollination.

Ward (1998) found an accession PI 510536 in the USDA-ARS collection, which had normal hermaphrodite and male sterile quinoa plants. This male sterility was of cytoplasmic nature and was characterized by small shrunken anthers and the absence of pollen. It was interesting that a dominant nuclear allele that interacted with this male sterile cytoplasm to restore male fertility was present in this accession. The ratio between male fertile and male sterile plants observed in F1 and F2 generations of the crosses between male sterile and the plants carrying restorer allele suggested either duplication of the restorer locus within the quinoa genome or tetraploid segregation. These characteristics may facilitate hybrid production in quinoa. Still, there is a need to obtain restorer lines, which can facilitate the easy production of hybrids and may overcome the extreme difficulty in the hybridization process.

10.4 Breeding in Quinoa

Breeding research started in the United Kingdom, Denmark and The Netherlands in the 1980s. The objectives were adaptation of quinoa to European climatic conditions and improvement of the agronomic performance as a seed crop with respect to uniformity, early maturity, seed yield, 1000-seed weight, and seed starch, protein and saponin content (Risi and Galwey, 1989; Galwey, 1993; Jacobsen and Stølen, 1993; Jacobsen et al., 1994; Mastebroek and van Soest, 1994; Limburg and Mastebroek, 1997; Mastebroek and

Limburg, 1997). Breeding for fodder quality has been a priority in Denmark and Sweden (Carlsson, 1980; Haaber, 1991). Breeding research in the Netherlands started in 1986 at the former DLO-Centre for Plant Breeding and Reproduction Research, since 1999 part of Plant Research International (Mastebroek *et al.*, 2002). About 100 different accessions derived from gene-banks, botanical gardens and universities were evaluated. Uniform lines adapted to the Western European climate were selected (Mastebroek and Limburg, 1997). The Indian subcontinent is another hotspot for breeding research in quinoa. After initial establishment of quinoa in India in the 1990s, extensive field trials were carried out at the National Botanical Research Institute, India on diverse parameters (Bhargava *et al.*, 2006). The crop has been evaluated as both a grain and fodder crop for different morphological and quality traits.

The studies on this crop include variability and diversity analysis, correlation and path coefficient studies, analysis of the genotype × environmental interactions (GEI) and hybridization experiments.

10.4.1 Genetic variability

Variability plays an important role in any crop-breeding programme and determines the limit of selection for yield improvement (Bhargava *et al.*, 2007a). Quantitative characteristics are those in which variation is continuous so that classification in discrete categories is not possible. These characteristics are governed by polygenes and environment plays an important role in the expression of such characteristics. Quantitative inheritance is primarily partitioned by magnitude, nature and interaction of genotypic and non-genotypic variation in various plant characteristics. This suggests the need to partition overall variability into its heritable and non-heritable components. Apart from this, knowledge of heritability and genetic gain is extremely important for plant improvement experiments. Heritability is defined as the transmissibility of a trait from the parents to the offspring (Falconer, 1981). Since heritability is the ratio of genotypic variance to phenotypic variance, it can be termed the heritable portion of phenotypic variance. Knowledge of heritability is important because it indicates the possibility and extent to which improvement can be brought about through selection (Robinson *et al.*, 1949; Holland *et al.*, 2003). Heritability is generally utilized in conjunction with selection differential, which indicates the expected genetic gain resulting from selection. Genetic advance is defined as improvement in the genotypic value of selected plants over the parental population. High heritability alone is not enough to make sufficient improvement through selection generally in advance generations unless accompanied by substantial amount of genetic advance. Thus, the knowledge of genetic variability present among populations and its quantitative assessment usually helps a plant breeder in choosing desirable parents for breeding programmes. The estimates of genetic variance and heritability play a major role in creating a pool of variable germplasm, selection of superior genotypes from the pool and utilization of selected individuals to create the desired variety (Dudley and Moll, 1969).

Genetic variability studies in quinoa

Based on the adaptations and some highly heritable morphological traits, quinoa is divided into five major groups (Lescano, 1989; Tapia, 1990):

1. Sea-level: Plants in this group grow 1–1.4 m tall, robust with branched growth, and produce transparent cream-coloured grains (Chullpi type). They are similar to *C. nuttalliae*, grown in Mexico (20°N latitude). Found in the areas of Linares and Concepción (Chile) at 36°S latitude.

2. Valley: Plants in this group grow up to 2.5 m or more, are greatly branching with lax inflorescences and are usually resistant to downy mildew. They are usually found at altitudes between 2500 and 3500 m.

3. Altiplano: Plants in this group grow to heights of 0.5–1.5 m, and have a compact main panicle. They are found at altitudes between 3600 and 3800 m in the Peruvian-Bolivian Altiplano, where the largest variability of characteristics is found and the grains with the most specialized uses are produced. This group contains the largest number of improved varieties that are susceptible to mildew when grown in moist areas.

4. Salt flat: Plants in this group are known as 'Royal Quinoa' and have the largest grain size (>2.2 mm in diameter). They are characterized by a thick pericarp and high saponin content. They grow in salt flat areas to the south of the Bolivian Altiplano.

5. Quinoas of the Yungas: A small group of quinoas that have adapted to the conditions of the Bolivian Yungas at altitudes between 1500 and 2000 m. These plants grow up to 2.20 m and the whole plant takes on an orange colour when in bloom.

Very few studies have been conducted on heritability and genetic advance in quinoa. Ballon *et al.* (1991) estimated heritability for quantitative traits like grain yield and grain size in *C. quinoa*. Broad sense heritabilities of 49 and 32% were estimated for grain yield and grain size, respectively.

Bhargava *et al.* (2003) evaluated some promising genotypes of *C. quinoa* to assess the suitability of selection on both normal and sodic soils. High heritability was obtained for all the traits on both types of soils. Heritability on sodic soil ranged from 91.09% (inflorescence length) to 98.27% (plant height), while on normal soil it ranged from 89.00% (grain yield) to 99.32% (stem diameter). Genetic gain on sodic soil was low in comparison with genetic gain on normal soil. Maximum genetic gain on sodic soil was observed for grain yield (34.44%), followed by inflorescence length (28.04%), while minimum genetic gain was noted for dry weight of plant (11.17%). Genetic gain on normal soil was most for stem diameter (92.31%), followed by primary branches/plant (52.16%) and number of inflorescence/plant (48.27%).

Bhargava *et al.* (2007a) analysed the morphological and nutritional variability of 27 germplasm lines of *C. quinoa* and two lines of *C. berlandieri* subsp. *nuttalliae* in subtropical North Indian conditions over a 2-year period. The analysis of variance for two separate years revealed significant differences among the strains for all the 16 characteristics (Table 10.1). The results indicated the presence of a high degree of morphological and qualitative variation among the lines studied. Phenotypic variance was most for inflorescence/plant among

Table 10.1. Range, variance, coefficient of variation, heritability and genetic gain for various morphological and quality traits in 27 germplasm lines of *C. quinoa* and two lines of *C. berlandieri* subsp. *nuttalliae*. [Reprinted from Bhargava *et al.* (2007a), with permission from Elsevier.]

Traits	Range	$\sigma^2 p$	$\sigma^2 g$	$\sigma^2 e$	PCV	GCV	Heritability (%)	Genetic gain (%)
Days to flowering	70.78–101.55	42.36	40.43	1.93	7.96	7.77	95.45	15.65
Days to maturity	109.33–163.33	185.75	182.29	3.46	10.53	10.43	98.14	21.27
Plant height (cm)	11.27–144.03	1,347.37	1,333.55	13.82	43.82	43.59	98.97	89.35
Leaf area (cm²)	4.42–30.91	62.47	59.16	3.31	43.53	42.36	94.69	84.94
Primary branches/plant	8.55–35.74	35.20	33.59	1.61	28.76	28.10	95.42	56.56
Inflorescence length (cm)	0.84–6.47	1.750	1.73	0.018	50.12	49.85	98.92	102.08
Inflorescence/plant	11.67–141.55	1,824.32	1,744.54	79.78	48.20	47.14	95.63	94.98
Seed size (mm)	1.34–2.21	0.053	0.045	0.008	12.53	11.49	84.90	21.88
1000-seed weight (g)	0.78–4.09	0.759	0.732	0.027	32.39	31.80	96.44	64.34
Dry weight/plant (g)	1.11–52.89	150.63	144.03	6.60	74.95	73.29	95.62	147.68
Harvest index	0.26–1.43	0.106	0.105	0.001	32.15	32.05	99.42	65.84
Seed yield (t/ha)	0.32–9.83	7.71	7.68	0.032	68.33	68.19	99.59	140.19
Total chlorophyll (mg/g)	0.55–2.04	0.122	0.109	0.013	24.48	23.15	89.34	44.95
Leaf carotenoid (mg/kg)	230.23–669.56	10,452.16	9,441.26	1,010.90	21.11	20.07	90.33	39.29
Seed carotenoid (mg/kg)	1.69–5.52	0.833	0.813	0.020	32.18	31.78	97.60	64.84
Seed protein (%)	12.55–21.02	6.73	6.68	0.052	15.98	15.92	99.23	32.69

$\sigma^2 p$ = phenotypic variance; $\sigma^2 g$ = genotypic variance; $\sigma^2 e$ = environmental variance; PCV = phenotypic coefficient of variation; GCV = genotypic coefficient of variation.

morphological traits and for leaf carotenoid among the quality traits. The phenotypic coefficient of variation (PCV) values for all the traits were higher than the corresponding genotypic values (GCV) values, though the differences were small (Bhargava et al., 2007a). Dry weight/plant, seed yield and inflorescence length recorded high coefficient of variation values. Broad sense heritability was high and exceeded 80% for all the traits. Genetic gain as percentage of mean was highest for seed yield (140.19%), followed by dry weight/plant (133.62%) and inflorescence length (102.08%). Seed protein showed lowest genetic gain among quality traits (32.69%), while for morphological traits the lowest value was for days to flowering (15.65%).

10.4.2   Genetic correlation and path coefficient analysis

Breeding response for one trait depends on the genotypic variations of that trait within the breeding population and on genotypic correlations between the traits. Therefore, determination of correlation coefficients between yield and other variables is necessary for the selection of favourable characteristics for an effective breeding programme. Correlation coefficient measures the extent of the relationship between two or more variables. The study of correlation is regarded as an important step in breeding programmes since the information obtained is useful for estimating the correlated response to directional selection for the formulation of selection indices.

The concept of correlation was initially presented by Galton (1889) and was later elaborated by Fisher (1918) and Wright (1921). Investigation of genotypic and phenotypic interrelationships between various agronomic/economic traits is of particular interest to plant breeders theoretically as well as practically, since selection is usually concerned with changing two or more characteristics simultaneously. The utility of estimates of correlation considerably increases by partitioning into phenotypic, genotypic and environmental components (Burton, 1952; Falconer and Mackay, 1996; Cervantes-Martinez et al., 2002). Genotypic correlation provides a measure of genetic association between characteristics and is used in selection for one characteristic as a means of improving the other. Such correlation coefficients may be helpful to the breeder since they are based on transmissible genetic variance (Jerome et al., 1956; Miller et al., 1958). Moreover, genotypic correlation is helpful in simplifying the approach for selection by formulating the most effective breeding methodology under particular environmental conditions.

Correlation coefficients generally show relationships among independent characteristics and the degree of linear relation between these characteristics. However, it is not sufficient to describe this relationship when the causal relationship among characteristics is needed (Korkut et al., 1993). Path analysis is used when a breeder wants to determine the amount of direct and indirect effects of the causal components on the effect component (Guler et al., 2001). The path coefficient is known as a standardized partial-regression coefficient, and separates the direct and indirect effects of a correlation coefficient. Hence, path analysis plays an important role in determining the degree of relationship between yield and yield components. The original concept of path analysis was put forward by Wright (1921), but was later elaborated and extensively used in plants by Dewey and Lu (1959).

Correlation and path studies in quinoa

Early studies found a lack of correlation between the variation of qualitative and quantitative descriptors in the core quinoa collections. Gandarillas-Santa Cruz and Espindola-Canedo (1981) found no relationship between inflorescence type and grain yield, or with its components such as earliness, plant height, inflorescence length or seed diameter, in quinoa cultivars.

Espindola and Gandarillas (1985) reported significant phenotypic correlations of plant height and inflorescence length with grain yield in 36 accessions of quinoa collected from Peru and Bolivia. The path analysis revealed that inflorescence length was the most important component influencing grain yield in *C. quinoa*.

Risi and Galwey (1989) calculated the correlation coefficients among various traits in 294 accessions of quinoa evaluated at Cambridge, UK. Plant height, stem diameter, inflorescence length and inflorescence diameter were all strongly correlated with each other. Also, there were significant associations between seed brightness and plant habit, and plant colour and inflorescence compactness. However, the associations between the duration of the developmental phases were weak, suggesting that there is great scope for manipulation of the pattern of development through breeding.

Risi and Galwey (1991) observed strong correlation between early flowering, early maturity and grain yield while studying genotype × environment interaction in quinoa.

Ortiz *et al.* (1998) noted high correlation between stem and inflorescence colour in 1029 quinoa ecotypes or landraces collected from farms in the Peruvian Andes and maintained in the gene bank of Universidad Nacional del Altiplano-Puno (UNAP). This confirmed the presence of partial common genetic control for pigmentation in the crop. There were significant phenotypic correlations between days to flowering with inflorescence length, plant biomass, plant height and grain yield.

Bhargava *et al.* (2003) calculated the correlation coefficients among various traits and their direct and indirect effects on grain yield in some exotic genotypes of quinoa grown on normal as well as sodic soils. Grain yield/plant on sodic soil was significantly and positively associated with all the traits except for primary branches/plant, which was negatively correlated (-0.797). Stem diameter and dry weight/plant were positively correlated with grain yield on normal as well as sodic soils. Stem diameter showed highest positive direct path (0.837) on sodic soil, while on normal soils dry weight/plant had highest positive path with grain yield. It was concluded that selection of thick-stemmed plants with a greater number of inflorescences and high dry weight would be more desirable for breeding for high-grain yield on sodic soils.

Spehar and Santos (2005) evaluated 26 quinoa lines, selected from individual plant progenies of hybrids among varieties Amarilla de Marangani, Blanca de Junín, Chewecca, Faro 4, Improved Baer, Kancolla, Real and Salares-Roja, for numerous agronomic characteristics in Planaltina, DF, Brazil. Grain yield was positively associated with plant height, length and diameter of inflorescence, and plant cycle. Inflorescence length and diameter were positively associated with grain yield, which indicated that selection for these characteristics may result in more productive genotypes. Positive correlation observed between plant height and inflorescence length suggested that high grain yield can be attained by selecting for stem/inflorescence ratio. Stem diameter was positively correlated with grain yield and biomass production, indicating that, under low population, plants increase their stem diameter and branching, and compensate for grain yield (Spehar and Santos, 2005).

The interrelationships among yield and yield components were elucidated in 27 germplasm lines of *C. quinoa* and two lines of *C. berlandieri* subsp. *nuttalliae* in subtropical North Indian conditions over a 2-year period using correlation analysis and the results were supplemented by path analysis (Bhargava *et al.*, 2007a). All morphological traits except days to flowering, days to maturity and inflorescence length exhibited significant positive association with seed yield. Significant associations between branches/plant, inflorescence length and inflorescence/plant indicate that plants with good branching habit tend to develop a large number of long inflorescences. Inflorescence length also correlated positively with plant height, indicating that lines with greater plant height also developed longer panicles, a fact also reported by Ochoa and Peralta (1988) and Rojas (2003). The non-significant correlation between seed yield and seed quality traits and the low values of direct path reported by Bhargava *et al.* (2007a) are beneficial because it would be possible to breed lines with both greater grain yield and high seed protein and carotenoid. Seed carotenoid seemed to be correlated with seed coat colour because six out of the seven dark-coloured accessions had high seed carotenoid. The path analysis revealed that 1000-seed weight had the highest positive direct relationship with seed yield (1.057), followed by total chlorophyll (0.559) and branches/plant (0.520). Traits showing a high negative direct effect on seed yield were leaf carotenoid (−0.749), seed size (−0.678) and days to flowering (−0.377). Comparing both correlation and path analysis, it was observed that seed yield and seed protein were the only traits exhibiting high positive direct path and significant positive association with harvest index, indicating a true relationship between these traits (Bhargava *et al.*, 2007a).

Fuentes and Bhargava (2010) noticed high correlation between stem diameter and plant weight (R = 0.83), stem diameter and plant height (R = 0.78), plant weight and plant height (R = 0.69), plant weight and inflorescence length (R = 0.65), plant height and inflorescence length (R = 0.62), and leaf length and leaf width (R = 0.62) while evaluating quinoa accessions in a lowland desert environment in Chile. Harvest index showed negative association with stem diameter, plant height, inflorescence length, inflorescence width and inflorescence branch number. Likewise, leaf tooth number exhibited a strong negative association with stem diameter, plant height and inflorescence length.

10.4.3 Genetic diversity

Study of genetic diversity is the process by which variation among individuals or groups is analysed using numerical data on different types of variables.

Efficient management and utilization of genetic potential held in a germplasm collection requires efficient characterization of genetic diversity of the collection (Karp and Edwards, 1995). Genetic diversity differs from variability in that variability has distinct phenotypic differences, whereas genetic diversity may or may not have such distinctness. Accurate assessment of the levels and the patterns of genetic diversity can be invaluable in crop breeding for diverse applications that include: (i) analysis of genetic variability in genotypes (Smith, 1984; Cox *et al.* 1986); (ii) identifying diverse parental combinations to create segregating progenies with maximum genetic variability for further selection (Barrett and Kidwell, 1998); and (iii) introgressing desirable genes from diverse germplasm into the available genetic base (Thompson *et al.*, 1998).

Analysis of genetic diversity in germplasm collections can facilitate reliable classification of accessions and identification of subsets of core accessions for possible utility for specific breeding purposes (Mohammadi and Prasanna, 2003). Diverse data sets have been used by researchers to analyse genetic diversity in crop plants, which include pedigree data (Cupic *et al.*, 2009; Geribello Priolli *et al.*, 2010), passport data-morphological data (Grenier *et al.*, 2000; Ferreira *et al.*, 2010; Brown *et al.*, 2012), isozymic data (Hamrick and Godt, 1997; Kurt *et al.*, 2008) and storage proteins (Smith *et al.*, 1997; Masoumi *et al.*, 2012). Genetic diversity in crop plants may be analysed at different levels, such as inbred lines/pure lines/clones, populations, germplasm accessions and at species level.

The use of established multivariate statistical algorithms is an important strategy for classifying germplasm, ordering variability for a large number of accessions or analysing genetic relationships among breeding materials (Mohammadi and Prasanna, 2003). Multivariate analytical techniques like cluster analysis, principal component analysis (PCA), principal coordinate analysis (PCoA) and multidimensional scaling (MDS) simultaneously analyse multiple measurements on each individual under investigation and have been widely used in the analysis of genetic diversity irrespective of the data set (Thompson *et al.*, 1998). Multivariate analysis has been successfully used to classify and order variation for both quantitative and qualitative traits in diverse crops like tomato (*Lycopersicon esculentum*) (Pratta *et al.*, 2000), mustard (Alemayehu and Becker, 2002), *Vigna* spp. (Bisht *et al.*, 2005), common bean (*Phaseolus vulgaris*) (Ceolin *et al.*, 2007) and amaranthus (*Amaranthus tricolor*) (Shukla *et al.*, 2010).

Genetic diversity analysis in quinoa

Cultivated quinoa displays genetic diversity, mainly represented in an ample range of characteristics like plant coloration, flower protein content, seed saponin content and leaf calcium oxalate content, which allows adaptability to a wide range of agroecological conditions (Anabalón-Rodriguez and Thomet-Isla, 2009).

Risi and Galwey (1989) assessed genetic diversity in 294 accessions of quinoa using principal component and canonical analysis. Accessions from near sea level in Chile formed a homogenous group. Ortiz *et al.* (1999) created a phenotypic distance matrix among 76 accessions from a Peruvian quinoa core collection.

Rojas (2003) analysed genetic diversity in *C. quinoa* using three multivariate methods. Multiple group discriminant function analysis resulted in six statistically significant functions, which separated the different groups. The assessments showed that phonological variables such as initiation of the flowering and midbloom date were stronger discriminants than the yield variables (grain diameter and 100-grain weight). The descriptor of plant emergence date was the least important, having the least variation within the germsplasm examined (Rojas, 2003). It was suggested that this descriptor can be excluded from any new morphological or phonological assessments within quinoa.

Twenty-nine germplasm lines of *C. quinoa* and two of *C. berlandieri* subsp. *nuttalliae* were evaluated for 19 traits in sub-tropical North Indian conditions using cluster and principal component analysis (Bhargava *et al.*, 2007b). The results revealed that an enormous amount of genetic variability existed in the quinoa germplasm lines. Multivariate analysis showed that most of the variations were accounted for by the first four principal components (PCs). The main traits that accounted for more variability in both PC1 and PC2 include days to maturity, primary branches/plant, chlorophyll content and seed yield/plant. Thus, these traits could be considered as important in distinguishing the material under study. Good vegetative growth and high seed yield/plant characterized the lines with high PC1 values, whereas lines with high PC2 values were characterized by late maturity and larger seed size. The germplasm lines were grouped into six clusters based on average linkage method (Fig. 10.1). The lines in cluster I were early maturing and high yielding but had low carotenoid content. Cluster II comprised lines having low seed quality but were higher in leaf-quality components. Cluster III had highest seed yield and high values for protein and carotenoids. The lines in cluster IV matured earliest and had high seed protein, while cluster V had high seed yield, dry weight/plant, stem diameter and maximum number of inflorescences. Cluster VI had low values for traits related to seed morphology and quality, except for carotenoids content. The present investigation based on morphological and quality characteristics also clustered two lines of *C. berlandieri* subsp. *nuttalliae* separately from the quinoa line that is phylogenetically correct.

Morphological diversity among 28 quinoa accessions from the Chilean highlands was assessed under desert lowland conditions using multivariate techniques to analyse measurements of 11 morphological descriptors (Fuentes and Bhargava, 2010). The first four principal components accounted for 70% of the total variation among the accessions. PC1, accounting for 36% of the total variation, included stem diameter, plant weight and plant height with high positive coefficients, and harvest index and leaf tooth number with high negative coefficients. PC2 contributed an additional 19% of the total variation and reflected the patterns of variation in leaf morphology, all of which had high positive values. A correlation matrix involving the complement of the Pearson coefficient was used to construct a dendrogram using the UPGMA algorithm. Cluster analysis allowed classification of the accessions into six discrete groups (Fig. 10.2). Group I was represented by two accessions having large inflorescences and fluctuating grain yield. Group II comprised three accessions characterized by medium plant height and medium grain yield. Group III comprised

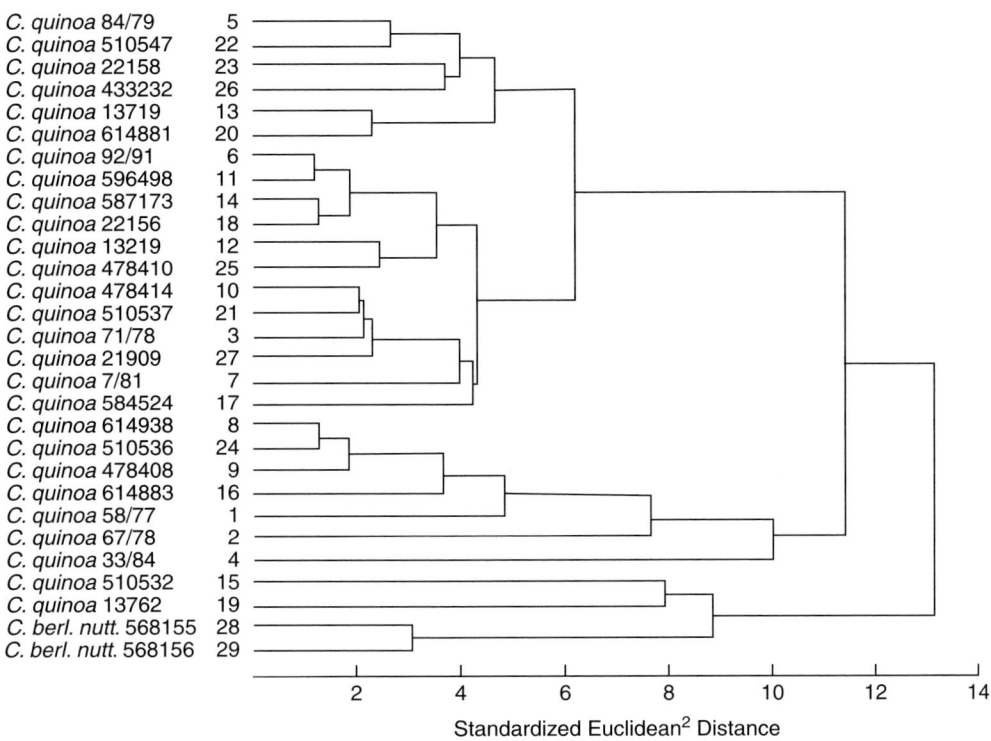

Fig. 10.1. Dendrogram of 29 chenopod lines derived from average linkage cluster analysis [Reprinted from Bhargava *et al.* (2007b), with permission from Springer.]

four accessions that were tall and had low grain yield. Group IV represented eight accessions having small plants with low grain yield. Group V included five accessions, all of which had medium height and medium-to-low grain yield. Group VI was represented by six accessions that comprised plants of medium plant height with medium-to-high grain yield.

The yield assessment in the Atacama Desert was considered low, probably due to the negative effects of the high-temperature stress around flowering.

10.4.4 Hybridization experiments

Intrageneric hybridization experiments involving a number of related species have explored the breeding value of the tetraploid species of the subsection Cellulata for transfer of genes to improve quinoa (Nelson, 1968; Heiser and Nelson, 1974; Wilson and Heiser, 1979; Wilson, 1980). A series of interspecific hybridization experiments have been performed to assess the relationships of quinoa with other chenopods (Wilson and Heiser, 1979; Wilson, 1980). Wilson and Heiser (1979) reported limited self-fertility in hybrids between *C. quinoa* and *C. berlandieri* subsp. *nuttalliae* or its wild North

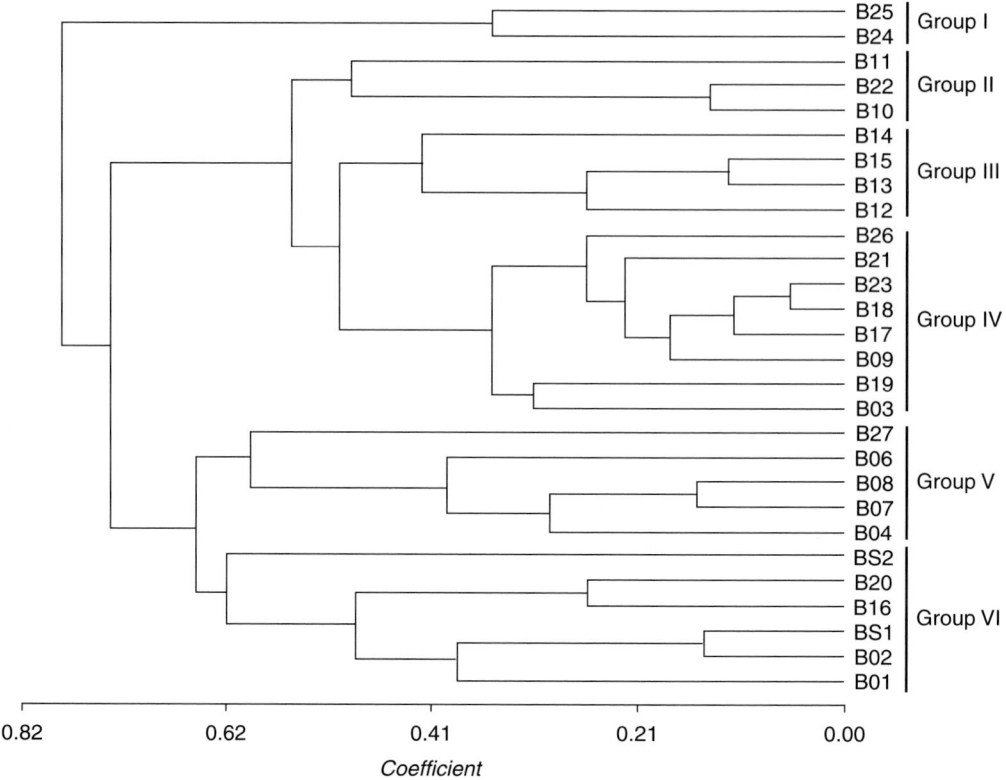

Fig. 10.2. Cluster analysis for characterization of quinoa germplasm grown under lowland desert conditions. Group I: Accessions having large inflorescences and fluctuating grain yields (GYs); group II: medium plant heights and medium GYs; group III: taller plants and low GYs (forage use); group IV: shorter plants with low GYs; group V: medium plant heights and medium-to-low GYs, and group VI: medium plant heights and medium-to-high GYs [Reprinted from Fuentes and Bhargava (2010), with permission from John Wiley and Sons.]

American relative *C. berlandieri*. Crosses between *C. berlandieri* subsp. *nuttalliae* and *C. quinoa* result in F1 plants with uniformly low fertility (Wilson and Heiser, 1979). Wilson (1980) observed that hybrids produced from the *C. quinoa* × *C. petiolare* and *C. quinoa* × *C. neomexicanum* crosses did not survive to flowering. However, hybrids between quinoa and *C. bushianum*, another tetraploid member of subsection Cellulata, produced partially fertile hybrids, although some backcross progeny could be produced. Bonifacio (1995) has reported viable, though sterile, hybrids from crosses between *Atriplex hortensis* (saltbush) and *C. quinoa*. Breeding strategies involving bridging crosses, colchicine-mediated chromosome doubling and/or embryo rescue may prove necessary to expand the breeding gene pool of quinoa to include Cellulata diploids and more distantly related *Chenopodium* and *Atriplex* species (Maughan *et al.*, 2007).

A half-diallelic crossing experiment with six diverse but uniform breeding lines was conducted for a better understanding of the inheritance of seed yield traits in quinoa (Mastebroek *et al.*, 2002). True hybrid plants were detected by means of differences in panicle colour in 14 out of the 15 crosses performed. The agronomic performance of 14 F2 populations and six parental lines was evaluated in field trials on fertile clay soil. General and specific combining ability of the lines (GCA and SCA) were estimated. Highly significant differences in GCA effects were found between the lines for plant height at maturity, early flowering, early maturity, 1000-seed weight and seed yield. Significant SCA effects were only found for plant height at maturity. However, SCA effects were rather small and accounted for 7% of the total variance. The finding of large GCA effects and low SCA effects suggested that best selection results could be expected by crossing best-performing lines, inbreeding and selection of pure recombinant lines that combine the agronomically positive traits from the different parental lines.

10.4.5   Genotype × environment interaction (GEI)

The interaction between cultivars and environmental factors is an important consideration for plant breeders. Plant breeders continuously strive to broaden the genetic base of a crop to prevent its vulnerability to changing environments. A material planted in different environments exhibits different responses because of environmental variation (Bhargava *et al.*, 2007c). Consistent performance across different sites and/or years is referred to as stability. A number of statistical models are available for evaluating the yield stability of a genotype in yield trials (Finlay and Wilkinson, 1963; Eberhart and Russel, 1966; Tai, 1971; Shukla, 1972; Shafii and Price, 1998). A study of GEI provides information about the suitability of genotypes over wide agroclimatic conditions. Partitioning of GEI into stability statistics assignable to each genotype is useful in selecting stable genotypes for any crop breeding programme. Genotypes showing considerable GEI effects are not suited for diverse environments and severely limit the selection of superior genotypes (Thillainathan and Fernandez, 2002; Bhargava *et al.*, 2007c). Thus, knowledge of GEI and yield stability is important for breeding new cultivars with improved adaptation to the environmental constraints prevailing in the target environment (Bhargava *et al.*, 2007c). Few reports on GEI and stability analysis in quinoa have been published to date (Risi and Galwey, 1991; Jacobsen *et al.*, 1996; Jacobsen, 1998; Bertero *et al.*, 2004).

A study of GEI by Risi and Galwey (1991) demonstrated that GEI differed among the variables measured. Grain yield was strongly dependent on the variety, but micronutrient deficiency and weed competition affected the varieties differently.

The stability of various descriptive characteristics was studied over a 5-year period in 14 lines of quinoa to determine the most appropriate time in a breeding programme when selection for these characteristics could be performed (Jacobsen *et al.*, 1996). Another objective was to ascertain which lines could

serve as potential parents. Various measures of stability were employed to analyse the data, appropriately modified for the purpose of this investigation. GE
interactions were significant for all characteristics, but for several traits of interest to the plant breeder, namely earliness, height and inflorescence size, some
of the best lines track the optimum response. From these results it was concluded that selection for height, inflorescence size and developmental stage
could be satisfactorily performed at an early stage of the breeding programme.
Potential parents were identified in this material for use in the development of
varieties suitable for North European conditions.

Studies on developmental patterns of quinoa for North European conditions were carried out in five groups of quinoa lines from different maturity
classes over 3 years, and measured on five occasions between bud formation
and seeds set (Jacobsen, 1997). The seed types originating from Chile were
found to be better adapted for growth under North European conditions.

Five quinoa lines, from different maturity groups, were assessed for their
adaptability to European conditions (Jacobsen, 1998). Ranking of the lines for
earliness was determined by the beginning of July and was consistent over
years, indicating that selection for this characteristic could take place at an early
stage in plant development. Discussion of the plant breeding implications of
these and other results suggested the selection of early, uniformly maturing
plants with more branches, low saponin content and high seed yield for raising
a model genotype (Jacobsen, 1998).

The size and nature of the GEI effects for grain yield, its physiological
determinants and grain size exhibited by the Andean grain crop quinoa were
comprehensively examined in a multi-environment trial involving a diverse set
of 24 cultivars tested in 14 sites under irrigation across three continents (Bertero
et al., 2004). The G × E interaction to G component of variance ratio was 4:1
and 1:1 for grain yield and grain size, respectively. Two-mode pattern analysis
of the environment-standardized matrix of grain yield revealed four genotypic
groups of different response pattern across environments. The clustering,
which separated cultivars from mid-altitude valleys of the northern Andes,
northern Altiplano, southern Altiplano and sea level showed a close corre-
spondence with adaptation groups previously proposed. The results can be
utilized to choose genotypes of contrasting relative performance across environments for further studies aimed at assessing the opportunity to select for
broad or specific adaptation. Classification of sites for grain yield grossly discriminated between cold highland sites, tropical valleys of moderate altitude
and warmer, low-altitude sites. No single genotype group showed consistently
superior grain yield across all environment groups. The genotype (G) and GEI
effects observed for the duration of the crop cycle had a major influence on the
average cultivar performance and on the form of GEIs observed for total above-
ground biomass and grain yield. Although different environment types showed
contrasting effects on the physiological attributes underlying grain yield variation among cultivars, it was noticed that good average performance and broad
adaptation could come from the combination of medium–late maturity and
high harvest index. Correlation analysis revealed no association between the
average cultivar responses for grain yield and grain size (Bertero et al., 2004).

Both observations indicated that simultaneous selection for grain yield and grain size can be expected from selection.

A high level of grain saponin is a major impediment in the diversification of this crop. Ward (2000) screened 10 South-American quinoa accessions for saponin content and performed three cycles of pedigree selection. The results indicated that dominant gene action is a major component of genetic variance for the trait. Fixed heterozygosity at loci controlling saponin content may also occur due to the allotetraploid nature of the crop.

10.5 Concluding Remarks

Because of the emerging potential of the crop and the limited amount of research work on breeding, there is an urgent need to initiate elaborate breeding programmes in quinoa (conventional as well as biotechnological) for its genetic improvement. Quinoa has a fairly versatile breeding system. The self-incompatible or male sterile lines and the lines with a very low frequency of hermaphrodite flowers can be used in breeding programmes because the quinoa flowers, being rather small, are not amenable to emasculation. Since breeding work in quinoa is still in its infancy, diverse strategies need to be formulated to improve the plant for edible, fodder and/or medicinal use.

References

Acquaah, G. (2007) *Principles of Plant Genetics and Breeding*. Blackwell, Oxford.

Alemayehu, N. and Becker, H. (2002) Genotypic diversity and patterns of variation in a germplasm material of Ethiopian mustard (*Brassica carinata* A. Braun). *Genetic Resources and Crop Evolution* 49, 573–582.

Anabalon-Rodriguez, L. and Thomet-Isla, M. (2009) Comparative analysis of genetic and morphologic diversity among quinoa accessions (*Chenopodium quinoa* Willd.) of the South of Chile and highland accessions. *Journal of Plant Breeding and Crop Science* 1, 210–216.

Ballon, E., Robison, D., Johnson, D., Nelson, S. and Claure, T. (1991) The adaptation of quinoa varieties (*Chenopodium quinoa* Willd.) and their implications for selection for yield and size of grain. *Anales V Congreso Internacional sobre Cultivos Andinos*.

Barrett, B.A. and Kidwell, K.K. (1998) AFLP-based genetic diversity assessment among wheat cultivars from Pacific Northwest. *Crop Science* 38, 1261–1271.

Bertero, H.D., Vega A.J. de la, Correa, G., Jacobsen, S.-E. and Mujica, A. (2004) Genotype and genotype-by-environment interaction effects for grain yield and grain size of quinoa (*Chenopodium quinoa* Willd.) as revealed by pattern analysis of international multienvironment trials. *Field Crops Research* 89, 299–318.

Bhargava, A., Shukla, S., Katiyar, R.S. and Ohri, D. (2003) Selection parameters for genetic improvement in *Chenopodium* grain on sodic soil. *Journal of Applied Horticulture* 5, 45–48.

Bhargava, A., Shukla, S. and Ohri, D. (2006) *Chenopodium quinoa*: an Indian perspective. *Industrial Crops and Products* 23, 73–87.

Bhargava, A., Shukla, S. and Ohri, D. (2007a) Genetic variability and interrelationship among various morphological and quality traits in quinoa (*Chenopodium quinoa* Willd.). *Field Crops Research* 101, 104–116.

Bhargava, A., Shukla, S., Rajan, S. and Ohri, D. (2007b) Genetic diversity for morphological and quality traits in quinoa (*Chenopodium quinoa* Willd.) germplasm. *Genetic Resources and Crop Evolution* 54, 167–173.

Bhargava, A., Shukla, S. and Ohri, D. (2007c) Evaluation of foliage yield and leaf quality traits in *Chenopodium* spp. in multiyear trials. *Euphytica* 153, 199–213.

Bisht, I.S., Bhat, K.V., Lakhanpaul, S., Latha, M., Jayan, P.K., Biswas, B.K. and Singh, A.K. (2005) Diversity and genetic resources of wild *Vigna* species in India. *Genetic Resources and Crop Evolution* 52, 53–68.

Bonifacio, A. (1995) Interspecific and intergeneric hybridization in Chenopod species. Master's thesis. Brigham Young University, Provo, Utah.

Brown, J.E., Bauman, J.M., Lawrie, J.F., Rocha, O.J. and Moore, R.C. (2012) The structure of morphological and genetic diversity in natural populations of *Carica papaya* (Caricaceae) in Costa Rica. *Biotropica* 44, 179–188.

Burton, G.W. (1952) Quantitative inheritance in grasses. In: *Proceedings of the VIth International Grassland Congress* 1, 227–283.

Carlsson, R. (1980) Quantity and quality of leaf protein concentrates from *Atriplex hortensis* L., *Chenopodium quinoa* Willd. and *Amaranthus caudatus* L., grown in Southern Sweden. *Acta Agriculturae Scandinavica* 30, 418–426.

Ceolin, A.C.G., Goncalves-Vidigal, M.C. and Vidigal-Filho, P.S. (2007) Genetic divergence of the common bean (*Phaseolus vulgaris* L.) group *Carioca* using morpho-agronomic traits by multivariate analysis. *Hereditas* 144, 1–9.

Cervantes-Martinez, C.T., Frey, K.J., White, P.J., Wesenberg, D.M. and Holland, J.B. (2002) Correlated responses to selection for greater β-glucan content in two oat populations. *Crop Science* 42, 730–738.

Cox, T.S., Murphy, J.P. and Rodgers, D.M. (1986) Changes in genetic diversity in the red winter wheat regions of the United States. *Proceedings of National Academy of Sciences (USA)* 83, 5583–5586.

Cupic, T., Tucak, M., Popovic, S., Bolaric, S., Grljusic, S. and Kozumplik, V. (2009) Genetic diversity of pea (*Pisum sativum* L.) genotypes assessed by pedigree, morphological and molecular data. *Journal of Food, Agriculture and Environment* 7, 343–348.

Dewey, D.R. and Lu, K.H. (1959) A correlation and path coefficient analysis of components of crested wheat grass seed production. *Agronomy Journal* 51, 515–518.

Dostalek, J. (1987) Influence of the mode of pollination on offsprings of some species of the genus *Chenopodium*. *Preslia* 59, 263–269.

Dudley, J.W. and Moll, R.H. (1969) Interpretation and use of estimates of heritability and genetic variance in plant introduction. *Crop Science* 9, 257–262.

Eberhart, S.A. and Russel, W.A. (1966) Stability parameters for comparing varieties. *Crop Science* 6, 36–40.

Espindola, G. and Gandarillas, H. (1985) Study of correlated characters and their effects on quinoa yield. *Boletin Genetico* 13, 47–54.

Falconer, D.S. (1981) *Introduction to Quantitative Genetics*, 2nd edn. Longman, New York.

Falconer, D.S. and Mackay, T.F.C. (1996) *Introduction to Quantitative Genetics*, 4th edn. Longman Technical and Scientific, Harlow Essex, UK.

Ferreira, J.J., Garcia-González, C., Tous, J. and Rovira, M. (2010) Genetic diversity revealed by morphological traits and ISSR markers in hazelnut germplasm from northern Spain. *Plant Breeding* 129, 435–441.

Finlay, K.W. and Wilkinson, G.N. (1963) The analysis of adaptation in a plant-breeding program. *Australian Journal of Agricultural Research* 14, 742–754.

Fisher, R.A. (1918) The correlation among relatives on the supposition of Mendelian inheritance. *Transactions of the Royal Society of Edinburgh* 52, 399–433.

Fleming, J.E. and Galwey, N.W. (1995) Quinoa (*Chenopodium quinoa*). In: Williams, J.T. (ed.) *Cereals and Pseudocereals*, Chapman and Hall, London.

Fuentes, F.F. and Bhargava, A. (2010) Morphological analysis of quinoa germplasm grown under lowland desert conditions. *Journal of Agronomy and Crop Science* 197, 124–134.

Galton, F. (1889) *Natural Inheritance*. Macmillan, London.

Galwey, N.W. (1993) The potential of quinoa as a multi-purpose crop for agricultural diversification: a review. *Industrial Crops and Products* 1, 101–106.

Gandarillas, H. (1969) Esterilidad genetica y citoplasmica en la quinua. *Turrialba* 19, 429–430.

Gandarillas, H. (1979) Mejoramiento genetico. In: Tapia, M.E. (ed.) *Quinua y Kaniwa. Cultivos Andinos, Serie Libros y Materiales Educativos*. Instituto Interamericano de Ciencias Agricolas, Bogota, Colombia, pp. 65–82.

Gandarillas-Santa Cruz, H. and Espindola-Canedo, G. (1981) Relacionentre el rendimiento y la forma de la panoja en quinua. *Turrialba* 31, 385–388.

Geribello Priolli, R.H., Pinheiro, J.B., Zucchi, M.I., Bajay, M.M. and Vello, N.A. (2010) Genetic diversity among Brazilian soybean cultivars based on SSR loci and pedigree data. *Brazilian Archives of Biology and Technology* 53, 519–531.

Grenier, C., Bramel-Cox, P.J., Noirot, M., Prasada Rao, K.E. and Hamon, P. (2000) Assessment of genetic diversity in three subsets constituted from the ICRISAT sorghum collection using random vs non-random sampling procedures A. Using morpho-agronomical and passport data. *Theoretical and Applied Genetics* 101, 190–196.

Guler, M., Adak, M.S. and Ulukan, H. (2001) Determining relationships among yield and some yield components using path coefficient analysis in chickpea (*Cicer arietinum* L.). *European Journal of Agronomy* 14, 161–166.

Haaber, J. (1991) *Chenopodium quinoa* Willd. as a green crop for the pelleting industry – the effect of heat treatment on the palatability in green pellets made of quinoa. *First European Symposium on Industrial Crops and Products*, Maastricht, The Netherlands.

Hamrick, J.L. and Godt, M.J.W. (1997) Allozyme diversity in cultivated crops. *Crop Science* 37, 26–30.

Heiser, C.B. and Nelson, D.C. (1974) On the origin of the cultivated chenopods (*Chenopodium*). *Genetics* 78, 503–505.

Holland, J.B., Nyquist, W. and Cervantes-Martinez, C.T. (2003) Estimating and interpreting heritability for plant breeding. *Plant Breeding Reviews* 22, 9–112.

Jacobsen, S.-E. (1997) Adaptation of quinoa (*Chenopodium quinoa*) to Northern European agriculture: studies on developmental pattern. *Euphytica* 96, 41–48.

Jacobsen, S.-E. (1998) Developmental stability of quinoa under European conditions. *Industrial Crops and Products* 7, 169–174.

Jacobsen, S.-E. and Stølen, O. (1993) Quinoa: morphology, phenology and prospects for its production as a new crop in Europe. *European Journal of Agronomy* 2, 19–29.

Jacobsen, S.-E., Jorgensen, I. and Stølen, O. (1994) Cultivation of quinoa (*Chenopodium quinoa*) under temperate climatic conditions in Denmark. *Journal of Agricultural Sciences* 122, 47–52.

Jacobsen, S.-E., Hill, J. and Stølen, O. (1996) Stability of quantitative traits in quinoa (*Chenopodium quinoa* Willd). *Theoretical and Applied Genetics* 93, 110–116.

Jerome, F.N., Henderson, C.R. and King, S.C. (1956) Heritabilities, gene interactions and correlations associated with certain traits in the domestic fowl. *Poultry Science* 35, 995–1013.

Karp, A. and Edwards, K.J. (1995) Techniques for the analysis, characterization and conservation of plant genetic resources. In: Ayad, W.G., Jaradat, A. and Rao, V.R. (eds) *Molecular Techniques in the Analysis of Extent of Genetic Diversity*. Reports of IPGRI Workshop, 9–11 October 1995, Rome, Italy, pp. 39–43.

Korkut, Z.K., Baser, I. and Bilir, S. (1993) Makarnalik bugdaylarda korelasyon ve path katsayilari uzerinde calismalar. *Makarnalik Bugday ve Mamulleri Simpozyumu*, Ankara, pp. 183–187.

Kurt, Y., Kaya, N. and Isik, K. (2008) Isozyme variations in four natural populations of *Cedrus libani* A.Rich. in Turkey. *Turkish Journal of Agriculture and Forestry* 32, 137–145.

Lescano, J.L. (1989) Recursos fitogenéticos altoandinos y bancos de germoplas. In: *Curso: 'Cultivos altoandinos'*. Potosí, Bolivia. 17–21 April 1989, pp. 1–18.

Lescano, R.J.L. (1980) Avances en la genetica de la quinua. In: *Primera Reunion de Genetica y Fitomejoramiento de la Quinua*. Universidad Nacional Tecnica del Altiplano, Instituto Boliviano de Tecnologia Agropecuaria, Instituto Interamericano de Ciencias Agricolas, Centro Internacional de Investigaciones para el Desarrollo, Puno, Peru, pp. B1–B9.

Limburg, H. and Mastebroek, H.D. (1997) Breeding high yielding lines of *Chenopodium quinoa* Willd. with saponin free seed. In: Stølen, O., Bruhn, K., Pithan, K. and Hill, J. (eds) *Small Grain Cereals and Pseudo-Cereals*. Proc. COST 814 workshop, 22–24 February 1996, Copenhagen, Denmark, pp. 103–114.

Lindhout, P. and Danial, D. (2006) Participatory genomics in quinoa. *Tailoring Biotechnologies* 2, 31–50.

Masoumi, S.M., Kahrizi, D., Rostami-Ahmadvandi, H., Soorni, J., Kiani, S., Mostafaie, A. and Yari, K. (2012) Genetic diversity study of some medicinal plant accessions belong to Apiaceae family based on seed storage proteins patterns. *Molecular Biology Reports* 39, 10361–10365.

Mastebroek, H.D. and Limburg, H. (1997) Breeding for harvest security in *Chenopodium quinoa*. In: Stølen, O., Bruhn, K., Pithan, K. and Hill, J. (eds) *Small Grain Cereals and Pseudo-Cereals*. Proc. COST 814 Workshop. 22–24 February 1996, Copenhagen, Denmark, pp. 79–86.

Mastebroek, H.D. and van Soest, L.J.M. (1994) Gierstmelde blijkt multi-purpose-gewas [*Chenopodium quinoa* proves multipurpose crop]. *Prophyta* 1, 15–17.

Mastebroek, H.D., van Loo, E.N. and Dolstra, O. (2002) Combining ability for seed yield traits of *Chenopodium quinoa* breeding lines. *Euphytica* 125, 427–432.

Maughan, P.J., Bonifacio, A., Coleman, C.E., Jellen, E.N., Stevens, M.R. and Fairbanks, D.J. (2007) Quinoa: *Chenopodium quinoa*. In: Kole, C. (ed.) *Genome Mapping and Molecular Breeding in Plants. Volume 3 Pulses, Sugar and Tuber Crops*. Springer, Berlin, pp. 147–158.

McCouch, S. (2004) Diversifying selection in plant breeding. *PLoS Biology* 2, e347.

Mehdi, M. and Anwar, A. (2009) Role of genetically engineered system of male sterility in hybrid production of vegetables. *Journal of Phytology* 1, 448–460.

Mohammadi, S.A. and Prasanna, B.M. (2003) Analysis of genetic diversity in crop plants: salient statistical tools and considerations. *Crop Science* 43, 1235–1248.

Miller, P.A., William, J.C., Robinson, H.F. and Comstock, R.E. (1958) Estimates of genotypic and environmental variances and covariances in upland cotton and their implication in selection. *Agronomy Journal* 50, 126–131.

Nelson, D.C. (1968) Taxonomy and origins of *Chenopodium quinoa* and *Chenopodium nuttalliae*. PhD. thesis, University of Indiana, Bloomington, Indiana.

Ochoa, J. and Peralta, E. (1988) Evaluación preliminar morfológica y agronomica de 153 entradas de quinua en Santa Catalina, Pichincha. *Actas del VI Congreso Internacional sobre Cultivos Andinos*. Quito, Ecuador, pp. 137–142.

Ortiz, R., Ruiz-Tapia, E.N. and Mujica-Sanchez, A. (1998) Sampling strategy for a core collection of Peruvian quinoa germplasm. *Theoretical and Applied Genetics* 96, 475–483.

Ortiz, R., Madsen, S., Ruiz-Tapia, E.N., Jacobsen, S.-E., Mujica-Sanchez, A., Christiansen, J.L. and Stolen, O. (1999) Validating a core collection of Peruvian quinoa germplasm. *Genetic Resources and Crop Evolution* 46, 285–290.

Pratta, G., Zorzoli, R. and Picardi, L. (2000) Multivariate analysis as a tool for measuring the stability of morphometric traits in *Lycopersicon* plants from in vitro culture. *Genetics and Molecular Biology* 23, 479–493.

Rea, J. (1969) Biologia floral de la quinoa (*Chenopodium quinoa*). *Turrialba* 19, 91–96.

Risi, J. and Galwey, N.W. (1984) The *Chenopodium* grains of the Andes: Inca crops for modern agriculture. *Advances in Applied Biology* 10, 145–216.

Risi, J. and Galwey, N.W. (1989) The pattern of genetic diversity in the Andean grain crop quinoa (*Chenopodium quinoa* Willd.). I. Association between characteristics. *Euphytica* 41, 147–162.

Risi, J. and Galwey, N.W. (1991) Genotype × environment interaction in the Andean grain crop quinoa (*Chenopodium quinoa*) in temperate environments. *Plant Breeding* 107, 141–147.

Robinson, H.F., Comstock, R.E. and Harvey, P.H. (1949) Estimates of heritability and degree of dominance in corn. *Agronomy Journal* 41, 353–359.

Rojas, W. (2003) Multivariate analysis of genetic diversity of Bolivian quinoa germplasm. *Food Reviews International* 19, 9–23.

Shafii, B. and Price, W.J. (1998) Analysis of genotype-by-environment interaction using the additive main effects and multiplicative interaction model and stability estimates. *Journal of Agricultural, Biological and Environmental Statistics* 3, 335–345.

Shukla, G.K. (1972) Some statistical aspects of partitioning genotype-environmental components of variability. *Heredity* 29, 237–245.

Shukla, S., Bhargava, A., Chatterjee, A., Pandey, A.C. and Mishra, B.K. (2010) Diversity in phenotypic and nutritional traits in vegetable amaranth (*Amaranthus tricolor*), a nutritionally underutilized crop. *Journal of the Sciences of Food and Agriculture* 90, 139–144.

Silvestri, V. and Gil, F. (2000) Alogamia en quinua. Tasa en Mendoza (Argentina). *Revista de la facultad de Ciencias Agrarias*. Universidad Nacional de Cuyo, pp. 71–76.

Simmonds, N.W. (1971) The breeding system of *Chenopodium quinoa*. I. Male sterility. *Heredity* 27, 73–82.

Smith, J.S.C. (1984) Genetic variability within U.S. hybrid maize: multivariate analysis of isozyme data. *Crop Science* 24, 1041–1046.

Smith, J.S.C., Paszkiewics, S., Smith, O.S. and Schaeffer, J. (1987) Electrophoretic, chromatographic and genetic techniques for identifying associations and measuring genetic diversity among corn hybrids. In: *Proc. 42nd Annual Corn and Sorghum Research Conference*, Chicago, Illinois. Am. Seed Trade Assoc., Washington, DC, pp. 187–203.

Spehar, C.R. and Santos, R.L.B. (2005) Agronomic performance of quinoa selected in the Brazilian Savannah. *Pesquisa Agropecuária Brasileira* 40, 609–612.

Tai, G.C.C. (1971) Genotypic stability analysis and its application to potato regional trials. *Crop Science* 11, 184–190.

Tapia, M. (1990) *Cultivos Andinos subexplotados y su aporte a la alimentación*. Instituto Nacional de Investigación Agraria y Agroindustrial INIAA–FAO, Oficina para América Latina y El Caribe, Santiago de Chile.

Thillainathan, M. and Fernandez, C.J. (2002) A novel approach to plant genotypic classification in multi-site evaluation. *HortScience* 37, 793–798.

Thompson, J.A., Nelson, R.L. and Vodkin, L.O. (1998) Identification of diverse soybean germplasm using RAPD markers. *Crop Science* 38, 1348–1355.

Ward, S.M. (1998) A new source of restorable cytoplasmic male sterility in quinoa. *Euphytica* 101, 157–163.

Ward, S.M. (2000) Response to selection for reduced grain saponin content in quinoa (*Chenopodium quinoa* Willd). *Field Crops Research* 68, 157–163.

Ward, S.M. and Johnson, D.L. (1993) Cytoplasmic male sterility in quinoa. *Euphytica* 66, 217–223.

Ward, S.M. and Johnson, D.L. (1994) A recessive gene determining male sterility in quinoa. *Journal of Heredity* 85, 231–233.

Wilson, H.D. (1980) Artificial hybridization among species of *Chenopodium* sect. *Chenopodium*. *Systematic Botany* 5, 253–263.

Wilson, H.D. and Heiser, C.B. (1979) The origin and evolutionary relationships of 'huauzontle' (*Chenopodium nuttaliae* Safford), domesticated chenopod of Mexico. *American Journal of Botany* 66, 198–206.

Wright, S. (1921) Correlation and causation. *Journal of Agricultural Research* 20, 557–558.

11 Molecular Studies

Francisco Fuentes and Andrés Zurita-Silva

11.1 Introduction

Recently, an increasing number of publications have highlighted the potential of quinoa as alternative crop species, especially in areas where climatic conditions are affecting food security and people's survival. In this context, many of quinoa's productive and environmental-adaptive attributes have received new attention from scientists in order to develop new strategies to address major production concerns (e.g. disease resistance, stress tolerance, saponin content, early maturity and high yields). Among these strategies, molecular techniques have emerged as an important link between agricultural and systems biology research to facilitate development of improved plant breeding methods and enhance our knowledge of previously poorly understood plant genomes. So far, one of the most important molecular genetic resources reported in quinoa has been molecular markers. This molecular approach has been widely used to characterize quinoa's genetic diversity along the Andes as well as to create the first generation of genetic linkage maps. Functional genomic techniques have also been important for identification and characterization of genomic regions involved in the expression of target traits such as salt stress tolerance, seed development and saponin synthesis. Taken together, the information emerging from these studies will be key to addressing the new questions related to germplasm characterization (e.g. *in situ* vs *ex situ* collections) and taxonomic gaps within the family Amaranthaceae, as well as genetic dissection of new target traits that could be useful for breeding purposes and comparative analyses with closely related species and other crops. This chapter presents the recent progress in developing molecular genetic resources for quinoa and the potential impact of their use for the next generation challenges and opportunities facing quinoa research.

11.2   First Generation of Molecular Studies in Quinoa

Hugh Wilson initiated the first molecular studies in quinoa during the 1980s. These works were focused on establishing genetic variability into a representative collection of germplasm grouped into domesticated quinoa germplasm (including the five quinoa ecotypes) and a free-living collection, namely *C. hircinum* and wild quinoa (ajara). By using a comparative morphometric and electrophoretic data analysis of 21 isozyme loci, the glutamate oxaloacetate transaminase (GOT), isocitrate dehydrogenase (IDH), leucine aminopeptidase (LAP), malic dehydrogenase (MDH), phosphoglucoisomerase (PGI) and phosphoglucomutase (PGM) systems were assessed. The results of these works highlighted for the first time, on the basis of molecular information, two distinct groups: a coastal type from southwestern Chile and an Andean type distributed at elevations above 1800 m from northwestern Argentina to southern Colombia. In addition, a low level of intraspecific variation was detected among Andean ecotypes based on patterns of molecular and leaf-shape variation, suggesting a co-evolutionary relationship between domesticated and free-living populations of the southern highlands possibly derived from human-mediated dispersal and genetic manipulation (Wilson, 1988a, 1988b).

After these pioneering studies, further efforts were later addressed to provide a new set of molecular tools. Thus, Fairbanks *et al.* (1990) developed a protein-based approach to characterize quinoa seed proteins on the basis of electrophoretic mobility, solubility fractionation and genetic variability, from a wide genetic base. The analysis revealed the presence of three polymorphic polypeptides from the globulin fraction of approximately 34.3, 35.6 and 36.2 kD in size, suggesting their use as an effective tool for cultivar identification and breeding programmes for improved protein quantity and quality based on the major seed storage proteins of quinoa. Similarly, Bhargava *et al.* (2005) described the seed protein profiles of 40 cultivated and wild taxa of *Chenopodium* compared by sodium dodecyl sulfate polyacrylamide gel electrophoresis (SDS-PAGE). In this study, eight accessions of *C. quinoa* were clustered together showing genetic similarity with closely related *C. bushianum* (39.3–76.2% genetic similarity) and *C. berlandieri* subsp. *nuttalliae* (26.5–64.5% genetic similarity). The findings reported from the above study were congruent with taxonomic position (subsect. Favosa of sect. *Chenopodium*) and crossability relationships and other biochemical characters reported in these species. Although these protein-based approaches developed in quinoa have revealed interesting issues related to its genetic relationships with the *Chenopodium* genus, their use as marker of choice has basically ceased due to its relatively small number of marker loci and its modest level of polymorphism representing only a small and not random part of the quinoa genome.

11.3   Sequence-Arbitrary Methods

11.3.1   Random amplified polymorphic DNA (RAPD) and directed amplification of minisatellite-region DNA (DAMD) markers

The first source of DNA-based markers developed in quinoa was reported by Fairbanks *et al.* (1993) using the RAPD technique. RAPD markers consist of

DNA fragments amplified by PCR using short decanucleotide primers of random sequence, which are usually able to amplify fragments from 1 to 10 genomic sites simultaneously, in a common range size between 0.5 and 5 kb. Thus, the preliminary data using RAPD markers revealed that of 30 selected primers, 26 produced polymorphic markers among 16 randomly selected quinoa accessions, indicating a relatively common presence of multiple polymorphic markers. The RAPD markers were also useful for identifying genetic variation among 19 accessions of six species of the genus *Chenopodium*, as reported by Ruas *et al.* (1999). A total of 33 10-mer primers generated 399 molecular markers with an average of 12 polymorphisms per RAPD primer, which discriminated the germplasm collection into five different clusters: (i) three cultivated varieties of *C. nuttalliae*; (ii) eight cultivars and two wild varieties of *C. quinoa*; (iii) *C. berlandieri* and *C. album*; (iv) two accessions of *C. pallidicaule*; and (v) two accessions of *C. ambrosioides*. Nevertheless, as in previous reports using different molecular approaches, it was detected that wild and crop populations of *C. quinoa* share a low level of molecular variation, without differentiation between sympatric domesticated and weedy populations, and a low level of intraspecific variation within the accessions of *C. quinoa*, *C. nuttalliae* and *C. pallidicaule*. Similarly, RAPD markers have also helped understanding of the hierarchical structure of ecotype populations of the highlands and inter-Andean valleys in Bolivia. The findings reported by Del Castillo *et al.* (2007), scoring 38 selected bands from ten RAPD primers on eight representative populations (*n* = 87) directly sampled in farmers' fields, revealed a marked geographical effect on the populations' structure, probably explained by climatic and orographic barriers in the studied zone rather than a distance effect. Thus population structure was related to the three major biogeographic zones present in Bolivia: northern and central highland, inter-Andean valley and southern Salar. Interestingly, the intra-population genetic diversity was higher than expected, because of the mainly autogamous reproduction as well as the limited seed exchange among these isolated regions. Genetic diversity was higher than that reported in studies based on germplasm collections, suggesting that current germplasm collections may not be representative of the genetic variation of the quinoa complex and that further sampling for *ex situ* conservation will also have to take into account the hierarchical structure of the genetic variation.

A combined study conducted by Rana *et al.* (2010) using RAPD and DAMD markers revealed the suitability and reliability of these markers to assess molecular diversity in the *Chenopodium* genus. Thus, the amplification of 12 decanucleotide primers of random sequence and four minisatellite or variable number of tandem repeat (VNTR) sequences yielded a total of 350 polymorphic markers on 55 accessions belonging to 14 species of chenopods. The taxa was divided into two main UPGMA clusters; the first cluster grouped all the accessions of *Chenopodium quinoa* and its related species *C. berlandieri* subsp. *nuttalliae*, one *C. album* (4×) from Mexico and three north Indian 2× accessions of *C. album*, while the other clusters comprised mainly 6× accessions of *C. album* and *C. giganteum* in addition to *C. strictum*, *C. bushianum*, *C. opulifolium* and *C. ficifolium*. Other wild species in the analysis were more or less distant to their taxonomic position, including representations of *C. ugandae*,

C. botrys, *C. foetidum*, *C. pallidicaule*, *C. murale* and *C. vulvaria*. Thus the analysis of these markers allowed intra- and interspecific variation within cultivated and non-cultivated species in this large genus to be assessed and taxonomic problems to be solved either at or below the species level.

Many different random sequence primers have been useful for assessing the quality and frequency of polymorphisms in several crop species (e.g. maize, soybean), permitting reliable scoring of polymorphisms in segregating populations. In spite of this, RAPD makers have been partially involved to generate genetic linkage maps in quinoa. The first genetic linkage map of quinoa reported by Maughan *et al.* (2004) included six RAPD primers, which generated an average of 3.8 prominent bands per RAPD amplification. Interestingly, the analysis revealed that one RAPD marker was scored in a co-dominant fashion, while the remaining five were scored as dominant. The data reported in this study demonstrated that single primers could be used to amplify DNA from dispersed polymorphic loci and thus increase the saturation of maps by filling in some genomic gaps. Despite its utility to detect polymorphism in quinoa, the RAPD technique can be less reproducible in comparison with other molecular approaches (e.g. isozyme, AFLP and/or SSR analysis) because of its sensitivity to the reaction conditions, DNA quality and PCR temperature profiles.

11.3.2 Amplified fragment length polymorphism (AFLP) markers

Another DNA-based marker successfully used in quinoa is AFLP. This technique is based on the detection of genomic restriction fragments by PCR amplification and has been efficiently utilized in generating the first genetic linkage map in quinoa. In this study, Maughan *et al.* (2004) reported a total of 230 polymorphic AFLP amplification products from 68 primer combinations scored on a F_2 population obtained from a Chilean lowland/coastal type (Ku-2) and a Peruvian highland type (0654) as female and male parent, respectively. The marker analysis revealed that 160 dominant markers were specific to 0654 and 144 were specific to Ku-2, while 26 of them were scored as co-dominant. Likewise, the chi-square test of segregation distortion showed that there was no statistically significant deviation from a disomic segregation model from the expected Mendelian ratio (3:1 or 1:2:1), in spite of the reported studies showing erratic and unpredictable meiotic pairing behaviour of some homoeologous chromosomes in quinoa affecting monogenic morphological traits (Ward, 2000). Taken together, these findings suggest that the AFLP markers are highly reliable for genetic analysis in *C. quinoa*. As a result, AFLP markers have been used to describe genetic relationships among 14 quinoa landraces from southern Chile, three landraces from the highland of northern Chile, a representation of *C. album* and *C. ambrosioides* (Anabalón-Rodriguez and Thomet-Isla, 2009). This study reported 150 AFLP bands generated by three *Eco*RI-*Mse*I primer combinations, of which 130 (86.6%) bands were polymorphic. Unfortunately the difficulties associated with AFLP marker technologies have limited its extensive use for genetic analysis in quinoa and little published information is available.

11.4   Sequence-Dependent Methods

11.4.1   Simple-sequence repeat (SSR) markers

In an attempt to make new molecular resources widely available, a Brigham Young University research team in the USA developed a collection of ~450 SSR markers for quinoa (Maughan *et al.*, 2004; Mason *et al.*, 2005; Jarvis *et al.*, 2008). In general, these markers, also called microsatellite or tandem repeat, consist of around 10–50 copies of motifs from 1 to 10 base pairs of DNA that can occur in perfect tandem repetition, as imperfect repeats (interrupted) or together with another repeat type. In this context, the SSR markers represent the best option for population studies, because of their co-dominant nature and their aptitude to detect a high level of polymorphism. Thus the reported SSR markers for quinoa have been successfully used to assess genetic diversity in different germplasm collections as well as to develop genetic linkage maps. Documented examples of the aforementioned are the assessment of genetic diversity performed by Christensen *et al.* (2007) in the USDA and CIP-FAO international nursery collections of quinoa; the assessment of genetic diversity patterns in Chilean quinoa germplasm reported by Fuentes *et al.* (2009, 2012); the characterization of genetic structure in cultivated quinoa of Northwest Argentina described by Costa Tártara *et al.* (2012) and the first genetic linkage maps developed by Maughan *et al.* (2004) and Jarvis *et al.* (2008) (Table 11.1).

The first large-scale development of quinoa SSR markers consisted of 208 polymorphic markers, which were validated and characterized in 31 cultivated quinoa accessions representative of the main growing areas of South America. This set of markers developed by Mason *et al.* (2005), on the basis of the Bolivian quinoa variety 'Surimi', was obtained by using a microsatellite-enriched library method, where the most common repeated motifs reported were CA, AAT, ATG, GA and CAA. The genetic analysis performed in the quinoa collection revealed a number of observed alleles ranging from 2 to 13, with an average of four alleles detected per locus. Heterozygosity values ranged from 0.2 to 0.9, with a mean value of 0.57. In addition, 67 SSR markers were highly polymorphic, with heterozygosity values equal to or greater than 0.70.

On the other hand, phenetic analysis showed similarity coefficients ranged from 0.22 to 0.65, with the least genetic similarity detected between a Peruvian highland ecotype accession (0654) and a Chilean coastal/lowland ecotype accession (G-205-95DK), and the greatest similarity occurring between the two sister-lines Chilean lowland/coastal accessions (KU-2 and RU-2). These findings correlated well with the aforementioned information based in different molecular approaches, which separated the quinoa germplasm into two distinct groups: highland and lowland/coastal type, with the highland group containing the Salar and inter-Andean valleys ecotypes.

Table 11.1. Overview of genetic diversity studies in different quinoa germplasm collections using DNA-based markers.

Marker type	# loci	Total alleles	Genetic diversity measure	# accessions	Ecotypes	Origin	References
AFLP	3	130	0.54–0.97[a]	18	LC-S	Ch	Anabalon-Rodriguez and Thomet-Isla (2009)
RAPD	10	38	0.10–0.22[b]	87	HL	B	Del Castillo *et al.* (2007)
SNP	427	854	0.02–0.50[c]	113	S-LC-HL-IAV	E, P, B, Ch, A	Maughan *et al.* (2012)
SSR	36	420	0.45–0.94[d]	151	S-LC-HL-IAV	E, P, B, Ch, A	Christensen *et al.* (2007)
SSR	216	888	0.12–0.90[d]	23	S-LC-HL-IAV	Ch, C, E, P, B, A	Jarvis *et al.* (2008)
SSR	20	150	0.07–0.90[d]	59	S-LC	Ch	Fuentes *et al.* (2009)
SSR	22	354	0.58–0.93[d]	35	S	A	Costa Tártara *et al.* (2012)
SSR	20	118	0.12–0.87[d]	34	S-LC-HL-IAV	Ch, C, E, P, B, A	Fuentes *et al.* (2012)

[a]Genetic similarity (Simple Matching coefficient); [b]He: averaged genetic diversity; [c]MAF: minor allele frequency; [d]heterozygosity. S: Salar; LC: lowland/coastal; HL: highlands; IAV: inter-Andean valley; C: Colombia; E: Ecuador; P: Peru; B: Bolivia; A: Argentina; Ch: Chile.

Interestingly, this set of SSR markers also revealed the potential utility for molecular studies across related species of the Chenopodioidae subfamily. Thus, the amplification of 202 of the 208 polymorphic SSR markers on a group consisting of two *C. pallidicaule* (Canihua, South America), two *C. berlandieri* subsp. *nuttaliae* (Huazontle, Central America) and two accessions of *C. giganteum* (Khan chi, Asia) revealed that 67% ($n = 136$) of markers amplified successfully in all groups assessed (Mason *et al.*, 2005). In addition, the highest level of PCR conservation was observed in *C. berlandieri* subsp. *nuttalliae*, with 99.5% reproducible amplifications, of which 81% were polymorphic on the basis of the two accessions assessed, confirming the close ancestry with quinoa. Moreover, the lowest level of conservation was for *C. pallidicaule*, with 74% amplifications, of which 10% were polymorphic between the two accessions. Finally, *C. giganteum* presented 81% specific PCR products, of which 34% were polymorphic. Likewise, 20 di-/tri-nucleotide loci microsatellites utilized by Fuentes *et al.* (2009) have also demonstrated the utility of SSR markers to analyse quinoa and related species. In this regard, these markers presented differential amplification pattern in a panel of *C. quinoa* cv. Regalona, *C. hircinum*, *C. carnosolum*, *C. murale* and *C. ambrosioides* accessions (F. Fuentes, unpublished data). Thus, these findings suggest the potential utility of this set of SSR markers for further genetic analysis across related species of the genus as well as covering the taxonomic gaps within the family Amaranthaceae (Fig. 11.1).

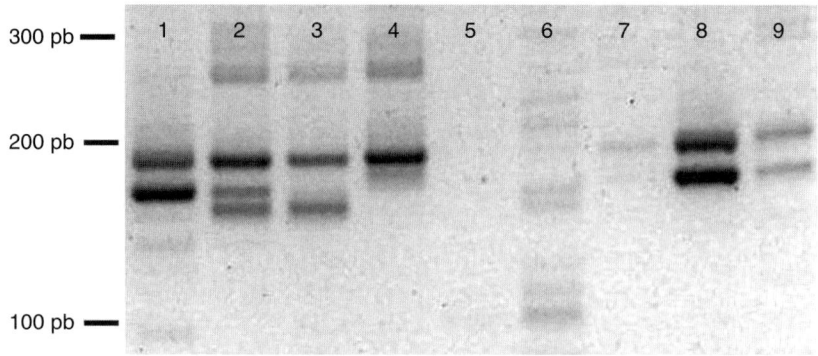

Fig. 11.1. Metaphor® gel profile of quinoa and wild relatives accessions amplified using SSR QCA120 locus. (1) *C. quinoa* cv. Regalona; (2) *C. hircinum* [32]; (3) *C. hircinum* [35]; (4) *C. hircinum* [36]; (5) *C. carnosolum* [41]; (6) *C. murale* [43]; (7) *C. hircinum* [43]; (8) *C. hircinum* [44]; and (9) *C. ambrosioides* [46].

Further efforts to develop a new set of polymorphic SSR markers and the construction of a new genetic linkage map of quinoa was reported by Jarvis *et al.* (2008). This work described the development of 216 new SSR markers through similar procedures as previously reported by Mason *et al.* (2005), as well as six SSR markers developed from bacterial artificial chromosome-end sequences (BES-SSRs). The most common repeats observed for the SSR markers in this study were GA, CAA and AAT. Likewise, repeats including CA, CGA, GAA and GGT also were observed, though in low frequency. The screening of these 216 SSR markers across 22 quinoa accessions representative of the South American Andes revealed an observed number of alleles per SSR ranging from 2 to 13, with an average of four alleles per SSR with *H* values from 0.12 to 0.90, with an average value of 0.57, similar to those previously reported by Mason *et al.* (2005) (Table 11.1). The 216 markers described by Jarvis *et al.* (2008) were considered polymorphic ($H \geq 0.10$) and 53 were considered highly polymorphic ($H \geq 0.70$). In addition, the polymorphism differences between di- and tri-nucleotide motifs confirmed the common observation that the development of highly polymorphic microsatellites markers should be focused to use tri-nucleotide motifs with a repeat of >20 bp.

11.4.2   Single nucleotide polymorphism (SNP) markers

In quinoa, the first source of SNP marker was reported by Coles *et al.* (2005). SNP markers consist of a single base change in a DNA sequence, with a usual alternative of two possible nucleotides at a given position. This kind of tool has quickly become the marker system of choice for many researchers because of its higher reproducibility, higher map resolution, higher throughput, lower cost and a lower error rate as reported in several economically important crops (e.g. corn, soybean, wheat). The SNPs markers reported by Coles *et al.* (2005)

were obtained from both an immature seed and floral expressed sequence tags (EST) libraries, which yielded a total of 51 SNPs markers in 20 EST sequences analysed, consisting of 38 single-base changes and 13 insertions–deletions (Indels), with an average of 1 SNP per 462 base pairs (bp) and 1 Indel per 1812 bp. In addition, when the *C. berlandieri* subsp. *jonesianum* accession was included in the analysis, 81 additional SNPs were identified, bringing the total number of SNPs discovered to 132 (1 per 179 bp).

Recently, a large-scale set of SNP markers has been reported by Maughan *et al.* (2012) to develop functional SNP assays for quinoa. In this work, the SNPs discovery was made using a genomic reduction protocol based on a representative set of eight genotypes including highland, Salar and lowland/coastal quinoa ecotypes. Of 511 functional SNP markers based on KASPar genotyping chemistry and Fluidigm dynamic array platform detection, 427 were utilized to analyse a set of 113 quinoa accessions that showed minor allele frequency (MAF) values between 0.02 and 0.50, with an average MAF of 0.28 per SNP. In this context, 46% of SNP loci were highly polymorphic (MAF $\geq$ 0.35) and 90% of them were polymorphic (MAF $\geq$ 0.10). As expected, the most frequent point mutation among all SNPs identified corresponded to transitions (A/G or C/T), being 1.6× higher than transversions (A/T, C/A, G/C, G/T). These results were higher than those reported previously by Coles *et al.* (2005), where 38 single-base changes presented a transition ($n = 20$) to transversion ($n = 18$) ratio of 1:1. Nevertheless, the higher than expected C/T (G/A) transition mutation rate has been well documented in other species (sugar beets) and is thought probably to result from hypermutability effects of CpG di-nucleotide sites and deamination of methylated cytosines.

The phenetic analysis reported in this study was consistent with all the previous molecular studies to separate the two major quinoa groups, namely the Andean and lowland/coastal ecotypes. In addition, the study also reported the potential transferability of SNP markers to related species, including two accessions of *C. hircinum*, four of *C. berlandieri* (subsp. *nuttalliae*, var. *macrocalycium*, var. *boscianum*, var. *zschackei*), and one accession of both *C. watsonii* and *C. ficifolium*. The test results showed that 34% of SNPs amplified in all eight accessions, while 17% of them failed to amplify in any of the related species. The analysis also revealed that the two *C. hircinum* accessions and the four *C. berlandieri* accessions presented 81% and 79% of successful amplification, respectively, confirming the close crop–weed sympatric relationship of these two tetraploid species with quinoa. Nevertheless, the inability to separate related species into discrete groups in the cluster analysis suggested the limited use of this set of SNP markers for phylogenetic studies at genus level.

11.5 Genetic Linkage Maps

The first quinoa genetic linkage map was constructed by means of molecular markers and was reported by Maughan *et al.* (2004). This map consisted of 35 genetic linkage groups containing a total of 230 amplified fragment length polymorphisms (AFLPs), 19 microsatellite or simple sequence repeats (SSRs), and six RAPD markers, spanning 1020 cM with an average density of 4.0 cM per marker. This map was based on 80 F_2 individuals derived from the cross of KU-2 and 0654 genotypes, a Chilean lowland type and a Peruvian Altiplano type, respectively. Four years later, a second version of the genetic linkage map was published by Jarvis *et al.* (2008), which was based on 275 polymorphic loci, including 200 SSR, 70 AFLP, two 11S seed storage protein loci, the nucleolar organizing region (NOR) and the morphological betalain colour locus (Fig. 11.2). The genetic map was comprised of 38 linkage groups in a recombinant inbred lines (RIL) population consisting of 82 F_6 individuals from a cross of Ku-2 and 0654. Nevertheless, this map just spanned 913 cM, namely approximately 54% of the predicted 1700-cM quinoa genome described by Maughan *et al.* (2004).

More recently, the first SNP-based linkage map has been developed from the large-scale set of SNP markers reported by Maughan *et al.* (2012). This map consisted of 29 linkage groups with 20 large linkage groups, spanning 1404 cM with a marker density of 3.1 cM per SNP marker. This linkage map was constructed employing a $F_{2:8}$ RIL population from two advanced quinoa mapping populations (Pop1 and Pop39) sharing a common paternal parent (0654, Altiplano type) whose molecular data were combined to construct an integrated linkage map based on 128 individuals. In this context, this SNP-based map consisted of approximately a twofold number of marker loci and spanned a greater genetic coverage than the previous reported maps, being closer to the 1700 cM total length of the quinoa map predicted (Maughan *et al.*, 2004). This latter result suggests that new molecular resources as well as a much larger RIL population are still needed to cover the remaining undetected areas of the quinoa genome.

11.6 Gene Discovery, Functional Genomics and Non-Coding Genotyping

The recent advances in developing genetic and genomic resources have led to an immense assembly of information in the area of plant genomics, through the sequencing project of different organisms. In spite of this, the identification of transcribed portions of the genomes by using the Expressed Sequence Tags (ESTs) approach provides a feasible alternative for analysing non-model systems and organisms with large genome sizes, wherein whole genome sequencing is still technically challenging and cost-prohibitive. The large scale of EST sequences in public databases enables access to functional information, which has been proven to be valuable for gene annotation and gene discovery. In this context, Coles *et al.* (2005) reported the development of the first EST libraries for quinoa from immature seed and floral tissues to characterize gene expression and its regulation during seed set. In this work, a total of 424 ESTs were found that corresponded to 331 sequences from the immature seed cDNA library and 83 sequences from the floral library, with an average length of 581 bp. All sequences reported in this study were deposited in the GenBank dbEST

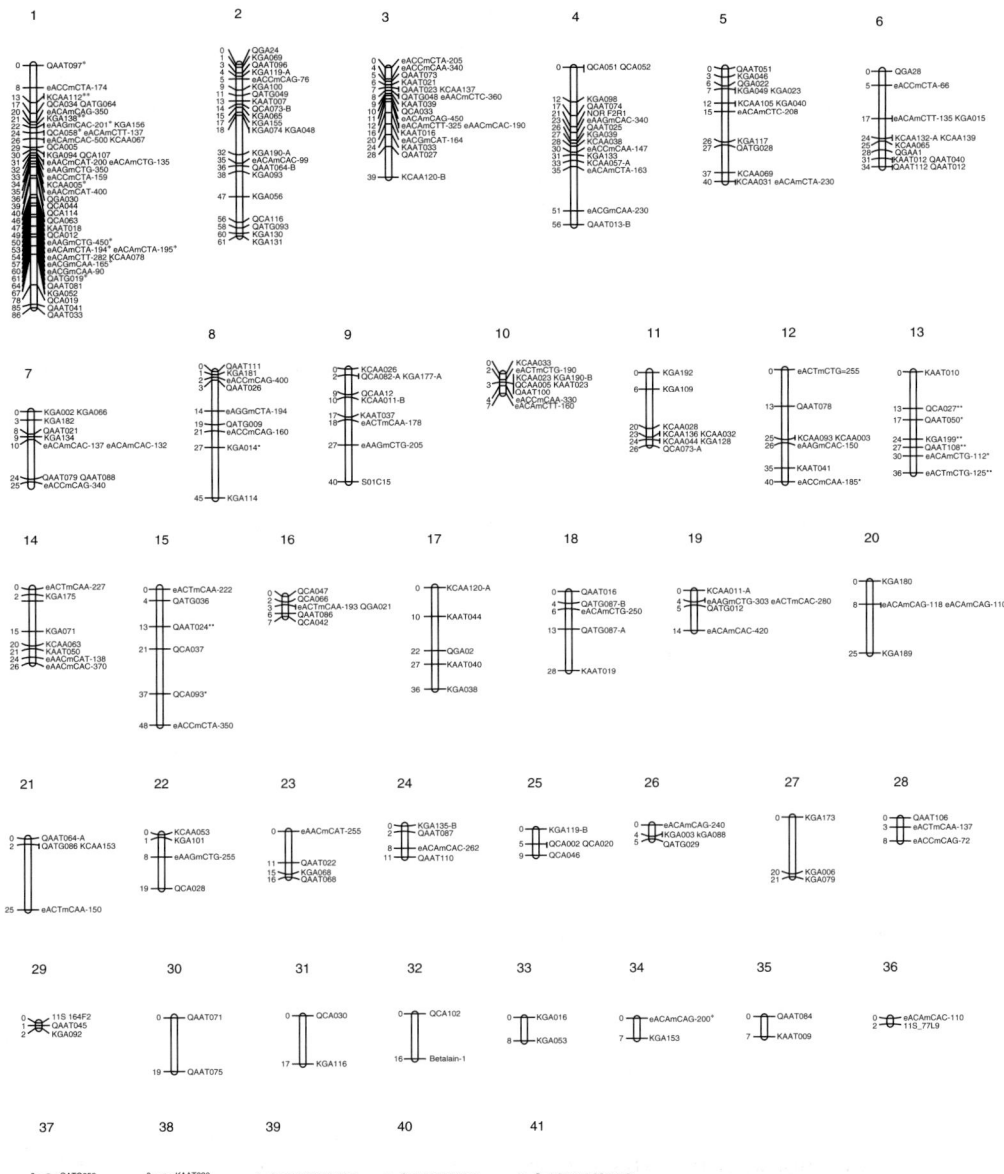

Fig. 11.2. Linkage map of the Ku-2/0654 population [Reprinted from Jarvis *et al.* (2008).] Distances in centiMorgans (cM) are indicated on the left side of each linkage group. All SSR markers begin with Q or K and AFLP markers begin with e. Markers skewed at P < 0.05 and P < 0.01 are indicated with * and **, respectively.

with accession numbers CN781906–CN782329. The homology analysis of quinoa sequences was performed using sequences present in the TIGR *Arabidopsis thaliana* and GenBank protein databases with BLASTX queries. The analysis yielded four groups of sequences, which corresponded to: (i) ESTs with significant homology to *Arabidopsis* proteins of known or putative function (67%); (ii) ESTs that match with *Arabidopsis* proteins of unknown function (9%); (iii) ESTs with homology to proteins from species other than *Arabidopsis* (6%); and (iv) ESTs not matching any protein in either database (18%). Although the main putative functions of the quinoa ESTs reported in this study corresponded to metabolism, protein synthesis, development and biogenesis of cellular components, among others, the major part of them were categorized as unclassified or not yet clear-cut. Interestingly, the most abundant ESTs from the two cDNA libraries were annotated with putative functions related to plant defence.

Subsequently, Reynolds (2009) reported the annotation of a large-scale EST collection from maturing quinoa seed tissues expressing saponins, in an attempt to elucidate the genetic components of its biosynthesis. Based on Sanger and 454 GS-FLX pyrosequencing technologies, a total of 39,366 unigenes were assembled, consisting of 16,728 contigs and 22,638 singletons. Additionally, the analysis of repeated sequences from unigenes sequences identified 291 new SSR markers (unpublished data). The assessment of transcriptional variation between sweet and bitter quinoa varieties at two different stages of development was developed using 102,834 oligonucleotide probes in a microarray assay. The microarray analysis allowed a set of candidate genes transcriptionally related with saponin biosynthesis to be identified, including genes with shared homology to cytochrome P450s, cytochrome P450 monooxygenases and glycosyltransferases, representing a potential new approach for quinoa grain improvement related to this economically important trait.

On the other hand, the sequence homology analysis on flanking sequences for polymorphic SSR marker reported by Mason *et al.* (2005) revealed that most homologies with GenBank sequences by using the BLASTN and BLASTX queries were from *Arabidopsis thaliana* or *Oryza sativa* (rice). Thus, 20% of sequences reported in this study presented a significant homology to sequences in GenBank, showing significant DNA- or protein-level homology as well as at both the nucleotide and amino acid level. The exhibited homology to annotated genes included sequences homologous to a mitochondrial pseudogene (orf764) (QAAT079), a fascilin-like arabinogalactan protein (QATG052 and QATG056), a putative mudrA transposon protein (QATG062), a putative auxin response factor IAA24 protein (QCA021), a protein kinase mRNA (QCA056), a chloroplast secA mRNA (QCA061), mRNA for a ubiquitin-conjugating enzyme UBC2 (QCA067), a GDP-fucose protein-*O*-fucosyltransferase 2 mRNA (QCA124) and a NADP-specific isocitrate dehydrogenase (QCA055). Additionally, the BLASTN and BLASTX searches in this SSR study reported by Jarvis *et al.* (2008) yielded the identification of 14 flanking sequences with significant DNA or protein homologies to sequences in the GenBank database. Thus, three clones presented homology to known sequences at the nucleotide level, while 11 were homologues to known sequences at the amino acid level only. The significant hits to annotated protein sequences on GenBank included proteins such as CCA (circadian clock-associated protein; KGA079), Nim1 (noninducible immunity-like protein; KCAA111), ICDH (isocitrate dehydrogenase; KGA026), succinyl CoA ligase (KGA171), sucrose transporter (KGA101) and oligosaccharyl transferase (KGA068). The sequence homologies identified in this study were from *Arabidopsis thaliana*, *Pinus pinaster*, *Oryza sativa japonica*, *Beta vulgaris*, *Plasmodium chabaudi*, *Allophyllum glutinosum*, *Mesembryanthemum crystallinum* and *Plantago major*.

Another approach to facilitate the exploration of the quinoa genome has been the development of Bacterial Artificial Chromosomes (BAC) libraries. This important genomic tool has been useful in many plant species for large-scale gene discovery and elucidation of gene function as well as for map-based cloning of quantitative trait loci, molecular cytogenetics and comparisons of specific regions with different species or ecotypes. Accordingly, the quinoa BAC library developed by Stevens *et al.* (2006) consisted of two separated libraries constructed by using *Bam*HI and *Eco*RI restriction endonucleases, which yielded 26,880 and 48,000 clones respectively, from the 'Real' quinoa type, with an average insert size of approximately 123 kb. These results allowed to determine the di-haploid genome (1C) of quinoa to be 967 Mbp, or 2C = 2.01 pg. In order to demonstrate the utility of this BAC library for gene identification, this study revealed the presence of two distinct genetic loci encoding 11S globulin seed storage proteins. Interestingly, the different pattern of the hybridized bands (Southern blot), occurring in a single copy, was consistent with the differences between quinoa genotypes from highland and lowland/coastal, suggesting the utility of this locus for improving the protein content and quality of quinoa grain.

In addition to its nutritional and functional features, quinoa has also been known for its ability to grow under harsh environmental conditions (e.g. soil, climate). Thus, the crop has demonstrated remarkable salt tolerance, even being able to grow in salt concentrations as high as those found in seawater. Nevertheless, this important characteristic in quinoa has been recently explored, in particular those molecular mechanisms involved in the ability to thrive in saline soils, namely those related to salt ion accumulation in specialized tissues and adjustment of leaf water potential. In an attempt to provide valuable information related to the molecular basis of salt tolerance in quinoa, Maughan *et al.* (2009) described the first molecular characterization of the *Salt Overly Sensitive 1* (*SOS1*) gene, which encodes a plasma membrane Na^+/H^+ antiporter that plays an important role in germination and growth of plants in saline environments. This study reported the complete genomic sequence of two homoeologous *SOS1* loci: *cqSOS1A* and *cqSOS1B*, which spanned 98,357 and 132,770 bp, respectively. Furthermore, when the *cqSOS1A* and *cqSOS1B* coding sequences were translated, these yielded proteins of 1158 and 1161 amino acids. The comparison of these translated genomic sequences revealed a high degree of similarity with *SOS1* sequence from species belonging to the Caryophyllales order.

On the other hand, the assessment of gene expression levels of *cqSOS1A* and *cqSOS1B* in quinoa plants grown hydroponically revealed that under saline conditions (450 mmol/l) relative gene expression of *SOS1* in roots was consistently 3–4-fold higher than in leaf tissue. Moreover, the *SOS1* expression was more strongly up-regulated by salt stress in leaves than in roots, suggesting a constitutive expression of *SOS1* genes in roots and a inducible expression in leaves under salt stress conditions. Likewise, Ruiz-Carrasco *et al.* (2011), using a similar approach, reported gene expression analyses for *CqSOS1* and cloning of *CqNHX1* genes confirming the different response to salt stress between leaf and root tissues, but also introduced the notion that they are differentially regulated in more salt-tolerant vs less salt-tolerant quinoa genotypes.

A recent revision of salt tolerance mechanisms studies in quinoa has been performed by Adolf *et al.* (2012) with the aim of elucidating the mechanisms used to cope with high salinity at various stages of plant development. They pointed to key traits that included an efficient control of Na^+ sequestration in leaf vacuoles, xylem Na^+ loading, higher ROS tolerance, better K^+ retention and an efficient control of stomatal development and aperture. The control of key transporters and channels mediating Na^+ and K^+ homeostasis in quinoa is considered pivotal for salt tolerance and reinforces the potential of quinoa for cultivation in salt-affected regions and as model for further research in the field of plant salt tolerance. Taken together, these findings open new challenges to provide a realistic view of mechanisms and signalling pathways that mediate salt stress responses in quinoa

The analysis of a non-coding segment of chloroplast DNA has provided useful information for genetic analysis in many angiosperms. Although a consensus has not been reached regarding which DNA sequences can be used as the best plant barcodes, the *psbA-trnH* spacer region has been tested extensively in recent years. The use of this barcode technique to characterize the genetic relationships among quinoa and wild relatives in quinoa fields in Chile has demonstrated evidence of natural hybridization of quinoa with *C. hircinum* based on similarities and differences between DNA sequences (Fig. 11.3; F. Fuentes, unpublished data). This finding agrees well with the current status of genetic diversity within coastal/lowlands germplasm reported by Fuentes *et al.* (2012). Based on this knowledge it has been suggested that coastal/ lowland ecotypes are representative of active crop/weed complexes having a monophyletic coevolving behaviour. This might also explain why lowland breeders find it difficult to obtain pure quinoa cultivars, because of its close affinity to weed populations of *C. album* and/or *C. hircinum* (Fuentes *et al.*, 2009).

11.7 Molecular Cytogenetics

The development of molecular genetic resources for quinoa has also involved the use of cytogenetic tools. Classical cytogenetic and recent molecular studies have revealed important information related to chromosome counts and estimation of quinoa genome size as well as cytological characterization of ribosomal RNA genes and repetitive sequences across the genome (Bhargava *et al.*, 2006; Maughan *et al.*, 2006; Kolano *et al.*, 2012). In this regard, these studies have consistently demonstrated that *C. quinoa* exhibits a tetraploid chromosome number $2n = 4x = 36$, with a basic number $x = 9$. Although the haploid genome size (1C value) of quinoa has been estimated initially between 1.005 and 1.500 pg, further determinations using Feulgen microdensitometry

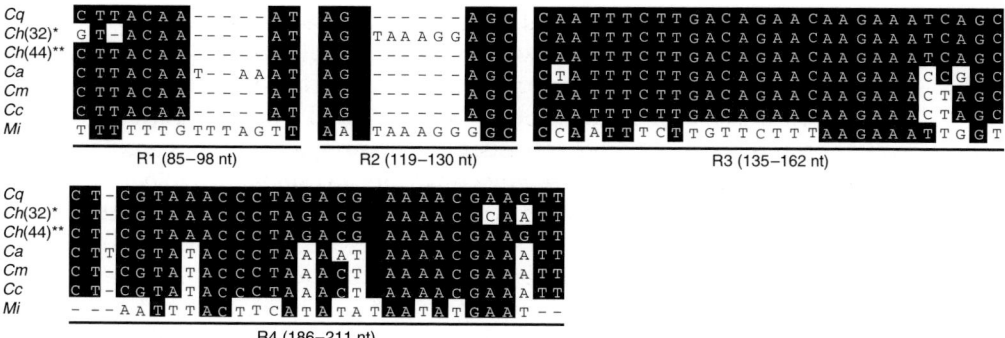

R1 (85–98 nt)   R2 (119–130 nt)   R3 (135–162 nt)

R4 (186–211 nt)

* *C. hircinum* collected from southern area in Chile, Temuco (38°40'S, 72°30'W).

** *C. hircinum* collected from northern area in Chile, Tarapaca canyon (19°56'S, 69°31'W).

Fig. 11.3. Interspecific variation of four variable regions of *psbA-trnH* non-coding sequence of chloroplast from six *Chenopodium* species (*Cq*: *C. quinoa* cv. Regalona; *Ch*: *C. hircinum*; *Ca*: *C. ambrosioides*; *Cm*: *C. murale*; *Cc*: *C. carnosolum*) and *Mossia intervallaris* as external reference. Divergent nucleotides are indicated with a white background.

have revealed values between 1.330 and 1.596 pg, suggesting a bigger genome than those reported by using flow cytometry (Bhargava *et al.*, 2007; Kolano *et al.*, 2012).

Molecular cytogenetic approaches have also provided valuable information related to localized coding and non-coding DNA sequences on chromosomes and interphase nuclei. In this context, the fluorescent *in situ* hybridization (FISH) technique has been useful to study the organization and genomic distribution of 45S (NOR, nucleolus organizer region) and 5S ribosomal RNA (rRNA) genes in quinoa. The use of this molecular approach has revealed that quinoa possesses an interstitial and terminal part of one pair of chromosomes with 5S rRNA loci and only one pair of NOR, suggesting a reduction in the number of rRNA loci during evolution of this species considering its allotetraploid origin. DNA sequence analysis of NOR intergenic spacers (IGS) confirmed the close relationship between *C. quinoa* and tetraploid *C. berlandieri* Moq. subsp. *zschackei*. Likewise, the characterization of a 5S rDNA spacer region revealed the existence of two different non-transcribed spacers (NTS) sequence classes that presumably originated from the two subgenomes of allopolyploid *C. quinoa*. Interestingly, one of these was very similar in sequence to the NTS present in *C. berlandieri*, suggesting that these two allotetraploid species have at least one common diploid ancestor (Maughan *et al.*, 2006).

Sequences obtained from the quinoa genomic DNA library have also been analysed using *in situ* hybridization. In this regard, homologues sequences to retrotransposons (15-5D, 21-5D and 22-19A) and to transposase genes (20-20I) have hybridized across the quinoa genome, without a specific chromosome or subgenome distribution pattern (Kolano, 2004). Subsequently, the chromosomal organization of a repetitive sequence (pTaq10) isolated from the *Taq*I digest of genomic DNA of quinoa was also described. Sequence analysis indicated that this 286-bp monomer was not homologous to any known retroelement sequence. Thus, *in situ* hybridization showed that this sequence presents an interspersed genomic organization pattern, which is excluded from 45S rRNA gene loci, but partly overlapped with 5S rDNA loci (Kolano *et al.*, 2008). Likewise, Kolano *et al.* (2011) using a repetitive DNA sequence (18–24J) reported abundant hybridization on 18 chromosomes. Interestingly, when this DNA sequence was re-probed with rDNA it was possible to observe in one subgenome, one 35S rRNA locus and at least half of the 5S rDNA loci present. In this same study, a second sequence was analysed (12–13P), which was localized exclusively in the pericentromeric regions of each chromosome of quinoa. However, the intensity of the FISH signals differed considerably among chromosomes. In addition, the pattern observed on quinoa chromosomes after FISH with 12–13P was very similar to genomic *in situ* hybridization (GISH) results, suggesting that the 12–13P sequence constitutes a major part of the repetitive DNA of quinoa.

11.8   Concluding Remarks

The emerging development of molecular genetic resources for quinoa offers important opportunities and challenges for applying advanced science for its improvement. Future applications of these molecular resources will probably include (i) bioinformatics-based comparative and predictive approaches at the genomic level; (ii) differential gene expression studies targeting multiple tissue types and different stress inductions; and (iii) identification of allele-specific markers for quantitative and qualitative traits (e.g. nutritional and anti-nutritional biosynthesis pathways, cytoplasmic-nuclear male-sterility system). The availability of a quinoa reference genome sequence will undoubtedly facilitate greater integration of biotechnological tools into quinoa breeding efforts to minimize the yield gap in farmer's fields in the Andes and new areas. Nevertheless, there is an urgent need to support the application of these genomic tools into traditional breeding programmes in order to achieve the desired results.

References

Adolf, V.I., Jacobsen, S.-E. and Shabala, S. (2012) Salt tolerance mechanisms in quinoa (*Chenopodium quinoa* Willd.). *Environmental and Experimental Botany* http://dx.doi.org/10.1016/j.envexpbot.2012.07.004.

Anabalon-Rodriguez, L. and Thomet-Isla, M. (2009) Comparative analysis of genetic and morphologic diversity among quinoa accessions (*Chenopodium quinoa* Willd.) of the South of Chile and highland accessions. *Journal of Plant Breeding and Crop Science* 1, 210–216.

Bhargava, A., Rana, T.S., Shukla, S. and Ohri, D. (2005) Seed protein electrophoresis of some cultivated and wild species of *Chenopodium* (Chenopodiaceae). *Biologia Plantarum* 49, 505–511.

Bhargava, A., Shukla, S. and Ohri, D. (2006) Karyotypic studies on some cultivated and wild species of *Chenopodium* (Chenopodiaceae). *Genetic Resources and Crop Evolution* 53, 1309–1320.

Bhargava, A., Shukla, S. and Ohri, D. (2007) Genome size variation in some cultivated and wild species of *Chenopodium* (Chenopodiaceae). *Caryologia* 60, 245–250.

Christensen, S.A., Pratt, D.B., Pratt, C., Nelson, P.T., Stevens, M.R., Jellen, E.N., Coleman, C.E., Fairbanks, D.J., Bonifacio, A. and Maughan, P.J. (2007) Assessment of genetic diversity in the USDA and CIP-FAO international nursery collections of quinoa (*Chenopodium quinoa* Willd.) using microsatellite markers. *Plant Genetic Resources* 5, 82–95.

Coles, N.D., Coleman, C.E., Christensen, S.A., Jellen, E.N., Stevens, M.R., Bonifacio, A., Rojas-Beltran, J.A., Fairbanks, D.J. and Maughan, P.J. (2005) Development and use of an expressed sequenced tag library in quinoa (*Chenopodium quinoa* Willd.) for the discovery of single nucleotide polymorphisms. *Plant Science* 168, 439–447.

Costa Tártara, S.M., Manifesto, M.M., Bramardi, S.J. and Bertero, H.D. (2012) Genetic structure in cultivated quinoa (*Chenopodium quinoa* Willd.), a reflection of landscape structure in Northwest Argentina. *Conservation Genetics* 13, 1027–1038.

Del Castillo, C., Winkel, T., Mahy, G. and Bizoux, J.P. (2007) Genetic structure of quinoa (*Chenopodium quinoa* Willd) from the Bolivian altiplano as revealed by RAPD markers. *Genetic Resources and Crop Evolution* 54, 897–905.

Fairbanks, D.J., Burgener, K.W., Robison, L.R., Andersen, W.R. and Ballon, E. (1990) Electrophoretic characterization of quinoa seed proteins. *Plant Breeding* 104, 190–195.

Fairbanks, D., Waldrigues, A., Ruas, C.F., Maughan, P.J., Robison, L.R., Andersen, W.R., Riede, C.R., Pauley, C.S., Caetano, L.G., Arantes, O.M., Fungaro, M.H.P., Vidotto, M.C. and Jankevicius, S.E. (1993) Efficient characterization of biological diversity using field DNA extraction and random amplified polymorphic DNA markers. *Revista Brasileira de Genética* 16, 11–22.

Fuentes, F., Martinez, E.A., Hinrichsen, P.V., Jellen, E.N. and Maughan, P.J. (2009) Assessment of genetic diversity patterns in Chilean quinoa (*Chenopodium quinoa* Willd.) germplasm using multiplex fluorescent microsatellite markers. *Conservation Genetics* 10, 369–377.

Fuentes, F., Bazile, D., Bhargava, A. and Martínez, E.A. (2012) Implications of farmers' seed exchanges for on-farm conservation of quinoa, as revealed by its genetic diversity in Chile. *Journal of Agricultural Science* 150, 702–716.

Jarvis, D.E., Kopp, O.R., Jellen, E.N., Mallory, M.A., Pattee, J., Bonifacio, A., Coleman, C.E., Stevens, M.R., Fairbanks, D.J. and Maughan, P.J. (2008) Simple sequence repeat marker development and genetic mapping in quinoa (*Chenopodium quinoa* Willd.). *Journal of Genetics* 87, 39–51.

Kolano, B.A. (2004) Genome analysis of a few *Chenopodium* species. PhD thesis, University of Silesia, Katowice, Poland.

Kolano, B., Plucienniczak, A., Kwasniewski, M. and Maluszynska, J. (2008) Chromosomal localization of a novel repetitive sequence in the *Chenopodium quinoa* genome. *Journal of Applied Genetics* 49, 313–320.

Kolano, B., Gardunia, B.W., Michalska, M., Bonifacio, A., Fairbanks, D., Maughan, P.J., Coleman, C.E., Stevens, M.R., Jellen, E.N. and Maluszynska, J. (2011) Chromosomal localization of two novel repetitive sequences isolated from the *Chenopodium quinoa* Willd. genome. *Genome* 54, 710–717.

Kolano, B., Siwinska, D., Gomez-Pando, L., Szymanowska-Pulka, J. and Maluszynska, J. (2012) Genome size variation in *Chenopodium quinoa* (Chenopodiaceae). *Plant Systematics and Evolution* 298, 251–255.

Mason, S.L., Stevens, M.R., Jellen, E.N., Bonifacio, A., Fairbanks, D.J., McCarty, R.R., Rasmussen, A.G. and Maughan, P.J. (2005) Development and use of microsatellite markers for germplasm characterization in quinoa (*Chenopodium quinoa* Willd.). *Crop Science* 45, 1618–1630.

Maughan, P.J., Bonifacio, A., Jellen, E.N., Stevens, M.R., Coleman, C.E., Ricks, M., Mason, S.L., Jarvis, D.E., Gardunia, B.W. and Fairbanks, D.J. (2004) A genetic linkage map of quinoa (*Chenopodium quinoa*) based on AFLP, RAPD, and SSR markers. *Theoretical and Applied Genetics* 109, 1188–1195.

Maughan, P.J., Kolano, B.A., Maluszynska, J., Coles, N.D., Bonifacio, A., Rojas, J., Coleman, C.E., Stevens, M.R., Fairbanks, D.J., Parkinson, S.E. and Jellen, E.N. (2006) Molecular and cytological characterization of ribosomal RNA genes in *Chenopodium quinoa* and *Chenopodium berlandieri*. *Genome* 49, 825–839.

Maughan, P.J., Turner, T.B., Coleman, C.E., Elzinga, D.B., Jellen, E.N., Morales, A.J., Udall, J.A., Fairbanks, D.J. and Bonifacio, A. (2009) Characterization of salt overly sensitive (SOS1) gene homoeologs in quinoa (*Chenopodium quinoa* Willd). *Genome* 52, 647–657.

Maughan, P., Smith, S., Rojas-Beltrán, J., Elzinga, D., Raney, J., Jellen, E., Bonifacio, A., Udall, J. and Fairbanks, D. (2012) Single nucleotide polymorphisms identification, characterization and linkage mapping in *Chenopodium quinoa*. *The Plant Genome* 5, 1–7.

Rana, T.S., Narzary, D. and Ohri, D. (2010) Genetic diversity and relationships among some wild and cultivated species of *Chenopodium* L. (Amaranthaceae) using RAPD and DAMD methods. *Current Science* 98, 840–846.

Reynolds, D.J. (2009) Genetic dissection of triterpenoid saponin production in *Chenopodium quinoa* using microarray analysis. MSc. thesis. Brigham Young University, Provo, Utah.

Ruas, P., Bonifacio, A., Ruas, C., Fairbanks, D. and Andersen, W. (1999) Genetic relationship among 19 accessions of six species of *Chenopodium* L., by random amplified polymorphic DNA fragments (RAPD). *Euphytica* 105, 25–32.

Ruiz-Carrasco, K., Antognoni, F., Coulibaly, A.K., Lizardi, S., Covarrubias, A., Martínez, E.A., Molina-Montenegro, M.A., Biondi, S. and Zurita-Silva, A. (2011) Variation in salinity tolerance of four lowland genotypes of quinoa (*Chenopodium quinoa* Willd.) as assessed by growth, physiological traits, and sodium transporter gene expression. *Plant Physiology and Biochemistry* 49, 1333–1341.

Stevens, M.R., Coleman, C.E., Parkinson, S.E., Maughan, P.J., Zhang, H.B., Balzotti, M.R., Kooyman, D.L., Arumuganathan, K., Bonifacio, A., Fairbanks, D.J., Jellen, E.N. and Stevens, J.J. (2006) Construction of a quinoa (*Chenopodium quinoa* Willd.) BAC library and its use in identifying genes encoding seed storage proteins. *Theoretical and Applied Genetics* 112, 1593–1600.

Ward, S.M. (2000) Allotetraploid segregation for single-gene morphological characters in quinoa (*Chenopodium quinoa* Willd.). *Euphytica* 116, 11–16.

Wilson, H.D. (1988a) Quinoa biosystematics I: domesticated populations. *Economic Botany* 42, 461–477.

Wilson, H.D. (1988b) Quinoa biosystematics II: free living populations. *Economic Botany* 42, 478–494.

12 Chemistry

12.1 Introduction

The increasing world population requires an increase in food production and the cultivation of crops that are nutritious and require minimum inputs. The demand for food has increased dramatically over the last 50 years due to the enormous increase in population as well as people's greater purchasing power (Aggarwal *et al.*, 2004). The increase in crop yields has been primarily achieved through better crop yields in the most agriculturally productive areas. However, a large proportion of the world's population, especially in the developing countries, still has little access to protein and a mineral-rich diet, since wheat and rice are the principal food crops. Each person must eat an adequate amount of good-quality and safe food throughout the year in order to meet all nutritional needs for body maintenance, work and recreation, and for growth and development. Malnutrition is a general term that indicates a lack of some or all nutritional elements necessary for human health. There are two basic types of malnutrition: (i) protein-energy malnutrition and (ii) micronutrient deficiency malnutrition.

It was estimated that about 925 million people were undernourished in 2010 (FAO, 2010). Developing countries account for 98% of the world's undernourished people and have a prevalence of undernourishment of 16%. There is an urgent need to take remedial action so that malnutrition and undernourishment can be reduced. To mitigate this problem, recently attention has centred on the exploitation and utilization of unusual and underexploited plant material for food. In recent years underutilized crops like quinoa have evoked interest as a potential food crop for diversification of agriculture into newer areas, for environmental sustainability and for combating the nutritional deficiency prevalent in many parts of the world (Jacobsen, 2003; Bhargava *et al.*, 2006, 2007). It belongs to the group of crops known as pseudocereals (Cusack, 1984; Koziol, 1993) that includes other domesticated chenopods, amaranths and buckwheat. The outstanding nutritional and functional properties of quinoa seeds are the higher protein content than that of the traditional cereals, the presence of the whole set of essential and non-essential amino acids, a range of vitamins, along with important minerals, isoflavones, and high-quality lipids (Repo-Carrasco *et al.*, 2003; Vega-Gálvez *et al.*, 2010).

12.2   Grain

Perisperm, embryo and endosperm are the three areas where reserve food is stored in quinoa seed (Prego *et al.*, 1998). Starch is stored in the perisperm, and lipids and protein in the endosperm and embryo. The FAO has observed that quinoa seeds have high-quality proteins, and higher levels of iron, calcium, phosphorus, fibre and vitamins (Dini *et al.*, 2005; FAO, 2011).

12.2.1   Nutritional components of quinoa seed

Protein

Proteins are the main structural constituents of the cells and tissues of the body, and they make up the greater portion of the substance of the muscles and organs. Proteins are necessary for growth and development of the body, for body maintenance and the repair and replacement of worn out or damaged tissues, to produce metabolic and digestive enzymes and as an essential constituent of certain hormones, such as thyroxine and insulin (Latham, 1997). A proteinaceous diet is an essential element of a healthy diet, allowing both growth and maintenance of the 25,000 proteins encoded within the human genome, as well as other nitrogenous compounds, which together form the body's dynamic system of structural and functional elements that exchange nitrogen with the environment.

The nutritional quality of a product depends on the quantity and the quality of the nutrients (Repo-Carrasco *et al.*, 2003). The protein content in quinoa seed ranges from 7 to 23% (Cardozo and Tapia, 1979; González *et al.*, 1989; Koziol, 1992; Ruales and Nair, 1994a, 1994b; Ando *et al.*, 2002; Karyotis *et al.*, 2003; Abugoch *et al.*, 2008) (Table 12.1).

Wright *et al.* (2002a) reported 14.8 and 15.7% of protein for sweet and bitter quinoa respectively. Compared to cereal grains, the total protein content in quinoa is higher than that of barley (11%), rice (7.5%) or maize (13.4%), and is comparable to that of wheat (15.4%) (USDA, 2005; Abugoch *et al.*, 2008). The concentration of proteins in quinoa seed is greater in plants grown in controlled environments compared with plants grown in field studies (Schlick and Bubenheim, 1996). This is probably because of increased nutrient availability and decreased stress in controlled environments.

Apart from the availability of nutrients, one must be able to digest, absorb and utilize the food and nutrients effectively. The nutritional value of a food is determined by its protein quality, which depends mainly on its amino acid content, influence of antinutritional factors, protein digestibility, and the

Table 12.1. Protein content in quinoa grain (%).

Reference	Range	Mean
Cardozo and Tapia (1979)	7.47–22.08	13.81
Koziol (1992)	13.8–16.5	15.0
Wright *et al.* (2002a)	14.8–15.7	15.25
Bhargava *et al.* (2007)	12.55–21.02	16.22
Repo-Carrasco-Valencia and Serna (2011)	13.96–15.47	14.78

tryptophan to large neutral amino acids ratio (Comai *et al.*, 2007). The essential amino acids are not synthesized by animals and hence must be provided in the diet. If one of these amino acids is limited, the others will be broken down and excreted, resulting in poor growth of livestock and humans and loss of nitrogen in the diet (Vega-Gálvez *et al.*, 2010). The importance of the proteins in quinoa is primarily due to their quality. The proteins of quinoa mainly belong to albumin and globulin, which have a balanced composition of essential amino acids similar to the composition of casein, the protein in milk (Ranhotra *et al.*, 1993). Koziol (1992) reported that albumin and globulins were the major protein fractions (44–77% of total protein), while the percentage of prolamines and glutelin was low. Brinegar and Goundan (1993) isolated and characterized the principal protein of quinoa, the chenopodina, which is a globulin 11S type protein, separated into two subgroups of chenopodina, A and B having molecular weights of 32,000–39,000 and 22,000–23,000, respectively, which are higher than for casein (Ranhotra *et al.*, 1993). Studies of the molecular structures of quinoa globulin and albumin have shown that both proteins are stabilized by disulphide-type bridges (Brinegar and Goundan, 1993; Brinegar *et al.*, 1996). The net protein utilization (NPU) or protein efficiency ratio (PER) are widely used as indicators for the nutritional quality of proteins. The PER for a food is determined by dividing the gain in weight of a test subject who had been given proteins by the intake of that particular protein during the test period. The PER has become a standard for evaluating the protein quality of food proteins, especially in the developed countries. Mahoney *et al.* (1975) reported the PER values for quinoa proteins while working with Bolivian quinoa. It was concluded that the protein quality of cooked quinoa was like that of casein with high amounts of lysine and methionine. In this study, the PER of the cooked quinoa was 30% greater than that of uncooked quinoa. Gross *et al.* (1989) reported a high apparent digestibility and a high PER of washed quinoa seeds; they found that the PER is almost equal to that of casein.

The quinoa protein products exhibited solubilities of 47.0–93.0%, depending on the material extracted and the method of protein concentration (Lindeboom, 2005). The solubilities of the quinoa protein products were significantly higher than that of soybean protein and similar to that of egg white, with the exception of the precipitated protein from saponin-extracted bran. The solubility of the precipitated protein from defatted/saponin-extracted bran was lower compared with other protein products. On precipitation, protein aggregates were formed that were hard to solubilise (Aluko and Monu, 2003;

Lindeboom, 2005), i.e. the solubility of protein decreased on saponin extraction. Therefore, the solubility of precipitated protein from bran that did not undergo the saponin-extraction step was high (93.0%). Saponins and additional substances that co-extracted with the saponins in 60% (v/v) ethanol, such as polyphenols, might influence the structure, denaturation, precipitation and resolubilization of protein. Lindeboom (2005) stated that quinoa protein foams better than egg white, but less well than soybean protein. The results were in sharp contrast to those obtained earlier by Aluko and Monu (2003), who concluded that quinoa protein had a very low foaming capacity, which was due to the globular nature of the protein. The globular nature reduced its ability to form interfacial membranes around air bubbles. The foam stabilities of quinoa protein products have been reported to be similar and significantly higher than that of soybean protein, and lower than that of egg white protein (Lindeboom, 2005).

The seed proteins of quinoa have a balanced amino-acid spectrum with high lysine, histidine and methionine (Van Etten *et al.*, 1963; Koziol, 1992). All the ten amino acids that are essential for good growth (phenylalanine, tyrosine, lysine, isoleucine, leucine, threonine, tryptophan, histidine, valine and methionine) are presented in quinoa (Table 12.2), providing it with a similar value to casein, the protein in milk.

It has been reported that quinoa protein can supply around 180% of the histidine, 212% of the methionine+cysteine, 228% of the tryptophan, 274%

Table 12.2. Comparative picture of the amino acid composition of quinoa, wheat and soybean (g/100 g).

Amino acid	Quinoa Ruales and Nair (1992)	Ranhotra *et al.* (1993)	Koziol (1992)	Repo-Carrasco *et al.* (2003)	Wheat Wright *et al.* (2002a)	Souci *et al.* (1986)	Soybean Janssen *et al.* (1979)	FAO/WHO recommended Friedman and Brandon (2001)
Lysine	5.10	6.30	6.10	5.60	6.10	2.90	6.40	1.60
Isoleucine	3.20	3.02	4.40	3.40	3.30	3.80	4.90	1.30
Leusine	7.90	6.88	6.60	6.10	5.80	6.80	7.60	1.90
Threonine	3.80	4.41	3.80	3.40	2.50	3.10	4.20	0.90
Methionine+ cystine	3.40	3.66	4.80	4.80	2.00	4.00	2.90	1.70
Phenylalanine+ tyrosine	6.30	8.18	7.30	6.20	6.20	7.60	8.40	1.90
Histidine	2.40	4.09	3.20	2.70	3.10	2.20	2.50	1.60
Valine	3.90	3.67	4.50	4.20	4.00	4.70	5.00	1.30
Tryptophan	1.50	Not determined	1.20	1.10	Not detected	1.10	1.30	0.50

of the isoleucine, 338% of the lysine, 320% of the phenylalanine+tyrosine, 323% of the valine and 331% of the threonine recommended in protein sources for adult nutrition (FAO/WHO/UNU, 1985; Abugoch, 2009). The total amino acid composition (protein-bound and free amino acids) of quinoa grain shows variation during seed development, especially at 12 and 16 days after anthesis (Prakash and Pal, 1998). The relative amounts of glutamic acid, glycine and arginine increased as the grain matured. In contrast, the amounts of aspartic acid, serine, proline, threonine, valine and lysine decreased. Glutamic acid showed major increase with 13.9% on the 4th day to 15.1% on the 12th and 21.7% at maturity, and was most abundant throughout the seed development (Prakash and Pal, 1998). Glycine varied from 8.0 to 8.7% on the 12th day, 9.0 on the 16th and 9.9% at maturity. Threonine (4.6–3.4%), serine (5.7–4.8%) and lysine (7.3–5.8%) showed major decrease from the 4th day to maturity. The variations in the relative amounts of different amino acids indicate that there might be a direct or indirect relationship between the utilization and synthesis or metabolism of these amino acids in grain. Further, the seed needs to be harvested at a specific maturity level to obtain a protein fraction with higher amounts of particular amino acids (Raina and Datta, 1992).

The content of essential amino acids in quinoa is higher than in common cereals (Ruales and Nair, 1992; Wright et al., 2002a). Quinoa proteins have higher histidine content than wheat and soy proteins, while the methionine+cystine content of quinoa is adequate for children and adults (Abugoch, 2009), it is similar to that soya and higher than the amounts in wheat (Table 12.2). Quinoa proteins have adequate levels of aromatic amino acids (phenylalanine and tyrosine) and histidine, isoleucine, threonine, phenylalanine, tyrosine and valine that are in conformance with FAO/WHO suggested requirements for 10-year-old children. In quinoa, lysine and leucine are the limited amino acids for 2–5-year-old children, while all the essential amino acids of this protein are sufficient according to FAO/WHO suggested requirements for 10–12-year-old children (Abugoch, 2009). Thus, the protein quality of quinoa grain is superior to most cereal grains, including wheat. Quinoa proteins may be one of the more promising food ingredients, capable of complementing cereal or legume proteins, and there is the potential for the production of protein concentrates from dehulled quinoa seeds, which could be used as raw materials in the food industry. The high protein quality and energy value of the grain can be utilized in the poultry industry. The good amino acid profile could be a good source of proteins for feeding infants and children. It is interesting to note that quinoa proteins have high concentrations of sulfur-containing amino acids compared with other plants. This could possibly be due to the type of land (volcanic) where this plant originated (Schlick and Bubenheim, 1996; Vega-Gálvez et al., 2010).

However, the digestibility of quinoa seed is the limiting factor in protein and energy utilization (Lopez de Romana et al., 1981), which can be significantly improved by milling. Lorenz and Coulter (1991) mixed corn grits with different concentrations of quinoa and found that the addition of quinoa produced extruded products that were higher in protein than corn grit products, but had lower in vitro digestibility. The importance of the non-protein tryptophan fraction is that it is the only one that can enter the brain and is more easily absorbed, so it guarantees a

greater amount available for uptake by the central nervous system. The tryptophan content of quinoa proteins is similar to that of wheat, but higher than that
of other cereals (Comai *et al.*, 2007). Free tryptophan in quinoa flour has values
higher than that in rice, maize and rye; similar to those of wheat and oat; and
lower than those of barley and pearl millet (Comai *et al.*, 2007). Enzymatic
hydrolysis studies have shown that short-chain peptides are more active in quinoa
seeds than long-chain peptides (Aluko and Monu, 2003). Low-molecular-weight
peptides possess higher potential than high-molecular-weight peptides as antihy-
pertensive agents or as compounds that reduce the amount of free radicals.

Carbohydrates

Carbohydrates play a basic nutritional function in living organisms and have different physiological health effects, such as sources of energy, effects on satiety/
gastric emptying, control of blood glucose and insulin metabolism, protein glycosylation, cholesterol and triglyceride metabolism (FAO, 1998). Carbohydrates
can be classified according to their degree of polymerization into three principal
groups: sugars (monosaccharides, disaccharides, polyols), oligosaccharides and
polysaccharides (starch and nonstarch). Starch, the major biopolymeric constituent of plants, occurs typically as granular forms of various shapes and sizes,
making up approximately 60–70% of the dry matter of most seeds. It provides
the major source of physiological energy in the human diet and accordingly it is
classified, in general, as the available carbohydrate (Tharanathan and
Mahadevamma, 2003). In quinoa, starch is the most important carbohydrate
comprising approximately 32–75% of the seed (Chauhan *et al.*, 1992a, 1992b;
Ranhotra *et al.*, 1993; Ahamed *et al.*, 1998; Oshodi *et al.*, 1999; Ando *et al.*,
2002; Wright *et al.*, 2002b; Lindeboom, 2005; USDA, 2005) (Table 12.3).

Starch consists of two polysaccharides: (i) amylose, which has long linear
chains of (1→4)-linked α-D-glucopyranose residues, some with a few (>10)
branches (Jane *et al.*, 1999) and (ii) amylopectin, which has a large molecular
weight and highly branched structures consisting of much shorter chains of
(1→4) α- D-glucose residues (Jane *et al.*, 1999).

Amylopectin is the major component of most starches, and its fine structure plays a critical role in the characteristics of starch. But very few results are
available on the molecular weight of cereal amylopectin since cereal starches
are difficult to dissolve in water and may be easily degraded (Inouchi *et al.*,
1999; Abugoch, 2009). The amylopectin and amylase content of quinoa and
its comparison with other food crops is provided in Table 12.4. Quinoa

Table 12.3. Carbohydrate content in quinoa grain (%).

Reference	Carbohydrate (%)	
	Range	Mean
DeBruin, 1964	58.10–64.20	60.34
Cardozo and Tapia, 1979	38.72–71.30	59.74
Repo-Carrasco-Valencia and Serna (2011)	68.84–75.82	73.07

Table 12.4. Amylose and amylopectin content in quinoa starch and its comparison with starch of other food crops.

Plant	Amylose (%)	Amylopectin (%)	Reference
Barley	27–32	68–73	Pieper *et al.* (2008)
Maize	20–30	70–80	Sharma *et al.* (2006)
Oats	23–27	73–77	Pieper *et al.* (2008)
Fagopyrum	15.6–17.9	NR	Yoshimoto *et al.* (2004)
Amaranthus	7.8	NR	Qian and Kuhn (1999)
Quinoa	7.10	NR	Tang *et al.* (2002)

NR: not reported.

amylopectin, like amaranth and buckwheat amylopectins, contains a large number of short chains containing 8–12 units and a small number of long chains of 13–20. This is in sharp contrast to the endosperm starches of commonly used cereals (Abugoch, 2009).

The digestibility of the starch depends on its native structure, physical encapsulation, proportion of damaged granules, crystallinity, degree of gelatinization and retrogradation (Ruales and Nair, 1994b). Starch digestibility may also be affected by the formation of amylose–lipid complexes, starch–protein interactions, the presence of such antinutritional factors as amylase inhibitors, modification of its structure and functional groups and processing techniques (Eyaru *et al.*, 2009; Wong *et al.*, 2009; Putseys *et al.*, 2010; Barrett and Udani, 2011). Quinoa starch is highly branched, with a minimum and maximum degree of polymerization of 4600 and 161,000 glucan units, and a weighted average of 70,000 (Praznik *et al.*, 1999). Chain length depends on the botanical origin of the starch and is likely to be of the order of 500–6000 glucose units. Considerable variability has also been reported in the amylase content in quinoa seed, which has been found to be around 3–20% (Lorenz, 1990; Praznik *et al.*, 1999; Tang *et al.*, 2002; Repo-Carrasco *et al.*, 2003; Watanabe *et al.*, 2007). However, detailed investigation is required to confirm whether this variation is truly a reflection of genetic variability or is due to variations in cultural practices or environment or to differences in the methods employed for amylose measurement (Lindeboom, 2005). The high performance size exclusion chromatography (HPSEC) method has been shown to give higher values for amylose content compared with the colorimetric method (Lindeboom *et al.*, 2005). Amylose content affects the functional and physicochemical properties of starch, including its pasting, gelatinization, retrogradation and swelling properties (Li *et al.*, 1994; Wootton and Panozzo, 1998; Baldwin, 2001; Bao *et al.*, 2001; Grant *et al.*, 2001; Svegmark *et al.*, 2002; Lindeboom, 2005).

Several reports have been published on the characteristics of quinoa starch (Atwell *et al.*, 1983; Lorenz, 1990; Inouchi *et al.*, 1999; Tang *et al.*, 2002). Quinoa starch is present in the form of small granules about 1–1.5 µm in diameter (Lorenz, 1990; Chauhan *et al.*, 1992a; Qian and Kuhn, 1999; Tang *et al.*, 2002; Wright *et al.*, 2002b; Tari *et al.*, 2003; Lindeboom, 2005) (Fig. 12.1).

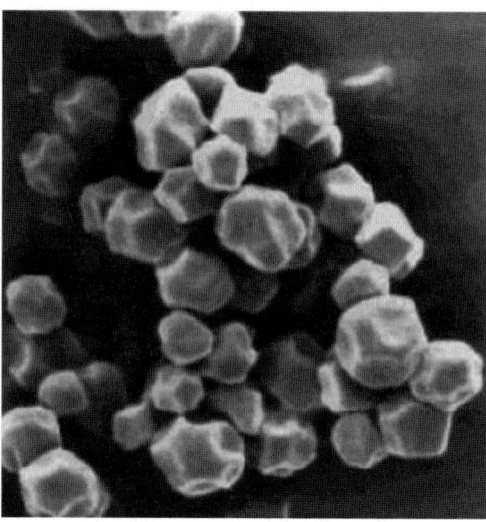

Fig. 12.1. Scanning micrograph of quinoa starch granules (10,000× magnification). [Reprinted from Qian and Kuhn (1999), with permission from John Wiley and Sons.]

Granule size affects the physicochemical characteristics of starch and is related to the biological source from which the starch is isolated. The shape of quinoa starch is similar to that of rice starch, but the particle size is less than that reported for wheat (0.7–39.2 μm), rice (0.5–3.9 μm), barley (1.0–39.2 μm) and maize (1.0–7.7 μm) (Tang *et al.*, 1998; Ando *et al.*, 2002). The extremely small size of the quinoa starch granule can be exploited by using it as a biodegradable filler in polymer packaging (Atwell *et al.*, 1988). Quinoa starch has an average molar mass of 11.3×10^6 g/mol, which is comparable to that of amaranth (11.8×10^6 g/mol) starch, higher than that of wheat starch (5.5×10^6 g/mol), but lower than that of waxy maize starch (17.4×10^6 g/mol) and rice starch ($0.52–1.96 \times 10^8$ g/mol) (Lindeboom, 2005; Park *et al.*, 2007). The shape of quinoa starch has been determined by scanning electron microscopy (SEM) and has been explained as polygonal (Lorenz, 1990; Ruales and Nair, 1994b; Qian and Kuhn, 1999; Wang *et al.*, 2003; Lindeboom, 2005), which is similar to that of amaranth and rice starch (Qian and Kuhn, 1999; Kong *et al.*, 2009). The starch also has polygonal granules and the particles can be present singly and as spherical aggregates (Fig. 12.2) packed in the quinoa perisperm (Atwell *et al.*, 1983; Ruales and Nair, 1994b; Ando *et al.*, 2002).

X-ray diffraction is a versatile, non-destructive technique that reveals detailed information about the crystallographic structure and chemical composition of natural as well as manufactured materials. A large number of studies have utilized the X-ray diffraction method to explain the structure of whole starch and amylose. Starch granules, depending on their botanical origin, amylose/amylopectin ratio and amylopectin branch length, show three types of X-ray diffraction patterns, associated with different crystalline polymorphic forms (Qian and Kuhn, 1999; Lopez-Rubio *et al.*, 2004): (i) A-type (cereal); (ii) B-type (tubers); and (iii) C-type (where A and B crystals coexist in the granule).

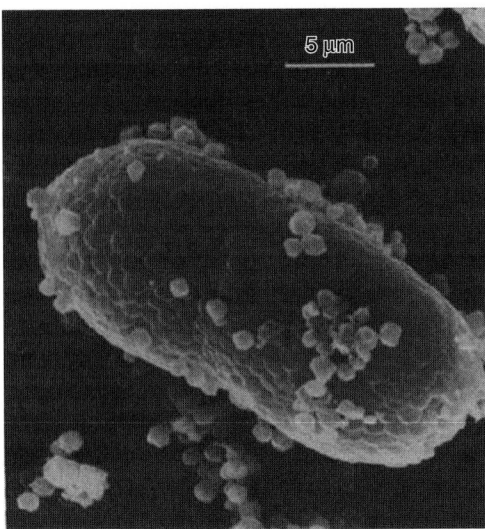

Fig. 12.2. Scanning micrograph of quinoa starch aggregate. [Reprinted from Atwell *et al.* (1983), with permission from AACC Intl.]

Quinoa starch has been shown to exhibit the typical A-type X-ray diffraction pattern (reflections at 15.3°, 17.0°, 18.0°, 20.0° and 23.4° 2θ angles) (Inouchi *et al.*, 1999; Watanabe *et al.*, 2007) that is a characteristic of cereal starches (Zobel, 1988; Lopez-Rubio *et al.*, 2004; Watanabe *et al.*, 2007), and a relative crystallinity of 35.0% (Tang *et al.*, 2002). The degree of relative crystallinity varies between 35% and 43% (Qian and Kuhn, 1999; Tang *et al.*, 2002; Watanabe *et al.*, 2007), which is lower than that of amaranth starch, higher than normal barley starch and similar to that of waxy barley starch (Qian and Kuhn, 1999; Tang *et al.*, 2002). Since amylose is known to disrupt the structure order within the amylopectin crystallites, the crystallinity of starch is associated with amylose content.

Starch gelatinization is a phenomenon that involves the disruption of the molecular order within the starch granule by heating it above its gelatinization temperature in excess of water. Apart from heat, high concentrations of an alkaline agent may also induce starch gelatinization (Lai *et al.*, 2002, Roberts and Cameron, 2002). Gelatinization is accompanied by an increase in the viscosity of the starch slurry, disentanglement of the linear amylose molecules and their solubilization in the surrounding medium. Upon continued heating, most unmodified starch pastes exhibit a decrease in viscosity after maximum viscosity has been attained. This breakdown of the starch paste results from extensive solubilization and fragmentation of granule structures such that they can no longer hold onto a large volume of water (Bean and Setser, 1992). The gelatinization properties of starch are related to a variety of factors that include the size, proportion and kind of crystalline organization, and the ultra-structure of the starch granule. The gelatinization temperatures are known to be positively associated with the amylose content (Lindeboom, 2005; Youa and Izydorczyk, 2007). The gelatinization temperature for different quinoa cultivars along with other thermal characteristic have been provided in Table 12.5.

Table 12.5. Thermal properties of starch of quinoa assessed by different researchers.

Parameter	Ando *et al.* (2002)	Wright *et al.* (2002b)	Lindeboom (2005)
T_0 (°C)	54	51.4–51.8	44.6–53.7
Tp (°C)	62.20	55.7–56.1	50.5–61.7
Tc (°C)	71	64.3–64.5	NR
Enthalpy (ΔH, J/g)	11	12.6	12.8–15
Retrogradation (%)	NR	NR	19.6–40.8

T_0: Gelatinization onset temperature (°C). Tp: Gelatinization peak temperature (°C).
Tc: Gelatinization conclusion temperature (°C). NR: not reported.

The quinoa starches exhibit higher gelatinization temperatures than rice starch, but lower gelatinization temperatures when compared with waxy barley and amaranth starches (Qian and Kuhn, 1999; Youa and Izydorczyk, 2007). Although quinoa starch gelatinizes at similar temperatures, its pasting behaviour is considerably different from that of wheat starch. At equal starch concentrations, quinoa starch exhibits a higher viscosity (Atwell *et al.*, 1983). A pasting temperature of 66.8°C was reported by Qian and Kuhn (1999), which is quite similar to the values reported by Wright *et al.* (2002b) (63.5–65.3°C) and Lindeboom (2005) (63–64°C). Quinoa starch also has excellent stability under freezing and retrogradation processes (Ahamed *et al.*, 1998). The term retrogradation refers to changes that occur on cooling and storage of gelatinized starch pastes, changes that often decrease the quality of starch-based foods (Lindeboom, 2005). Retrogradation of quinoa starch ranges from 19.6 to 40.8% of the initial gelatinization enthalpy. Amylose content is considered as one of the most influential factors in starch retrogradation, with a higher level of amylase resulting in greater association of starch molecules and a higher degree of retrogradation (Gudmundsson and Eliasson, 1990; Chang and Lui, 1991; Fan and Marks, 1998; Kaur *et al.*, 2002). The fact that quinoa starch is resistant to retrogradation suggests that it can be useful in frozen food products, sauces, cream soups, pie fillings and in emulsion types of food products such as salad dressings. Quinoa starch also has non-edible potential for utilization as biodegradable fillers in low-density polyethylene (LDPE) films (Ahamed *et al.*, 1996). This aspect needs more investigation for its effective utilization in the food, pharmaceutical and textile industry (Bhargava *et al.*, 2006). Because of better mechanical properties, quinoa starch can be utilized in the manufacture of carrier bags, where tensile strength is important (Bhargava *et al.*, 2006).

Lipids
There is a relatively high quantity of oil in quinoa, making it a potential source for oil extraction. Quinoa is referred to as a pseudo-oilseed crop (Cusack, 1984) because of its lipid fraction, i.e. it has an exceptional balance between oil, protein and fats. The concentration of fat in quinoa as reported by various workers has been provided in Table 12.6.

Table 12.6. Fat content in quinoa grain (%).

Reference	Range	Mean
DeBruin (1964)	5.5–6.7	6.2
Cardozo and Tapia (1979)	1.8–9.3	5.0
Romero (1981)	1.8–8.2	4.6
Koziol (1990)	4.3–9.5	7.2
Prakash *et al.* (1993)	3.3–4.2	3.75

The seeds have an interesting lipid composition of about 1.8–9.5%, with an average of 5.0–7.2% (Koziol, 1993; Ranhotra *et al.*, 1993; Wood *et al.*, 1993; Oshodi *et al.*, 1999; USDA, 2005; Ryan *et al.*, 2007). Quinoa has a higher oil content than maize (4.9% dry basis) and a lower content than soya (20.9% dry basis) (Koziol, 1993; USDA, 2005). As with maize, the oil is concentrated in the germ, which in quinoa represents 25–30% of the weight of the grain (Cardozo and Tapia, 1979; Koziol, 1993). In quinoa the germ encircles the endosperm and can be easily removed by a modified polishing procedure to give a fraction containing 19% oil (Koziol, 1993). Although special breeding programmes were necessary to achieve oil content of 6–8% in maize, several cultivars of quinoa initially showed oil contents in this range (Alvarez *et al.*, 1990). Also unlike maize, where an increase in oil content resulted in a decrease in starch content, increased oil content in the quinoa grain showed no significant correlation with total carbohydrate content and was negatively correlated with protein content (Koziol, 1993). The seed lipids of quinoa contain high amounts of neutral lipids, of which triglycerides are the major fraction accounting for over 50% of the neutral lipids (Przybylski *et al.*, 1994). Diglycerides are present in the whole seeds and contribute 20% of the neutral lipid fraction. Lysophosphatidyl ethanolamine and phosphatidyl choline are the most abundant (57%) of the total polar lipids (Przybylski *et al.*, 1994).

Fatty acids constitute the main components of the lipid entities and are required in human nutrition as a source of energy, and for metabolic and structural activities. Numerous fatty acids are synthesized by the human body and are known as 'non-essential fatty acids' because they are not essentially needed in the diet. However, since the body cannot produce all the types of fatty acids required by it, some fatty acids must come from the diet; these fatty acids are called 'essential fatty acids' (EFAs) which belong to a class of fatty acids called polyunsaturated fatty acids (PUFAs). PUFAs have several positive effects on cardiovascular disease (Abeywardena *et al.*, 1991) and improved insulin sensitivity (Lovejoy, 1999). Natural PUFA with methylene-interrupted double bonds and all of the *cis* configuration can be divided into 12 families, ranging from double bonds located at the n-1 position to the n-12 position (Gunstone, 1999). The most important families, in terms of extent of occurrence and human health and nutrition, are the ω-3 and ω-6 families (Moyad, 2005; FAO, 2008). These names refer to the chemical structure of the fatty acid; both types are unsaturated, i.e. they contain carbon–carbon double bonds; the type is determined by the final double bond being either at the n-3 or n-6 position. ω-6 and ω-3 fatty acids are essential fatty acids because they cannot be synthesized by humans, who must obtain them from foods. Linoleic acid is the parent fatty acid of the n-6 family. It has 18 carbon atoms and two double bonds and the first double bond is six carbon atoms from the methyl end of the fatty acid chain, hence the name n-6. Linoleic acid can be desaturated and elongated in humans to form a series of n-6 PUFA. α-linolenic acid is the parent fatty acid of the n-3 family. It also has 18 carbon atoms and three double bonds. In contrast to linoleic acid, the first double bond in α-linolenic acid is three carbon atoms from the methyl end of the fatty acid chain, hence the name n-3. α-linolenic acid can be desaturated and elongated to form a series of n-3 PUFA as linoleic acid. Linoleic acid and α-linolenic acid occur in almost all dietary fats and form the major proportions in most vegetable oils (White, 2008). Although the human body cannot synthesize either of these fatty acids from scratch, it can use them to synthesize other essential fatty acids. Table 12.7 presents the fatty acid composition in quinoa and its comparison with other commonly used grains. The fatty acid profile of quinoa clearly shows that it is rich in linoleic acid and α-linolenic acid, which are necessary substrates in animal metabolism. Linoleic acid is metabolized to arachidonic acid, and linolenic acid to eicosapentaenoic acid (EPA) and docosahexaenoic acid (DHA). EPA and DHA play important roles in prostaglandin metabolism, thrombosis and atherosclerosis, immunology and inflammation, and membrane function (Youdim *et al.*, 2000; Abugoch, 2009). Thus quinoa oil appears to be a high-quality edible oil, similar in fatty-acid composition to maize and soybean oil. The oil fraction of quinoa seed is of high quality and is highly nutritious because it has a high degree of unsaturation, with a polyunsaturation index of 3.9–4.7 (Abugoch, 2009). Given the high quality of its oil,

Table 12.7. Fatty acid composition in quinoa and its comparison with other common grains.

	Quinoa (Ando *et al.*, 2002)	Maize (USDA, 2005)	Soybean (USDA, 2005)	Wheat (Marton *et al.*, 2010)	Sunflower (Lusas, 2000)	Safflower (Lusas, 2000)
Myristic (C14:0)	0.2	Traces	Traces	0.8	0.1	0.1
Palmitic (C16:0)	10.3	10.7	10.7	31.2	6.5	6.7
Stearic (C18:0)	0.8	2.8	3.6	1.9	4.8	2.4
Oleic (C18:1)	25.6	26.1	22	10.7	26.5	14.9
Linoleic (C18:2)	52	57.7	56	25.6	57	75.5
Linolenic (C18:3)	9.8	2.2	7	2	0.2	0.1

and the fact that some varieties show fat concentrations up to 9.5%, quinoa has been considered a potentially valuable new oil crop (Koziol, 1993).

Minerals

Minerals are important for various physiological functions in the human body and are important as constituents of teeth, bones, muscles, soft tissues, blood and nerve cells (Sardesai, 1998). Minerals are required for metabolic reactions, transmission of nerve impulses, rigid bone formation and regulation of water and salt balance among other functions (Dini *et al.*, 2008). The human body requires more than 100 mg per day of each major mineral (Na, Mg, K, Ca, P, S and Cl) and less than 100 mg per day of trace elements (Cr, Mn, Fe, Co, Cu, Zn, Se, Mo and I) (Insel *et al.*, 2004). The mineral composition of quinoa seed and quinoa flour is shown in Table 12.8. Quinoa grain contains more calcium, magnesium, iron, copper and zinc than common cereals, and its iron content is particularly high (Table 12.8).

Phosphorus and magnesium are localized in the embryonic tissue of quinoa seed, while calcium and potassium are present in the pericarp (Konishi *et al.*, 2004). In general, the mineral content of quinoa seed is higher than that of cereals like oat and barley, especially for potassium, magnesium and calcium (Abugoch, 2009). According to the National Academy of Sciences (2004), the amount of iron, copper, magnesium and manganese present in 100 g of quinoa seeds is sufficient to cover the daily needs of infants and adults, while the phosphorus and zinc content in 100 g is sufficient for children. Polishing and washing quinoa seeds reduce the mineral content to some extent: 12–15% loss in the concentration of iron, zinc and potassium, 27% loss of copper and 3% loss of magnesium (Jancurova *et al.*, 2009). It is interesting to note that quinoa exhibits considerable diversity in mineral content (Oshodi *et al.*, 1999; Ogungbenle, 2003; Repo-Carrasco *et al.*, 2003), which might be due to the fact that the mineral content may vary depending

Table 12.8. Mineral content (mg/100 g) in quinoa and its comparison with some commonly used cereals.

	K	Mg	P	Ca	Fe	Cu	Zn
Whole quinoa seed (Koziol, 1992)	926.7	249.6	383.7	148.7	13.2	5.1	4.4
Quinoa flour (Ranhotra *et al.*, 1993; Oshodi *et al.*, 1999)	714–855	161–232	22–462	70–86	2.6–6.3	0.7–7.6	3.2–3.8
Quinoa leaves (Bhargava *et al.*, 2008)	6778	817	nd	1112	84.64	12.14	19.48
Wheat (Koziol, 1992)	578.3	169.4	467.7	50.3	3.8	0.7	4.7
Rice (Koziol, 1992)	118.3	73.5	137.8	6.9	0.7	0.2	0.6
Barley (USDA, 2005)	280	79	221	29	2.5	0.4	2.1
Oats (USDA, 2005)	566	235	734	58	5.4	0.4	3.11

on factors such as ripeness, variety, soil type, the use of fertilizers, intensity and exposure time to sunlight, temperature and rainfall (Zielisnki and Koslowska, 2000; Miranda *et al.*, 2009, 2010).

Vitamins

Table 12.9 shows the vitamin content of quinoa seed. Quinoa is a good source of thiamin, folic acid and vitamin C. The seeds contain twice as much γ-tocopherol (5.3 mg/100 g) than α-tocopherol (2.6 mg/100 g) (Jancurova *et al.*, 2009). The amount of vitamin B_6 and folic acid in 100 g of seeds can cover the requirements of children and adults. The vitamin B_2 content in 100 g quinoa seeds can contribute up to 80% of the daily needs of children and 40% of those of adults (National Academy of Sciences, 2004). The vitamin B_3 content in quinoa seeds is beneficial to the diet, although its amount does not cover the daily requirement (Abugoch, 2009). Vitamin B_1 values in quinoa are lower than those in oats or barley, but those of vitamin B_2, vitamin B_3, vitamin B_6 and total folate are higher (Ranhotra *et al.*, 1993; USDA, 2005). The antioxidant activity of quinoa might be of particular interest to medical researchers and needs more attention for its utilization as a potent antioxidant.

Fibre

Quinoa is generally considered to be high in fibre (Ranhotra *et al.*, 1993; Ruales and Nair, 1994), although crude fibre percentage as low as 1.1% has also been reported (Becker and Hanners, 1990). Grains have an average of about 4.1% fibre with a range from 1.1% to 16.32% (Cardozo and Tapia, 1979). DeBruin (1964) reported 3.4% fibre content, which is much higher than that of rice (0.4%), wheat (2.7%) and maize (1.7%). According to Ranhotra *et al.* (1993), quinoa contained 8.9% total dietary fibre, of which more than 80% was insoluble. Ruales and Nair (1994) reported a total dietary fibre content of 13.4% in quinoa, consisting of 11.0% insoluble fibre and 2.4% soluble fibre. Soluble and insoluble fibre content was higher than that reported for wheat and was comparable to that of rye. Removal of the outer layers of the seed by scrubbing and washing, in order to remove saponins, did not affect its dietary fibre content (Ruales and Nair, 1994).

Table 12.9. Vitamin content in quinoa and its comparison with commonly used crops.

	Quinoa (Koziol, 1992; USDA, 2005)	Rice (Koziol, 1992)	Wheat (Koziol, 1992)	Barley (Koziol, 1992)	Buckwheat (Wijngaard and Arendt, 2006)
β-carotene (Pro-vitamin A)	0.39	NR	0.02	0.01	0.21
Vitamin B_1	0.38	0.47	0.55	0.49	0.46
Vitamin B_2	0.39	0.10	0.16	0.20	0.14
Vitamin B_3	1.06	5.98	5.88	5.44	1.80
Vitamin C	4	0	0	0	5
α-Tocopherol	5.37	0.18	1.15	0.35	5.46

Polyphenols

Polyphenols are bioactive secondary plant metabolites that are widely present in commonly consumed foods of plant origin. The three main types of polyphenol are flavonoids, phenolic acids and tannins, which act as powerful antioxidants *in vitro* (Hirose *et al.*, 2010; Repo-Carrasco-Valencia *et al.*, 2010). Polyphenols are ubiquitous in plant foods and have many potential beneficial effects on human health, such as reduction of the risk of cardiovascular diseases, neurodegenerative diseases, cancers, diabetes, osteoporosis and Alzheimer's disease (Youdim *et al.*, 2004; Arts and Hollman, 2005; Kurosawa *et al.*, 2005; Repo-Carrasco-Valencia *et al.*, 2010). In food, polyphenols may contribute to bitterness, astringency, flavour, colour and oxidative stability of products (Shahidi and Naczk, 1995; Scalbert *et al.*, 2005; Han *et al.*, 2007). Until recently, little information was available on the polyphenolic content of quinoa (De Simone *et al.*, 1990; Rastrelli *et al.*, 1995; Zhu *et al.*, 2001), but recent research has dealt with this aspect in depth. Zhu *et al.* (2001) isolated and characterized six flavonol glycosides: four kaempferol glycosides and two quercetin glycosides. Among them kaempferol 3-O-[β-D-apiofuranosyl(1‴-2″)]-β-D-galactopyranoside, kaempferol 3-O-[2,6-di-α-L-rhamnopyranosyl)-β-D-galactopyranoside and quercetin 3-O-[2,6-di-α-L-rhamnopyranosyl)-β-D-galactopyranoside were the main flavonoid glycosides found in quinoa seeds. These compounds exhibited good antioxidant capacity, suggesting that quinoa seeds can serve as a good source of free radical scavenging agents. Gorinstein *et al.* (2008) reported 251.5 µg/g of ferulic acid, 0.8 µg/g of p-coumaric acid and 6.31 µg/g of caffeic acid. Pásko *et al.* (2009) analysed the total polyphenol content and antioxidant activity in quinoa and amaranth (*Amaranthus cruentus*). The results obtained showed significant correlation between total polyphenols content in seed and sprouts. Repo-Carrasco-Valencia *et al.* (2010) determined the levels of flavonoids, phenolic acids and betalains in three Andean grains of family Amaranthaceae, viz. *C. quinoa*, *C. pallidicaule* and *Amaranthus caudatus*. Both the soluble and total phenolic acid contents in the Andean cereals were quantified as aglycones. Soluble phenolic acids (free and bound soluble forms) were extracted with methanolic acetic acid, whereas the total phenolic acid content (i.e. sum of bound soluble, insoluble and free phenolic acids) was obtained after alkaline and acid hydrolyses. Table 12.10 shows the total contents (mg/100 g) and percentage share of soluble phenolic acids and flavonoids in the three cereals obtained by Repo-Carrasco-Valencia *et al.* (2010). The total content of phenolic acids varied from 16.8 to 59.7 mg/100 g and the percentage share of soluble phenolic acids varied from 7% to 61%. The Andean cereals contained lower levels of phenolic acids compared with common cereals like wheat and rye. In these cereals the phenolic acids accumulate in bran, where their levels are as high as 419 and 453 mg/100 g in rye and wheat bran, while whole grain flours of these grains contain 137 and 134 mg/100 g, respectively (Mattila *et al.*, 2005). The flavonoid content of *Chenopodium* species was exceptionally high, varying from 36.2 to 144.3 mg/100 g (Repo-Carrasco-Valencia *et al.*, 2010). The predominant flavonoids in quinoa samples were quercetin and kaempferol, while in some varieties myricetin and

Table 12.10. Soluble phenolic acids and flavonoid content in quinoa and its comparison with other seeds. [Reprinted from Repo-Carrasco-Valencia *et al.* (2010), with permission from Elsevier.]

	Soluble phenolic acids					Flavonoids			
	Caffeic acid	Ferulic acid	p-Coumaric acid	p-OH-benzoic acid	Vanillic acid	Myricetin	Quercetin	Kaempferol	Isorhamnetin
C. quinoa	0.7	15	8	2.9	11	0.5	36	20	0.4
C. pallidicaule	3	23	1	1.7	4	0.04	60	2	30
A. caudatus	0.9	6.9	0.89	3	5.4	ND	ND*	ND*	ND*

ND: not detected.

isorhamnetin were also found. Quinoa seeds can thus be considered a very good source of flavonoids. In contrast, the common cereals (wheat, rye, oat, barley, etc.) do not contain any flavonols (Shahidi and Naczk, 1995). Since dietary flavonoids are thought to have health benefits, possibly due to antioxidant and anti-inflammatory properties, it can be safely said that Andean indigenous crops have excellent potential as sources of health-promoting bioactive compounds such as flavonoids. Hirose *et al.* (2010) also evaluated antioxidative properties and flavonoid composition in quinoa seeds. The high performance liquid chromatography analysis of the crude glycosidic fraction yielded four flavonol glycosides 1–4 (Fig. 12.3). The results confirmed the presence of large amounts of quercetin and kaempferol glycosides. The amount of quercetin in Japanese samples was 150–225 µmol/100 g fresh weight, which was about three times higher than that of kaempferol (52.3–71.0 µmol/100 g fresh weight). The amount of quercetin in quinoa seeds from Japan was higher than those grown in South America and than in buckwheat (Hirose *et al.*, 2010).

12.2.2 Antinutritional components of quinoa seed

Antinutritional factors (ANF) are compounds that have the potential to adversely affect health and growth by preventing the absorption of nutrients from food and play a great role in limiting the wider use of many plants (Osagie, 1998; Soetan and Oyewole, 2009). These are natural compounds capable of precipitating deleterious effects in man and animals (Osagie, 1998). The levels of toxic substances in plants vary with the species, cultivar and postharvest treatment, such as soaking, drying, autoclaving, and seed germination (Soetan, 2008). Several antinutritional substances have been found in quinoa, such as saponins, phytic acid, tannins and protease inhibitors (González *et al.*, 1989; Improta

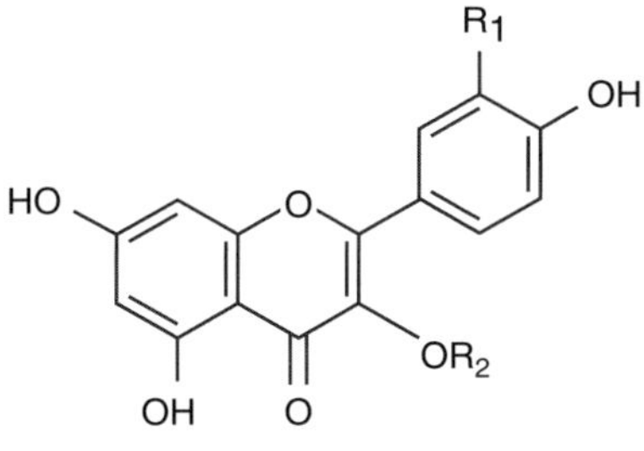

	R_1	R_2
1:	OH	2,6-di-*O*-α-rham nopyranosyl-β-galactopyranosyl
2:	OH	2,6-di-*O*-α-rham nopyranosyl-β-glucopyranosyl
3:	OH	2-*O*-β-apiofuranosyl-6-*O*-α-rham nopyranosyl-β-galactopyranosyl
4:	H	2,6-di-*O*-α-rham nopyranosyl-β-galactopyranosyl

Fig. 12.3. Structure of phytic acid.

and Kellems, 2001), that can have a negative effect on performance and survival of monogastric animals when quinoa is used as the primary dietary energy source (Improta and Kellems, 2001).

Phytic acid
Phytic acid, a hexaphosphorylated myo-inositol commonly known as inositol hexaphosphate, is an abundant constituent in plants, which comprises 1–3% of whole grains, cereals, legumes, nuts and oil seeds (Abdel-Hamid *et al.*, 2007). The molecular formula of phytic acid is $C_6H_{18}O_{24}P_6$ and its molecular weight is 660.03. The chemical structure of phytic acid is provided in Fig. 12.4. Phytic acid is considered an antinutrient and forms insoluble complexes with a wide range of minerals. In addition to depressing protein digestibility *in vitro* (Shimelis and Rakshit, 2005), phytic acid also forms insoluble complexes with polycations, decreasing their availability and intestinal absorption, mostly due to the reactive phosphate groups attached to the inositol ring (Ahamed *et al.*, 1998; Hayashi *et al.*, 2001; Ekholm *et al.*, 2003; Khattak *et al.*, 2007). Phytic acid is present in beans, seeds, nuts and grains, especially in the bran or outer hull; phytates are also found in tubers and trace amounts occur in certain fruits and vegetables like berries and green beans (Coulibaly *et al.*, 2011). In quinoa, phytic acid is not only present in the outer layers of seed, as in rye and wheat (Ahamed *et al.*, 1998), but is also evenly distributed in the endosperm. Koziol; (1992) reported a phytic acid content ranging from 10.5 to 13.5 mg/g for five different varieties of quinoa, which was similar to the range of 7.6–14.7 mg/g for other cereals. Chauhan *et al.* (1992a) obtained 174.4 mg/100g phytate phosphate in the whole quinoa seed and 76.9 mg/100 g in the hulls. The phytic acid in quinoa seed is located in the external layers as well as in the endosperm, which is in sharp contrast to cereals where phytic acid is concentrated in the germ. In quinoa, 60% of the total phytate is concentrated in the embryo, 35% in the perisperm and 5% in the bran (Ando *et al.*, 2002). Since quinoa is high in iron content compared with other cereals, the phytates could markedly decrease the bioavailability of this element. Valencia *et al.* (1999) reported that soaking, germination and lactic fermentation of quinoa resulted in improved iron solubility and reduced phytate content. However, the most effective treatment for reducing phytate was

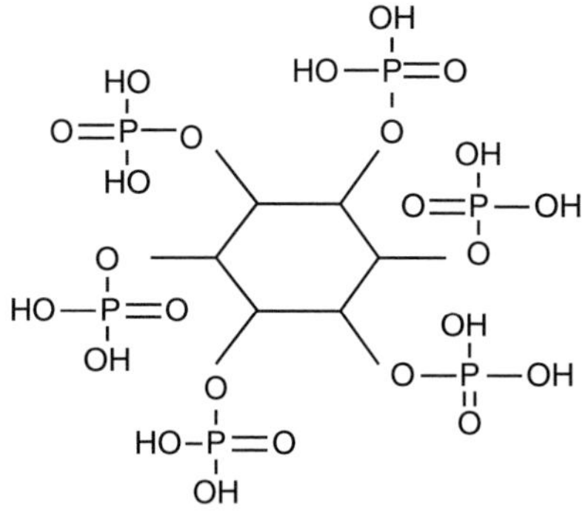

Fig. 12.4. Structures of flavonol glycosides isolated from quinoa seeds. [Reprinted from Hirose *et al.* (2010) with permission from Elsevier.]

fermentation of germinated quinoa flour, whereby phytate was almost completely hydrolysed and the iron solubility increased five to eight times compared with its unfermented counterpart (Lindeboom, 2005).

Protease inhibitors

Protease inhibitors are proteins that form very stable complexes with proteolytic enzymes broadly distributed in nature (Aguirre *et al.*, 2004). Traditionally, protease inhibitors belong to two major classes (Huisman and Tolman, 2001; Pusztai *et al.*, 2004): (i) Kunitz trypsin inhibitor, mainly present in soybeans, and (ii) Bowman-Birk trypsin/chymotrypsin inhibitor, mainly present in grain legumes.

Protease inhibitors act by inhibition of the proteolytic enzymes trypsin and chymotrypsin, secreted into the intestinal lumen, by forming stable inactive complexes (Lallès and Jansman, 1998). Moreover, an increased pancreatic secretion of trypsin and chymotrypsin due to trypsin inhibitor activity (which is used as a measure to determine protease inhibitor activity) may lead to an enhanced loss of endogenous methionine and cystine, because these pancreatic enzymes are rich in the sulfur-containing amino acids (Gatel, 1994). As a result, losses of endogenous methionine and cystine via enhanced secretion of trypsin and chymotrypsin may increase growth depression (Belitz and Weder, 1990). The concentration of protease inhibitors in quinoa seeds is <50 ppm. Ahamed *et al.* (1998) and Improta and Kellems (2001) reported that quinoa contains small amounts of trypsin inhibitors that are much lower than those in commonly consumed grains and hence do not pose any serious concern.

Tannins

Tannins are complex polyphenolics found widely in the plant kingdom in leaves, flowers, fruits and tree bark. Tannins have a large influence on the nutritive value of many foods eaten by humans and feedstuff eaten by animals. Tannins cause decreased feed consumption in animals, bind dietary protein and digestive enzymes to form complexes that are not readily digestible (Aletor, 1993; Soetan and Oyewole, 2009). They also cause decreased palatability and reduced growth rate (Roeder, 1995). The content of tannins measured as flavonols in whole raw quinoa seeds was 0.5% (Ahamed *et al.*, 1998). Chauhan *et al.* (1992a) obtained 0.53% tannins in the whole quinoa seed and 0.92% in the hulls. However, tannins were not detected in raw quinoa seeds that had been polished and washed (Chauhan *et al.*, 1992a).

Saponins

These are the main antinutritional factors present in the seed coat of quinoa. The saponin content in seeds of sweet and bitter genotypes vary from 0.2 to 0.4 and 4.7 to 11.3 g/kg dry matter respectively (Mastebroek *et al.*, 2000). Saponins in quinoa are basically glycosidic triterpenoids, with glucose constituting about 80% of the weight. Saponin content is affected by growth stages and soil water deficit. Saponin content is low in the branching stage and high in the blooming stage; a high water deficit lowers the saponin content (Soliz-Guerrero *et al.*, 2002; Bhargava *et al.*, 2006). Saponins in quinoa can be removed either by the wet method, i.e. washing and rubbing in cold water, or by the dry method, i.e. toasting and subsequent rubbing of the grains to remove the outer layers (Risi and Galwey, 1984). Saponin removal by the dry method reduces the vitamin and mineral content to some extent, the loss being significant in the case of potassium, iron and manganese (Ruales and Nair, 1992). Saponins will be dealt with in detail in Chapter 13.

12.3 Leaves

Quinoa leaves are widely used as food for humans and livestock and constitute an inexpensive source of vitamins and minerals (Weber, 1978; Ahamed *et al.*, 1998). Young leaves have been used like spinach since pre-Columbian times. The correlation between the nutrient content of a leaf and its age (as shown by its position on the plant) is an important factor in choosing leaves for harvesting (Ahamed *et al.*, 1998). Initial reports suggested a protein content of 2.79–4.17% and lipid content of 1.9–2.3% in fresh leaves (Cornejo, 1976). Quinoa leaves contain ample amounts of ash (3.3%), fibre (1.9%), nitrates (0.4%), vitamin E (2.9 mg α-TE/100g) and Na (289 mg/100g) (Koziol, 1992). Prakash *et al.* (1993) reported that leaves have about 82–190 mg/kg of carotenoids, 1.2–2.3 gm/kg of vitamin C and 27–30 gm/kg of protein. However, some reports also confirm a number of antinutrient factors such as nitrate and oxalates (Prakash *et al.*, 1993). Quinoa is considered to be among the 10 most productive species with respect to amount of extractable leaf protein out of 200 members of the families Chenopodiaceae and Amaranthaceae (Lexander *et al.*, 1970). Studies on fresh leaves (Bhargava *et al.*, 2007) revealed abundant moisture (83.92–89.11%), chlorophyll a (0.48–1.82 mg/g), chlorophyll b (0.25–0.07 mg/g) and much higher amount of leaf carotenoid (230.23–669.57 mg/kg) than reported previously. The leaf carotenoid content was higher than that reported for spinach, amaranth and *Chenopodium album* (Gupta and Wagle, 1988; Prakash and Pal, 1991; Shukla *et al.*, 2003). The results obtained by Bhargava *et al.* (2008) have shown that potassium, calcium, sodium, iron and copper contents in quinoa were particularly high and moderate consumption of the foliage would satisfy the 'Recommended Dietary Allowance' (National Research Council, 1989). The proximate value of the mineral composition of quinoa leaves compares favourably with other leafy vegetables reported by Aletor *et al.* (2002) and Kawashima and Valente-Soares (2003). However, a comparison with vegetable amaranth, a widely consumed vegetable crop, suggests that *Chenopodium* spp. is a richer source of potassium and copper, but is comparatively inferior in magnesium and zinc (Aletor and Adeogun, 1995). However, the lesser consumption of quinoa as a leafy vegetable compared with other chenopods is due to the fact that quinoa leaves contain considerable amounts of saponin (0.13–0.17 g/kg dry matter) that are produced rather late during plant development (Mastebroek *et al.*, 2000). Thus, the presence of saponin, an antinutritional factor, acts as a deterrent in the use of quinoa leaves for fodder and vegetable purposes (Bhargava *et al.*, 2008).

12.4 Concluding Remarks

Quinoa has recently gained prominence all over the world for its nutritional benefits. The chief features of the grain include excellent amino acid composition of protein and high content of oil, minerals and vitamins. It is also considered a good source of dietary fibre and other bioactive compounds such as phenolics and antioxidants. The higher concentrations of flavonoid derivatives in quinoa enhance its neutraceutical value in terms of health-promoting effects. Quinoa could serve as an important crop to complement the diet in rural/marginal regions of developing countries where energy-protein malnutrition affects most of the population. It may present a new viable crop option for low-income areas and also provide a new ingredient for specific foods for particular target populations with potential health benefits. Sincere efforts are needed to promote quinoa consumption by informing consumers of the good properties of quinoa so that they may incorporate it into their daily diet as a healthy, nutritious, good tasting and versatile food. Quinoa has rightly been selected by FAO as one of the crops destined to offer food security in the 21st century.

References

Abdel-Hamid, N.M., Faddah, L.M., Al-Rehany, M.A., Ali, A.H. and Bakeet, A.A. (2007) New role of antinutritional factors, phytic acid and catechin in the treatment of CCl_4 intoxication. *Annals of Hepatology* 6, 262–266.

Abeywardena, M., McLeannan, P. and Charnock, J. (1991) Differential effects of dietary fish oil on myocardial prostaglandin 12 and thromboxane A2 production. *American Journal of Physiology* 260, 379–385.

Abugoch, L.E. (2009) Quinoa (*Chenopodium quinoa* Willd.): composition, chemistry, nutritional, and functional properties. *Advances in Food Nutrition Research* 58, 1–31.

Abugoch, L., Romero, N., Tapia, C., Silva, J. and Rivera, M. (2008) Study of some physico-chemical and functional properties of quinoa (*Chenopodium quinoa* Willd.) protein isolates. *Journal of Agricultural and Food Chemistry* 56, 4745–4750.

Aggarwal, P.K., Joshi, P.K., Ingram, J.S.I. and Gupta, R.K. (2004) Adapting food systems of the Indo-Gangetic plains to global environmental change: key information needs to improve policy formulation. *Environmental Science and Policy* 7, 487–498.

Aguirre, C., Valdéz-Rodríguez, S., Mendoza-Hernández, G., Rojo-Domínguez, A. and Blanco-Labra, A. (2004) A novel 8.7 kDA protease inhibitor from chan seeds (*Hyptis suaveolens* L.) inhibits proteases from the larger grain borer *Prostephanus truncatus* (Coleoptera: Bostrichidae). *Comparative Biochemistry and Physiology Part B* 138, 81–89.

Ahamed, N.T., Singhal, R.S., Kulkarni, P.R., Kale, D.D. and Pal, M. (1996) Studies on *Chenopodium quinoa* and *Amaranthus paniculatas* starch as biodegradable fillers in LDPE films. *Carbohydrate Polymers* 31, 157–160.

Ahamed, N.T., Singhal, R.S., Kulkarni, P.R. and Mohinder, P. (1998) A lesser-known grain, *Chenopodium quinoa*: review of the chemical composition of its edible parts. *Food and Nutrition Bulletin* 19, 61–70.

Aletor, O., Oshodi, A.A. and Ipinmoroti, K. (2002) Chemical composition of common leafy vegetables and functional properties of their leaf protein concentrates. *Food Chemistry* 78, 63–68.

Aletor, V.A. (1993) Allelochemicals in plant food and feeding stuffs: nutritional, biochemical and physiopathological aspects in animal production. *Veterinary and Human Toxicology* 35, 57–67.

Aletor, V.A. and Adeogun, O.A. (1995) Nutrient and anti-nutrient components of some tropical leafy vegetables. *Food Chemistry* 53, 325–329.

Aluko, R. and Monu, L. (2003) Functional and bioactive properties of quinoa seed protein hydro-lysates. *Journal of Food Science* 68, 1254–1258.

Alvarez, M., Pavón, J. and von Rütte, S. (1990) Caracterización. In: Wahli, Ch. (ed.) Quinua: hacia su cultivo comercial. Latinreco S.A., Casilla 17-110-6053, Quito, Ecuador, pp. 5–30.

Ando, H., Chen, Y., Tang, H., Shimizu, M., Watanabe, K. and Miysunaga, T. (2002) Food components in fractions of quinoa seed. *Food Science and Technology Research* 8, 80–84.

Arts, I.C.W. and Hollman, P.C.H. (2005) Polyphenols and disease risk in epidemiologic studies. *American Journal of Clinical Nutrition* 81, 317S–325S.

Atwell, W.A., Patrick, B.M., Johnson, L.A. and Glass, R.W. (1983) Characterization of quinoa starch. *Cereal Chemistry* 60, 9–11.

Atwell, W.A., Hyldon, W.G., Godfret, P.D., Galle, E.I., Sperber, W.H., Pedersen, D.C., Evans, W.D. and Rabe, G.O. (1988) Germinated quinoa flour to reduce the viscosity of starchy foods. *Cereal Chemistry* 65, 508–509.

Baldwin, P.M. (2001) Starch granule-associated proteins and polypeptides: a review. *Starch* 53, 475–503.

Bao, J.S., Cai, Y.Z. and Corke, H. (2001) Prediction of rice starch quality parameters by near infrared reflectance spectroscopy. *Journal of Food Science* 66, 936–939.

Barrett, M.L. and Udani, J.K. (2011) A proprietary alpha-amylase inhibitor from white bean (*Phaseolus vulgaris*): a review of clinical studies on weight loss and glycemic control. *Nutrition Journal* 10, 24.

Bean, M.M. and Setser, C.S. (1992) Starch. In: Bowers, J. (ed.) *Food Theory and Applications*. Macmillan, New York, pp. 70–118.

Becker, R. and Hanners, G.D. (1990) Composition and nutritional evaluation of quinoa whole grain flour and mill fractions. *Lebensmittel-Wissenschaft und Technologie* 23, 441–444.

Belitz, H.D. and Weder, J.K.P. (1990) Protein inhibitors of hydrolases in plant foodstuffs. *Food Reviews International* 6, 151–211.

Bhargava, A., Shukla, S. and Ohri, D. (2006) *Chenopodium quinoa*: an Indian perspective. *Industrial Crops and Products* 23, 73–87.

Bhargava, A., Shukla, S. and Ohri, D. (2007) Genetic variability and interrelationship among various morphological and quality traits in quinoa (*Chenopodium quinoa* Willd.). *Field Crops Research* 101, 104–116.

Bhargava, A., Shukla, S., Srivastava, J., Singh, N. and Ohri, D. (2008) Genetic diversity for mineral accumulation in foliage of *Chenopodium* spp. *Scientia Horticulturae* 118, 338–346.

Brinegar, C. and Goundan, S. (1993) Isolation and characterization of chenopodin, the 11S seed storage protein of quinoa (*Chenopodium quinoa*). *Journal of Agricultural and Food Chemistry* 41, 182–185.

Brinegar, C., Sine, B. and Nwokocha, L. (1996) High-cysteine 2S seed storage proteins from quinoa (*Chenopodium quinoa*). *Journal of Agricultural and Food Chemistry* 44, 1621–1623.

Cardozo, A. and Tapia, M.E. (1979) Valor nutritivo. Quinua y Kaniwa. Cultivos Andinos. In: Tapia, M.E. (ed.) *Serie Libros y Materiales Educativos*. Instituto Interamericano de Ciencias Agricolas, Bogota, Colombia, 49, 149–192.

Chang, S.M. and Lui, L.C. (1991) Retrogradation of rice starches studied by differential scanning calorimetry and the influence of sugars, NaCl and lipids. *Journal of Food Science* 56, 564–566, 570.

Chauhan, G., Eskin, N. and Tkachuk, R. (1992a) Nutrients and antinutrients in quinoa seed. *Cereal Chemistry* 69, 85–88.

Chauhan, G., Zillman, R. and Eskin, N. (1992b) Dough mixing and breadmaking properties of quinoa–wheat flour blends. *International Journal of Food Science and Technology* 27, 701–705.

Comai, S., Bertazzo, A., Bailoni, L., Zancato, M., Costa, C. and Allegri, G. (2007) The content of proteic and nonproteic (free and protein-bound) tryptophan in quinoa and cereal flours. *Food Chemistry* 100, 1350–1355.

Cornejo, de Z.G. (1976) Hojas de quinua (*Chenopodium quinoa* Willd.) fuente de protein. In: *Segunda Convencion Internacional de Quenopodiaceas*, Universidad Boliviana Tomas Frias, Comite Departamental de Obras Publicas de Potosi, Instituto Interamericano de Ciencias Agricolas, Potosi, Bolivia, pp. 177–180.

Coulibaly, A., Kouakou, B. and Chen, J. (2011) Phytic acid in cereal grains: structure, healthy or harmful ways to reduce phytic acid in cereal grains and their effects on nutritional quality. *American Journal of Plant Nutrition and Fertilization Technology* 1, 1–22.

Cusack, D. (1984) Quinoa: grain of the Incas. *Ecologist* 14, 21–31.

De Bruin, A. (1964) Investigation of the food value of quinoa and canihua seed. *Journal of Food Science* 29, 872–876.

De Simone, F., Dini, A., Pizza, C., Saturnino, P. and Scettino, O. (1990) Two flavonol glycosides from *Chenopodium quinoa*. *Phytochemistry* 29, 3690–3692.

Dini, I., Tenore, G.C. and Dini, A. (2005) Nutritional and antinutritional composition of Kancolla seeds: an interesting and underexploited andine food plant. *Food Chemistry* 92, 125–132.

Dini, I., Tenore, G.C. and Dini, A. (2008) Chemical composition, nutritional value and antioxidant properties of *Allium cepa* L. Var. Troperana (red onion) seeds. *Food Chemistry* 107, 613–621.

Ekholm, P., Virkki, L., Ylinen, M. and Johansson, L. (2003) The effect of phytic acid and some natural chelating agents on the solubility of mineral elements in oat bran. *Food Chemistry* 80, 165–170.

Eyaru, R., Shrestha, A.K. and Arcot, J. (2009) Effect of various processing techniques on digestibility of starch in Red kidney bean (*Phaseolus vulgaris*) and two varieties of peas (*Pisum sativum*). *Food Research International* 42, 956–962.

Fan, J. and Marks, B.P. (1998) Retrogradation kinetics of rice flours as influenced by cultivar. *Cereal Chemistry* 75, 153–155.

FAO (1998) *Carbohydrates in Human Nutrition*. Food and Nutrition Papers, Food and Agriculture Organization, Rome, Italy.

FAO (2008) *Fats and Fatty Acids in Human Nutrition: Report of an Expert Consultation*. Food and Agriculture Organization, Rome, Italy.

FAO (2010) *The State of Food Insecurity in the World 2010*. Food and Agriculture Organization, Rome, Italy.

FAO (2011) *Quinoa: An Ancient Crop to Contribute to World Food Security*. Food and Agriculture Organization, Rome, Italy.

FAO/WHO/UNU (1985) *Energy and protein requirements*. Report of a joint FAO/WHO/ UNU meeting. Food and Agriculture Organization of the United Nations/World Health Organization/United Nations University, World Health Organization, Geneva, Switzerland.

Friedman, M. and Brandon, D.L. (2001) Nutritional and health benefits of soy proteins. *Journal of Agricultural and Food Chemistry* 49, 1069–1086.

Gatel, F. (1994) Protein quality of legume seeds for non-ruminant animals: a literature review. *Animal Feed Science and Technology* 45, 317–348.

González, J.A., Roldán, A., Gallardo, M., Escudero, T. and Prado, F.E. (1989) Quantitative determination of chemical compounds with nutritional value from Inca crops: *Chenopodium quinoa* ('quinoa'). *Plant Foods for Human Nutrition* 39, 331–337.

Gorinstein, S., Lojek, A., Cîz, M., Pawelzik, E., Delgado-Licon, E., Medina, O., Moreno, M., Salas, I. and Goshev, I. (2008) Comparison of composition and antioxidant capacity of some cereals and pseudocereals. *International Journal of Food Science and Technology* 43, 629–637.

Grant, L.A., Vignaux, N., Doehlert, D.C., McMullen, M.S., Elias, E.M. and Kianian, S. (2001) Starch characteristics of waxy and nonwaxy tetraploid (*Triticum turgidum* L. var. *durum*) wheat. *Cereal Chemistry* 78, 590–595.

Gross, R., Roch, F., Malaga, F., De Mirenda, A., Scoeneberger, H. and Trugo, L.C. (1989) Chemical composition and protein quality of some Andean food sources. *Food Chemistry* 30, 25–34.

Gudmundsson, M. and Eliasson, A.C. (1990) Retrogradation of amylopectin and the effects of amylose and added surfactants emulsifiers. *Carbohydrate Polymers* 13, 295–315.

Gunstone, F.D. (1999) What else besides commodity oils and fats. *European Journal of Lipid Science and Technology* 101, 124–130.

Gupta, K. and Wagle, D.S. (1988) Nutritional and antinutritional factors of green leafy vegetables. *Journal of Agricultural and Food Chemistry* 36: 472–474.

Han, X., Shen, T. and Lou, H. (2007) Dietary polyphenols and their biological significance. *International Journal of Molecular Sciences* 8, 950–988.

Hayashi, K., Hara, H., Asvarujanon, P., Aoyama, Y. and Luangpituksa, P. (2001) Ingestion of insoluble dietary fiber increased zinc and iron absorption and restored growth rate and zinc absorption suppressed by dietary phytate in rats. *British Journal of Nutrition* 86, 443–451.

Hirose, Y., Fujita, T., Ishii, T. and Ueno, N. (2010) Antioxidative properties and flavonoid composition of *Chenopodium quinoa* seeds cultivated in Japan. *Food Chemistry* 119, 1300–1306.

Huisman, J. and Tolman, G.H. (2001) Antinutritional factors in the plant proteins of diets for non-ruminants. In: Garnsworthy, P.C. and Wiseman, J. (eds) *Recent Developments in Pig Nutrition*. Nottingham University Press, Nottingham, UK, pp. 261–322.

Improta, F. and Kellems, R.O. (2001) Comparison of raw, washed and polished quinoa *Chenopodium quinoa* Willd. to wheat, sorghum or maize based diets on growth and survival of broiler chicks. *Livestock Research for Rural Development* 13, 10.

Inouchi, N., Nishi, K., Tanaka, S., Asai, M., Kawase, Y., Hata, Y., Konishi, Yue, S. and Fuwa, H. (1999) Characterization of amaranth and quinoa starches. *Journal of Applied Glycoscience* 46, 233–240.

Insel, P.M., Turner, E. and Ross, D. (2004) *Nutrition*, 2nd edn. Jones and Bartlett, Sudbury, Massachusetts.

Jacobsen, S.-E. (2003) The worldwide potential of quinoa (*Chenopodium quinoa* Willd.). *Food Reviews International* 19, 167–177.

Jancurova, M., Minarovicova, L. and Dandar, A. (2009) Quinoa – a review. *Czech Journal of Food Science* 27, 71–79.

Jane, J., Chen, Y.Y., Lee, L.F., McPherson, A.E., Wong, K.S., Radosavljevic, M. and Kasemsuwan, T. (1999) Effects of amylopectin branch chain length and amylose content on the gelatinization and pasting properties of starch. *Cereal Chemistry* 76, 629–637.

Janssen, W.M.M.A., Terpstra, K., Beeking, F.F.E. and Bisalsky, A.J.N. (1979) *Feeding Values for Poultry*, 2nd edn. Spelderholt Center for Poultry Research and Information Services, Beekbergen, The Netherlands.

Karyotis, T., Iliadis, C., Noulas, C. and Mitsibonas, T. (2003) Preliminary research on seed production and nutrient content for certain quinoa varieties in a saline-sodic soil. *Journal of Agronomy and Crop Science* 189, 402–408.

Kaur, L., Singh, N. and Singh-Sodhi, N. (2002) Some properties of potatoes and their starches II. Morphological, thermal and rheological properties of starches. *Food Chemistry* 79, 183–192.

Kawashima, L.M. and Valente-Soares, L.M. (2003) Mineral profile of raw and cooked leafy vegetables consumed in southern Brazil. *Journal of Food Composition and Analysis* 16, 605–611.

Khattak, A.B., Zeb, A., Bibi, N., Khalil, S.A. and Khattak, M.S. (2007) Influence of germination techniques on phytic acid and polyphenols content of chickpea (*Cicer arietinum* L.) sprouts. *Food Chemistry* 104, 1074–1079.

Kong, X., Bao, J. and Corke, H. (2009) Physical properties of *Amaranthus* starch. *Food Chemistry* 113, 371–376.

Konishi, Y., Hirano, S., Tsuboi, H. and Wada, M. (2004) Distribution of minerals in quinoa (*Chenopodium quinoa* Willd.) seeds. *Bioscience, Biotechnology and Biochemistry* 68, 231–234.

Koziol, M.J. (1990) Composicion quimica. In: Wahli, C. (ed.) *Quinua, Hacia su Cultivo Commercial*. Latinreco S.A., Casilla 17-110-6053, Quito, Ecuador, pp. 137–159.

Koziol, M.J. (1992) Chemical composition and nutritional value of quinoa (*Chenopodium quinoa* Willd.). *Journal of Food Composition and Analysis* 5, 35–68.

Koziol, M.J. (1993) Quinoa: a potential new oil crop. In: Janick, J. and Simon, J.E. (eds) *New Crops*. Wiley, New York, pp. 328–336.

Kurosawa, T., Itoh, F., Nozaki, A., Nakano, Y. and Katsuda, S. (2005) Suppressive effects of cacao liquor polyphenols (CLP) on LDL oxidation and the development of atherosclerosis in Kurosawa and kusanagi hypercholesterolemic rabbits. *Atherosclerosis* 179, 237–246.

Lai, L.N., Abd Karim, A., Norziah, M.H. and Seow, C.C. (2002) Effects of Na_2CO_3 and NaOH on DSC thermal profiles of selective native cereal starches. *Food Chemistry* 78, 355–362.

Lallès, J.P. and Jansman, A.J.M. (1998) Recent progress in the understanding of the mode of action and effects of antinutritional factors from legume seeds in non-ruminant farm animals. In: Jansman, A.J.M., Hill, G.D., Huisman, J. and van der Poel, A.F.B. (eds) *Recent Advances of Research in Antinutritional Factors in Legume Seeds and Rapeseed*. Centre for Agricultural Publishing and Documentation (PUDOC), Wageningen, The Netherlands, pp. 219–232.

Latham, M.C. (1997) *Human Nutrition in the Developing World*. Food and Nutrition Series No. 29. Food and Agriculture Organization, Rome, Italy.

Lexander, K., Carlsson, R., Schalen, V., Simonsson, A. and Lundborg, T. (1970) Quantities and qualities of leaf protein concentrates from wild species and crop species under controlled conditions. *Annals of Applied Biology* 66, 193–216.

Li, J., Berke, T.G. and Glover, D.V. (1994) Variation for thermal properties of starch in tropical maize germplasm. *Cereal Chemistry* 71, 87–90.

Lindeboom, N. (2005) Studies on the characterization, biosynthesis and isolation of starch and protein from quinoa (*Chenopodium quinoa* Willd.). Doctor of Philosophy thesis, University of Saskatchewan, Canada.

Lindeboom, N., Chang, P.R., Falk, K.C. and Tyler, R.T. (2005) Characteristics of starch from eight quinoa lines. *Cereal Chemistry* 82, 216–222.

Lopez, de Romana, G., Graham, G., Rojas, M. and MacLean, W. (1981) Digestibilidad y calidad proteinica de la quinua: estudio comparativo, en ninos, entre semilla y harina de quinua. *Archivos Latinoamericana de Nutrición* 31, 485–497.

Lopez-Rubio, A., Flanagan, B., Gilbert, E. and Gidley, M. (2004) A novel approach for calculating starch crystallinity and its correlation with double helix content: a combined XRD and NMR study. *Biopolymers* 89, 761–768.

Lorenz, K. (1990) Quinoa (*Chenopodium quinoa*) starch: physico-chemical properties and functional characteristics. *Starch* 42, 81–86.

Lorenz, K. and Coulter, L. (1991) Quinoa flour in baked products. *Plant Foods for Human Nutrition* 41, 213–224.

Lovejoy, J. (1999) Dietary fatty acids and insulin resistance. *Current Atherosclerosis Reports* 1, 215–220.

Lusas, E.W. (2000) Oilseeds and oil bearing materials. In: Kulp, K. and Ponte, J.G. Jr (eds) *Handbook of Cereal Science and Technology*. Marcel Dekker, New York, pp. 297–362.

Mahoney, A., Lopez, J. and Hendricks, D. (1975) Evaluation of the protein quality of quinoa. *Journal of Agricultural and Food Chemistry* 23, 190–193.

Marton, M., Mandoki, Zs. and Csapo, J. (2010) Evaluation of biological value of sprouts. I. Fat content, fatty acid composition. *Acta Universitatis Sapientiae, Alimentaria* 3, 53–65.

Mastebroek, H.D., Limberg, H., Gilles, T. and Marvin, H.J.P. (2000) Occurrence of sapogenins in leaves and seeds of quinoa (*Chenopodium quinoa* Willd.). *Journal of the Sciences of Food and Agriculture* 80, 152–156.

Mattila, P., Pihlava, J.-M. and Hellstrom, J. (2005) Contents of phenolic acids, alkyl and alkenyl resorcinols and avenanthramides in commercial grain products. *Journal of Agricultural and Food Chemistry* 53, 8290–8295.

Miranda, M., Maureira, H., Rodriguez, K. and Vega-Gálvez, A. (2009) Influence of temperature on the drying kinetic, physicochemical properties, and antioxidant capacity of *Aloe Vera* (*Aloe Barbadensis* Miller) gel. *Journal of Food Engineering* 91, 297–304.

Miranda, M., Vega-Gálvez, A., Lopez, J., Parada, G., Sanders, M., Aranda, M., Uribe, E. and Scala, K.D. (2010) Impact of air-drying temperature on nutritional properties, total phenolics content and antioxidant capacity of quinoa seeds (*Chenopodium quinoa* Willd.). *Industrial Crops and Products* 32, 258–263.

Moyad, M.A. (2005) An introduction to dietary/supplemental omega-3 fatty acids for general health and prevention. Part I. *Urologic Oncology* 23, 28–35.

National Academy of Sciences (2004) *Comprehensive DRI Table for Vitamins, Minerals and Macronutrients, Organized by Age and Gender.* Institute of Medicine, Food and Nutrition Board, Beltsville, Maryland.

National Research Council, Food and Nutrition Board (1989) *Recommended Dietary Allowances*, 10th edn. National Academy of Sciences, Washington, DC.

Ogungbenle, H.N. (2003) Nutritional evaluation and functional properties of quinoa (*Chenopodium quinoa*) flour. *International Journal of Food Sciences and Nutrition* 54, 153–158.

Osagie, A.U. (1998) Antinutritional factors. In: Osagie, A.U. and Eka, O.U. (eds) *Nutritional Quality of Plant Foods*. Ambik Press, Benin City, Nigeria, pp. 1–40, 221–244.

Oshodi, A., Ogungbenle, H. and Oladimeji, M. (1999) Chemical composition, nutritionally valuable minerals and functional properties of benniseed, pearl millet and quinoa flours. *International Journal of Food Sciences and Nutrition* 50, 325–331.

Park, I., Ibáñez, A. and Shoemaker, C. (2007) Rice starch molecular size and its relationship with amylose content. *Starch/Starke* 59, 69–77.

Pásko, P., Bartón, H., Zagrodzki, P., Gorinstein, S. and Folta Mand Zachwieja, Z. (2009) Anthocyanins, total polyphenols and antioxidant activity in amaranth and quinoa seeds and sprouts during their growth. *Food Chemistry* 115, 994–998.

Pieper, R., Jha, R., Rossnagel, B., Van Kessel, A.G., Souffrant, W.B. and Leterme, P. (2008) Effect of barley and oat cultivars with different carbohydrate compositions on the intestinal bacterial communities in weaned piglets. *FEMS Microbiology Ecology* 66, 556–566.

Prakash, D. and Pal, M. (1991) Nutritional and anti nutritional composition of vegetable and grain amaranth leaves. *Journal of the Sciences of Food and Agriculture* 57, 573–583.

Prakash, D. and Pal, M. (1998) *Chenopodium*: seed protein, fractionation and amino acid composition. *International Journal of Food Science and Nutrition* 49, 271–275.

Prakash, D., Nath, P. and Pal, M. (1993) Composition, variation of nutritional content in leaves, seed protein, fat and fatty acid profile of *Chenopodium* species. *Journal of the Sciences of Food and Agriculture* 62, 203–205.

Praznik, W., Mundigler, N., Kogler, A., Pelzl, B. and Huber, A. (1999) Molecular background of technological properties of selected starches. *Starch* 51, 197–211.

Prego, I., Maldonado, S. and Otegui, M. (1998) Seed structure and localization of reserves in *Chenopodium quinoa*. *Annals of Botany* 82, 481–488.

Przybylski, R., Chauhan, G. and Eskin, N. (1994) Characterization of quinoa (*Chenopodium quinoa*) lipid. *Food Chemistry* 51, 187–192.

Pusztai, A., Bardocz, S. and Martín-Cabrejas, M.A. (2004) The mode of action of ANFs on the gastrointestinal tract and its microflora. In: Muzquiz, M., Hill, G.D., Cuadrado, C., Pedrosa, M.M.

and Burbano, C. (eds) *Recent Advances of Research in Antinutritional Factors in Legume Seeds and Oilseeds*. Wageningen Academic Publishers, Wageningen, The Netherlands, pp. 87–100.

Putseys, J.A., Derde, L.J., Lamberts, L., Östman, E., Björck, I.M. and Delcour, J.A. (2010) Functionality of short chain amylase-lipid complexes in starch-water systems and their impact on in vitro starch degradation. *Journal of Agriculture and Food Chemistry* 58, 1939–1945.

Qian, J.Y. and Kuhn, M. (1999) Characterization of *Amaranthus cruentus* and *Chenopodium quinoa* starch. *Starch* 51, 116–120.

Raina, A. and Datta, A. (1992) Molecular cloning of a gene encoding seed-specific protein with nutritionally balanced amino acid composition from *Amaranthus*. *Proceedings of the National Academy of Sciences (USA)* 89, 11,774–11,778.

Ranhotra, G., Gelroth, J., Glaser, B., Lorenz, K. and Johnson, D. (1993) Composition and protein nutritional quality of quinoa. *Cereal Chemistry* 70, 303–305.

Rastrelli, L., Saturnino, P., Schettino, O. and Dini, A. (1995) Studies on the constituents of *Chenopodium pallicaule* (Canihua) seeds. Isolation and characterization of two new flavonol glycosides. *Journal of Agriculture and Food Chemistry* 43, 2020–2024.

Repo-Carrasco-Valencia, R. and Serna, L.A. (2011) Quinoa (*Chenopodium quinoa*, Willd.) as a source of dietary fiber and other functional components. *Ciencia e Tecnologia de Alimentos* 31, 225–230.

Repo-Carrasco, R., Espinoza, C. and Jacobsen, S.-E. (2003) Nutritional value and use of the Andean crops quinoa (*Chenopodium quinoa*) and kañiwa (*Chenopodium pallidicaule*). *Food Reviews International* 19, 179–189.

Repo-Carrasco-Valencia, R., Hellstrom, J.K., Pihlava, J.M. and Mattila, P.H. (2010) Flavonoids and other phenolic compounds in Andean indigenous grains: quinoa (*Chenopodium quinoa*), kaniwa (*Chenopodium pallidicaule*) and kiwicha (*Amaranthus caudatus*). *Food Chemistry* 120, 128–133.

Risi, J. and Galwey, N.W. (1984) The *Chenopodium* grains of the Andes: Inca crops for modern agriculture. *Advances in Applied Biology* 10, 145–216.

Roberts, S.A. and Cameron, R.E. (2002) The effects of concentration and sodium hydroxide on the rheological properties of potato starch gelatinisation. *Carbohydrate Polymers* 50, 133–143.

Roeder, E. (1995) Medicinal plants in Europe containing pyrrolidizine alkaloids. *Pharmazie* 50, 83–98.

Romero, J.A. (1981) Evaluación de las características físicas, químicas y biológicas de ocho variedades de quinua (*Chenopodium quinoa* Willd.). Tesis de Maestro, Universidad de San Carlos de Guatemala, Ciudad de Guatemala, Guatemala.

Ruales, J. and Nair, B.M. (1992) Nutritional quality of the protein in quinoa (*Chenopodium quinoa* Willd.) seeds. *Plant Foods for Human Nutrition* 42, 1–12.

Ruales, J. and Nair, B.M. (1994a) Effect of processing on in vitro digestibility of protein and starch in quinoa seeds. *International Journal of Food Science and Technology* 29, 449–456.

Ruales, J. and Nair, B. (1994b) Properties of starch and dietary fibre in raw and processed quinoa (*Chenopodium quinoa* Willd.) seeds. *Plant Foods for Human Nutrition* 45, 223–246.

Ryan, E., Galvin, K., O'Connor, T., Maguire, A. and O'Brien, N. (2007) Phytosterol, squalene, tocopherol content and fatty acid profile of selected seeds, grains, and legumes. *Plant Foods for Human Nutrition* 62, 85–91.

Sardesai, V.M. (1998) *Introduction to Clinical Nutrition*. Marcel Dekker, New York.

Scalbert, A., Manach, C., Morand, C. and Remesy, C. (2005) Dietary polyphenols and the prevention of diseases. *Critical Reviews in Food Science and Nutrition* 45, 287–306.

Schlick, G. and Bubenheim, D.L. (1996) Quinoa: candidate crop for NASA's controlled ecological life support systems. In: Janick, J. (ed.) *Progress in New Crops*. ASHS Press, Arlington, Virginia.

Shahidi, F. and Naczk, M. (1995) *Food Phenolics*. Technomic, Lancaster, USA,

Sharma, V., Rausch, K.D., Tumbleson, M.E. and Singh, V. (2006) *Starch Fermentation Characteristics for Different Proportions of Amylase and Amylopectins*. American Society of Agricultural and Biological Engineers, St Joseph, Michigan.

Shimelis, E.A. and Rakshit, S.K. (2005) Antinutritional factors and *in vitro* protein indigestibility of improved haricot bean (*Phaseolus vulgaris* L.) varieties growing in Ethiopia. *International Journal of Food Science and Nutrition* 56, 377–387.

Shukla, S., Pandey, V., Pachauri, G., Dixit, B.S., Bannerji, R. and Singh, S.P. (2003) Nutritional contents of different foliage cuttings of vegetable amaranth. *Plant Foods for Human Nutrition* 58, 1–8.

Soetan, K.O. (2008) Pharmacological and other beneficial effects of antinutritional factors in plants: a review. *African Journal of Biotechnology* 7, 4713–4721.

Soetan, K.O. and Oyewole, O.E. (2009) The need for adequate processing to reduce the antinutritional factors in plants used as human foods and animal feeds: a review. *African Journal of Food Science* 3, 223–232.

Soliz-Guerrero, J.B., Jasso de Rodriguez, D., Rodriguez-Garcia, R., Angulo-Sanchez, J.L. and Mendez-Padilla, G. (2002) Quinoa saponins: concentration and composition analysis. In: Janick, J. and Whipkey, A. (eds) *Trends in New Crops and New Uses*. ASHS Press, Alexandria, Virginia, pp. 110–114.

Souci, S.W., Fachmann, W. and Kraut, H. (1986) *Food Composition and Nutrition Tables 1986/87*. Wissenschaftliche Verlagsgesellschaft, Stuttgart, Germany.

Svegmark, K., Helmersson, K., Nilsson, G., Nilsson, P.O., Andersson, R. and Svensson, E. (2002) Comparison of potato amylopectin starches and potato starches: influence of year and variety. *Carbohydrate Polymers* 47, 331–340.

Tang, H., Yoshida, T., Watanabe, K. and Mitsunaga, T. (1998) Some properties of starch granules in various plants. *Bulletin of the Institute for Comprehensive Agricultural Sciences, Kinki University* 6, 83–89.

Tang, H., Watanabe, K. and Mitsunaga, T. (2002) Characterization of storage starches from quinoa, barley and adzuki seeds. *Carbohydrate Polymers* 49, 13–22.

Tari, T., Annapure, U., Singhal, R. and Kulkarni, P. (2003) Starch-based spherical aggregates: Screening of small granule sized starches for entrapment of a model flavouring compound, vanillin. *Carbohydrate Polymers* 53, 45–51.

Tharanathan, R.N. and Mahadevamma, S. (2003) Grain legumes: a boom to human nutrition. *Trends in Food Science and Technology* 14, 507–518.

USDA (U.S. Department of Agriculture) (2005) USDA National Nutrient Database for Standard Reference, Nutrient Data Laboratory.

Valencia, S., Svanberg, U., Sandberg, A.S. and Ruales, J. (1999) Processing of quinoa (*Chenopodium quinoa* Willd.): effects on in vitro iron availability and phytate hydrolysis. *International Journal of Food Science and Nutrition* 50, 203–211.

Van Etten, C.H., Miller, R.W., Wolff, I.A. and Jones, Q. (1963) Amino acid composition of seeds from 200 angiosperm plants. *Journal of Agricultural and Food Chemistry* 11, 399–410.

Vega-Gálvez, A., Miranda, M., Vergara, J., Uribe, E., Puente, L. and Martínez, E.A. (2010) Nutrition facts and functional potential of quinoa (*Chenopodium quinoa* Willd.). An ancient Andean grain: a review. *Journal of the Sciences of Food and Agriculture* 90, 2541–2547.

Wang, Y., Liu, W. and Sun, Z. (2003) Effects of granule size and shape on morphology and tensile properties of LDPE and starch blends. *Journal of Material Science Letters* 22, 57–59.

Watanabe, K., Peng, L., Tang, H. and Mitsunaga, T. (2007) Molecular structural characteristics of quinoa starch. *Food Science and Technology Research* 13, 73–76.

Weber, E.J. (1978) The Inca's ancient answer to food shortage. *Nature* 272, 486.

White, P.J. (2008) Fatty acids in oilseeds (vegetable oils). In: Chow, K.C. (ed.) *Fatty Acids in Foods and their Health Implications.* CRC Press, New York, pp. 227–262.

Wijngaard, H.H. and Arendt, E.K. (2006) Buckwheat. *Cereal Chemistry* 83, 391–401.

Wong, J.H., Lau, T., Cai, N., Singh, J., Pedersen, J., Vensel, W., Hurkman, W.J., Wilson, J.D., Lemaux, P.G. and Buchanan, B.B. (2009) Digestibility of protein and starch from sorghum (*Sorghum bicolor*) is linked to biochemical and structural features of grain endosperm. *Journal of Cereal Science* 49, 73–82.

Wood, S., Lawson, L., Fairbanks, D., Robinson, L. and Andersen, W. (1993) Seed lipid content and fatty acid composition of three quinoa cultivars. *Journal of Food Composition and Analysis* 6, 41–44.

Wootton, M. and Panozzo, J.F. (1998) Differences in gelatinization behaviour between starches from Australian wheat cultivars. *Starch* 50, 154–158.

Wright, K., Pike, O., Fairbanks, D. and Huber, C. (2002a) Composition of *Atriplex hortensis*, sweet and bitter *Chenopodium quinoa* seeds. *Journal of Food Science* 67, 1380–1383.

Wright, K., Huber, K.C., Fairbanks, D.J. and Huber, C.S. (2002b) Isolation and characterization of *Atriplex hortensis* and sweet *Chenopodium quinoa* starches. *Cereal Chemistry* 79, 715–719.

Yoshimoto, Y., Egashira, T., Hanashiro, I., Ohinata, H., Takase, Y. and Takeda, Y. (2004) Molecular structure and some physicochemical properties of buckwheat starches. *Cereal Chemistry* 81, 515–520.

Youa, S. and Izydorczyk, M. (2007) Comparison of the physicochemical properties of barley starches after partial α-amylolysis and acid/alcohol hydrolysis. *Carbohydrate Polymers* 69, 489–502.

Youdim, K., Martin, A. and Joseph, J. (2000) Essential fatty acids and the brain: possible health implications. *International Journal of Developmental Neuroscience* 18, 383–399.

Youdim, K.A., Shukitt-Hale, B. and Joseph, J.A. (2004) Flavonoids and the brain: interaction at the blood–brain barrier and their physiological effects on the central nervous system. *Free Radical Biology and Medicine* 37, 1683–1693.

Zhu, N., Sheng, S., Li, D., Lavoie, E.J., Karwe, M.V. and Rosen, R.T. (2001) Antioxidative flavonoid glycosides from quinoa seeds (*Chenopodium quinoa* Willd). *Journal of Food Lipids* 8, 37–44.

Zielisnki, H. and Koslowska, H. (2000) Antioxidant activity and total phenolics in selected cereal grains and their different morphological fractions. *Journal of Agricultural and Food Chemistry* 48, 2008–2016.

Zobel, H.F. (1988) Molecules to granules: a comprehensive starch review. *Starch* 40, 44–50.

13 Saponins

13.1 Introduction

Saponins are a large and structurally diverse group of bioactive natural products that are found in both wild plants and cultivated crops (Connolly and Hill, 2005; Phillips *et al.*, 2006; Vincken *et al.*, 2007; Yendo *et al.*, 2010; Osbourn *et al.*, 2011). Saponins are widely distributed in the plant kingdom, being reported in nearly 100 angiospermic plant families, as well as in ferns (Hanus *et al.*, 2003). Plants producing saponins are widely distributed in various geographical and climatic zones around the world, from tropical forests to polar tundras. They include trees, bushes, grasses, perennial evergreen and deciduous shrubs, biennial and annual herbs (Szakiel *et al.*, 2011). They are usually found in the roots, tubers, leaves, flowers or seeds. The word 'saponin' is derived from the soapwort plant *Saponaria* (Family: Caryophyllaceae), the root of which was once used as a soap ('Sapo' is the Latin word for soap). The names of some saponin-producing plant species – for example, soapwort (*Saponaria officinalis*), soapberry or soapnut (*Sapindus* spp.) and soapbark (*Quillaja saponaria*) – reflect their original use as sources of natural soaps (Osbourn *et al.*, 2011). Saponins are abundant in the botanical family Sapindaceae, and in the closely related families, viz. Hippocastanaceae and Aceraceae. They are also found in *Gynostemma pentaphyllum* (Family: Cucurbitaceae) in a form called 'gypenosides' (Kim and Han, 2011), and in *Panax* (ginseng or red ginseng) (Family: Araliaceae) in a form called 'ginsenosides' (Tang *et al.*, 2009). Legumes such as soybeans and chickpeas are major sources of saponins in the human diet (Oakenfull, 1981). Sources of nondietary saponins include alfalfa, sunflower, horse chestnut and a wide variety of herbs (Price *et al.*, 1987). Commercial formulations of plant-derived saponins are obtained from *Q. saponaria*, a tree native to the Andean region, and *Yucca schidigera* (Family: Asparagaceae). Saponins have also been reported in some marine animals (Riguera, 1997; Kalinin *et al.*, 2005).

The content of saponins is very high in some plants, reaching several per-cent of the dry weight. However, like other classes of plant secondary metabo-lites, the level of saponins can be significantly influenced by many intrinsic and external factors (Szakiel *et al.*, 2011). The intrinsic factors reflect the physio-logical status of the plant, which depends mainly on the stage of growth and development. The external factors comprise environmental stimuli, both abi-otic and biotic, including light, temperature, humidity and soil fertility, feeding of phytophagous insects or other herbivorous animals, competition with neigh-bouring plants, and interactions with pathogens and parasites such as bacteria, fungi, viruses and nematodes. These factors affect both the quantitative amount and qualitative composition of saponins (Szakiel *et al.*, 2011).

13.2 Chemical Structure of Saponins

Saponins are a highly diverse group of glycosides containing either a steroidal or triterpenoid aglycone (sapogenin) to which one or more sugar chains are attached (Francis *et al.*, 2002). The sugars may be glucose, galactose, glucuronic acid, xylose, rhamnose or methylpentose. These are synthesised from mevalonate via farnesyl diphosphate and squalene (Fig. 13.1) (Osbourn *et al.*, 2011). This biosynthetic pathway is primarily cytosolic, in contrast to the plastid-localized methylerythritol-phosphate pathway, which is the source of monoterpenes, diterpenes, tetrater-penes (carotenoids) and polyprenols (Chappell, 2002). Saponins are of two types:

- Steroidal saponins: They are C-27 with five methyl groups. Steroidal sapo-nins are less widely distributed in nature compared with the triterpenoidal saponins. These are used mainly as precursors for the partial synthesis of sex hormones and corticosteroids.
- Triterpenoidal saponins: They are C-30 compounds having a pentacyclic skeleton with eight methyl groups.

13.3 Quinoa Saponins

Triterpene saponins have been found in all parts of the quinoa plant, such as leaves, flowers, fruits, seeds and seed coats (Simmonds, 1965; Varriano-Marston and De Francisco, 1984; Cuadrado *et al.*, 1995; Mizui *et al.*, 1988, 1990; Mastebroek *et al.*, 2000; Kuljanabhagavad *et al.*, 2008) (Fig. 13.2). The plant produces saponins primarily in the papillose cells of the outer seed hull as a defence against bird predation (van Raamsdonk *et al.*, 2010). These saponins are a drawback for quinoa as a food and feed application because saponins possess a bitter taste and exhibit toxic effects.

The saponins of *Chenopodium quinoa* are bitter-tasting and represent the major antinutritional factor found in the grain (Ma *et al.*, 1989). In fact, quinoa can be classified into the following two types, according to its saponin content:

- Sweet: saponin absent or less than 0.11% on a fresh weight basis.
- Bitter: saponin greater than 0.11% on a fresh weight basis.

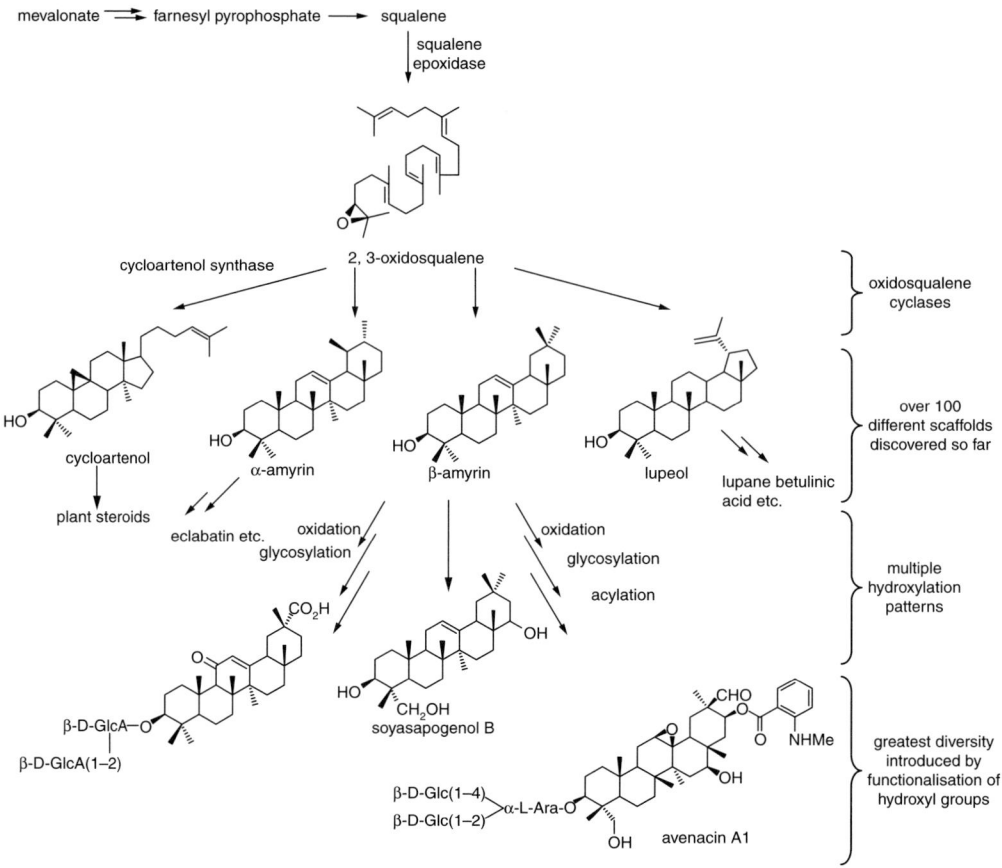

Fig. 13.1. An overview of the routes to saponin biosynthesis and structural diversification (Ara – arabinose; Glc – glucose; GlcA – glucuronic acid). [Reprinted from Osbourn *et al.* (2011), with permission from the Royal Society of Chemistry.]

The saponin content in seeds of sweet genotypes varies from 0.2 to 0.4 g/kg dry matter and in bitter genotypes from 4.7 to 11.3 g/kg dry matter (Mastebroek *et al.*, 2000). These values are higher than those in soybean and oat, but lower than in green pea and yucca (Güçlü-Üstündag and Mazza, 2007). The concentration of saponin in quinoa varies with variety and environmental conditions (Koziol, 1992; Repo-Carrasco *et al.*, 2003). Saponin content in the crop is also significantly affected by soil-water deficit. Saponin concentration and composition were studied by Soliz-Guerrero *et al.* (2002) in two quinoa cultivars 'Sajama' and 'Chucara' during plant development under three soil-water deficit treatments. Analysis of variance of saponin content showed significant differences only for water-deficit treatments, stages and stage × deficit interaction. Saponin content for the low soil-water deficit plants was 0.456%, whereas that for the high water-deficit plants was 0.386%. The lowest saponin content (0.309%) was found at the branching stage and the highest (0.608%) at blooming. It was concluded that soil water-deficit treatment affected the saponin content; high deficit promoted

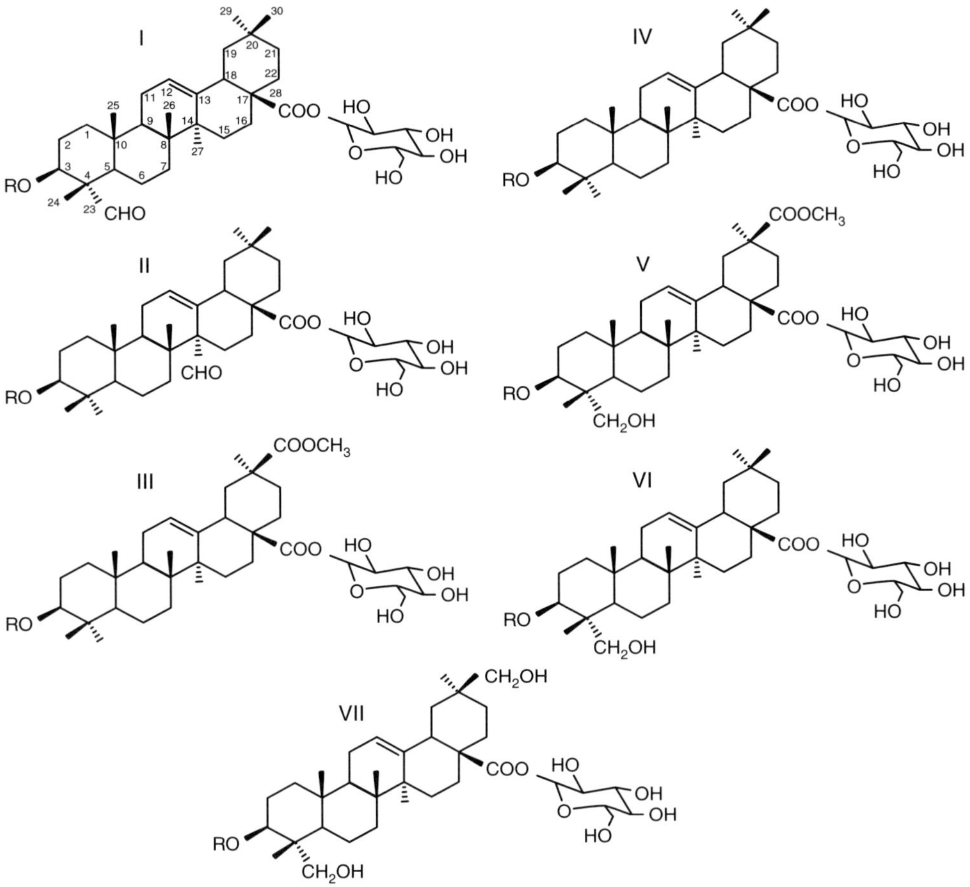

Compound	Aglycone	R	Flower	Fruit	Seed	Seed coat
1	I	β-D-Glc(1→3)-α-L-Ara	+	+	+	+
2	II	β-D-Glc(1→3)-α-L-Ara	+	+	+	+
3	III	α-L-Ara	+	+	++	++
4	III	β-D-GlcA	+	+	++	++
5	III	β-D-Glc(1→2)-β-D-Glc(1→3)-α-L-Ara	+	+	++	++
6	III	β-D-Glc(1→3)-α-L-Ara	+	+	+	++
7	IV	β-D-Xyl(1→3)-β-D-GlcA	+	+	+	+
8	IV	β-D-Glc(1→2)-β-D-Glc(1→3)-α-L-Ara	+	+	+	+
9	IV	β-D-Glc(1→3)-α-L-Ara	+	+	+	+
10	IV	β-D-GlcA	+	+	++	++
11	V	β-D-Glc(1→3)-α-L-Ara	++++	++++	++++	++++
12	V	α-L-Ara	+++	+++	++++	++++
13	V	β-D-Glc(1→3)-β-D-Gal	++	++	++	++
14	V	β-D-Glc(1→2)-β-D-Glc(1→3)-α-L-Ara	++	++	+	++
15	V	β-D-Glc(1→4)-β-D-Glc(1→4)-β-D-Glc	++	++	+	++
16	VI	β-D-Glc(1→3)-β-D-Gal	+	+	+	+
17	VI	α-L-Ara	+	+	+	+
18	VI	β-D-GlcA	++	++	++	++
19	VI	β-D-Glc(1→3)-α-L-Ara	+++	+++	++	+++
20	VII	β-D-Glc(1→3)-α-L-Ara	+	+	++	++++

(+) trace quantities, (++) low quantities, (+++) moderate quantities, (++++) high quantities

Fig. 13.2. Structures of saponins 1–20 (aglycones I–VII, 3β-hydroxy-23-oxo-olean-12-en-28-oic acid, I; 3β-hydroxy-27-oxo-olean-12-en-28-oic acid, II; serjanic acid, III; oleanolic acid, IV; phytolaccagenic acid, V; hederagenin, VI; 3β,23,30-trihydroxy olean-12-en-28-oic acid, VII). [Reprinted from Kuljanabhagavad *et al.* (2008), with permission from Elsevier.]

low saponin content in quinoa cultivars (Soliz-Guerrero *et al.*, 2002). Martinez *et al.* (2009) studied the influence of drought on the saponin level in two quinoa cultivars, one with a high initial saponin content (1.2% of seed dry weight) and the other with a low content (0.46%), growing in conditions with high or low irrigation. The saponin level in the cultivar with the high initial saponin content was found to decrease by 9% with high irrigation and increase by 25% with low irrigation, whereas in the cultivar with low initial saponin content it declined by 26 and 57% in high and low irrigation conditions, respectively. Gómez-Caravaca *et al.* (2012) studied the effects of different agronomic variables, such as irrigation and salinity, on the saponin profiles of quinoa. The amount of saponins decreased when samples were exposed to drought and saline regimens. In situations of severe water deficit, the saponin content decreased by 45% and 50% when a salt stress was added. The saponin yield in quinoa is also influenced by tillage system and fertilization. High saponin content (0.42–0.45%) has been found in quinoa with cow manure and compost treatments (Bilalis *et al.*, 2012). The highest saponin yield was found under the minimum tillage system (Bilalis *et al.*, 2012). Saponin yield had a positive and significant correlation with total nitrogen, root density and leaf area index.

13.4   Biosynthesis of Quinoa Saponins

Saponins from quinoa are a complex mixture of triterpene glycosides derived from β-amirin, being phytolaccagenic, oleanolic and serjanic acids, hederagenin, 3β,23,30 trihydroxy olean-12-en-28-oic acid, 3β-hydroxy-27-oxo-olean-12-en-28-oic acid and 3β,23,30 trihydroxy olean-12-en-28-oic acid, the most common aglycones found in seeds (Madl *et al.*, 2006; Kuljanabhagavad *et al.*, 2008; Kuljanabhagavad and Wink, 2009). The major carbohydrates are arabinose, glucose and galactose, while glucoronic acid and xylose have been found to be less common (Mizui *et al.*, 1988, 1990; Mastebroek *et al.*, 2000; Woldemichael, 2000; Dini *et al.*, 2001a, 2001b; Zhu *et al.*, 2002; Madl *et al.*, 2006; Kuljanabhagavad *et al.*, 2008).

The biosynthesis of quinoa saponins proceeds via the isoprenoid pathway in which three isoprene units are first linked in a head-to-tail manner to each other, resulting in 15 carbon atoms, called farnesyl pyrophosphate (FPP) (Kuljanabhagavad and Wink, 2009) (Fig. 13.3). Two FPPs are subsequently linked in a tail-to-tail manner, giving rise to a 30-carbon atom compound called squalene (Holstein and Hohl, 2004). Squalene is oxidized to oxidosqualene, which is the common starting point for cyclization reactions in triterpenoid biosynthesis. The first committed step in the pathway towards triterpene saponins is the cyclization of 2,3-oxidosqualene catalysed by β-amyrin synthase (Kuljanabhagavad and Wink, 2009). Judging from the structure of the saponins that accumulate in quinoa, the next steps in the pathway presumably include:

- oxidation of β-amyrin at carbon positions 23, 27 and/or 30;
- glycosylation at C3 and, for the major bisdesmosides, esterification at C28;
- in the oleanane skeleton a single double bond is present at position C12–C13.

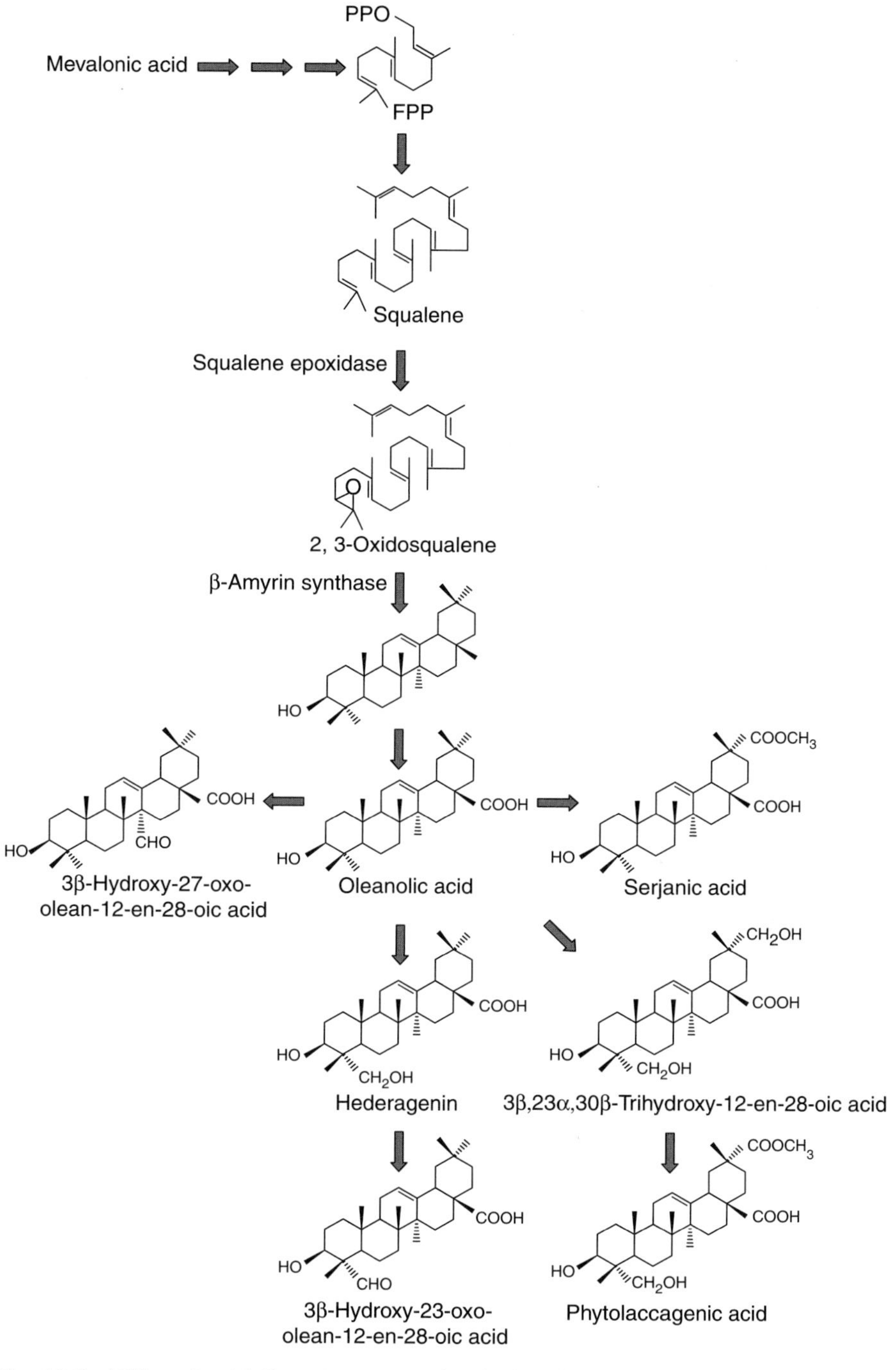

Fig. 13.3. Different oxidation steps occurring in the oleanolic acid and serjanic acid aglycones from the β amyrin leading to hederagenin, 3β-hydroxy-23-oxo-12-en-28-oic acid, serjanic acid, phytolaccagenic acid, and 3β, 23α, 30β-trihydroxy-12-en-28-oic acid. [Reprinted from Kuljanabhagavad and Wink (2009), with kind permission from Springer Science + Business Media B.V.]

13.5   Genetic Aspects

There has been little investigation of the genetic basis of saponin content in quinoa. Gandarillas (1979) proposed that the trait was controlled by two alleles at a single locus, the 'bitter' (high saponin) being dominant to 'sweet' (low saponin). Later researchers observed that grain saponin content in quinoa was a continuously distributed variable and was therefore more likely to be polygenically controlled and quantitatively inherited (Galwey et al., 1990; Jacobsen et al., 1996). Ward (2001) used cytoplasmic male sterile (CMS) quinoa lines as female parents in four different crosses, two of which used low saponin 'sweet' quinoa cultivars as pollen parents to characterize genetic control of grain saponin content in quinoa, especially in low-saponin lines. Subsequent F_1, F_2 and F_3 generations were screened for saponin content using Koziol's standardized afrosimetric test. It was observed that saponin production in quinoa required at least one dominant allele at the bitter saponin locus; quinoa with a fully recessive allele at the bitter saponin locus had no detectable amounts of saponin (Ward, 2001). However, the amount of saponin was determined by an unknown number of quantitative trait loci (Ward, 2001). It appeared that grain saponin level in quinoa was both qualitatively and quantitatively controlled, with saponin production requiring at least one dominant allele at the Sp locus and the amount of grain saponin determined by an unknown number of additional quantitative loci. Additionally, the bitter saponin locus has not been tightly (<5 cM) linked to molecular markers, making marker-assisted selection very difficult (Ricks, 2005). Genetic mapping has produced some linked markers, the most tightly linked being an AFLP marker linked 9.4 cM from the bitter saponin locus (Ricks, 2005). However, the exact nature of the gene responsible for bitter saponin in quinoa remains unknown. Reynolds (2009) reported the development and annotation of the first large-scale expressed sequence tags collection for quinoa containing 39,366 unigenes and the development of a custom microarray to assay gene expression in developing seeds of quinoa. Sanger sequencing produced 18,325 reads with an average read length of 693 nucleotides, while 454 GS-FLX pyrosequencing generated 295,048 reads with an average read length of 202 nucleotides. A hybrid assembly of all sequences generated 39,366 unigenes, consisting of 16,728 contigs and 22,638 singletons. Repeat sequence analysis of the unigene set identified 291 new microsatellite markers. Several candidate genes that could be involved in the production of bitter saponin in quinoa were also identified. From the unigene set, a custom microarray was developed and used to assay transcriptional changes in developing seeds of saponin-containing and saponin-free quinoa lines. The microarray consisted of 102,834 oligonucleotide probes representing 37,716 sequences of the unigenes set. Three different statistical comparisons, based on comparisons of 'sweet' vs 'bitter' seed tissue at two developmental stages, were assayed on the custom array. A list of candidate genes was identified that were known to be associated with identified triterpenoid (saponin) biosynthetic pathways. This list included candidate genes that shared homology to cytochrome P450s (20), cytochrome P450 monooxygenases (10) and glycosyltransferases (49), suggesting that transcriptional differences in the saponin biosynthesis pathway possibly responsible for the absence or presence of saponin in quinoa are determined after the formation of the β-amyrin skeleton. The candidate genes were suggested for use in future studies in the production of saponin in quinoa.

13.6 Methods for Studying Saponins in Quinoa

A number of methods have been used for isolation and characterization of saponins in plants (Marston *et al.*, 2000; Muir *et al.*, 2000; Schopke, 2000; Oleszek, 2002). The early determination of saponins was based on gravimetry or on methods taking advantage of some of their chemical or biological features. The simplest method is the afrosimetric method that is based on the foaming/frothing property of saponins after shaking in water. Other methods can be categorized into biological and non-biological methods (Oleszek, 2002). Of these, the use of conventional methods like solvent extraction, column chromatography and preparative thin layer chromatography (TLC) is quite limited since in many cases it is difficult to isolate individual saponins. An overview of the different methods used for saponin estimation in quinoa is provided in Table 13.1.

Table 13.1. Different methods used for estimation of quinoa saponins.

Study	Method
Aguilar *et al.* (1979); Reichert *et al.* (1986)	Haemolytic
Burnouf-Radosevich *et al.* (1985)	Gas chromatography-mass spectrometry
Mizui *et al.* (1988, 1990)	High performance liquid chromatography
Cuadrado *et al.* (1995)	Gas chromatography
Koziol (1991)	Afrosimetric
Meyer *et al.* (1990)	Gas chromatography and gas chromatography-mass spectrometry; thin layer chromatography
Chauhan *et al.* (1992)	Gravimetric
Ng *et al.* (1994)	Thin layer chromatography
Woldemichael and Wink (2001)	Thin layer chromatography
Zhu *et al.* (2002)	Thin layer chromatography
Mastebroek *et al.* (2000)	Gas chromatography and gas chromatography-mass spectrometry
Dini *et al.* (2001a, 2001b)	1D- and 2D-NMR (HMQC and HMBC) and electrospray ionization multi-stage analyses; 1D- and 2D-NMR (HMQC and HMBC), electrospray ionization multi-stage and fast atom bombardment-mass spectrometry analyses
Solíz-Guerrero *et al.* (2002)	Infrared spectroscopy
Madl *et al.* (2006)	Nano-HPLC electrospray ionization multi-stage tandem mass spectrometry approach
Gomez-Caravaca *et al.* (2011)	Liquid chromatography-diode array detection-electrospray ionization-time-of-flight mass spectrometry

Burnouf-Radosevich *et al.* (1985) developed a gas chromatography-mass spectrometry (GC-MS) method for the resolution and identification of the TMSi ether derivatives of seven oleanane- and two ursane-type triterpenes that are commonly distributed in the form of saponins in *C. quinoa*. Oleanolic acid and hederagenin were confirmed to be major triterpenes of *C. quinoa* seed saponins.

The saponin content of 17 quinoa cultivars, estimated using a virtual haemolytic assay, ranged from 0.13 to 0.74% (Reichert *et al.*, 1986). The 1000-seed weight correlated significantly with saponin content and the abrasive hardness index indicated that larger seeds were harder and contained more saponins. Abrasive dehulling to a flour extraction ranging from 85.2 to 98.9% reduced the saponin content to a low level. It was also observed that the higher the initial saponin content of the cultivar, the greater the loss of bran required before dehulling. It was concluded that cultivars with a low level of saponin content provided optimum yields with abrasive dehulling.

The aqueous extracts of the whole quinoa seeds were fractionated chromatographically to isolate the toxic/bitter principles (Meyer *et al.*, 1990). These undesirable compounds appeared to be a mixture of saponins whose acidic hydrolysis gave oleanolic acid and hederagenin (3:1) as the only detectable saponin aglycons by GC-MS analysis. One of these saponins (quinoside A) was identified as a tetraglycoside of hederagenin; the proposed structure (1) was named olean-12-ene-28-oic acid, 3,23-bis(*O*-β-D-glucopyranosyloxy)-*O*-β-D-glucopyranosyl-(1→3)-*O*-α-L-arabinopyranosyl ester (3β,4α).

Chauhan *et al.* (1992) estimated the total saponin content in quinoa seeds by gravimetric method. The results showed that approximately 34% of the saponins were present in the hull, indicating that dehulling can remove a large proportion of the saponins. The saponin content of the bran was double that found in the flour and represented a major source of saponin in the plant. Washing the seed during the moisture extraction process reduced the saponin content threefold compared with manual dehulling.

Ng *et al.* (1994) developed a TLC method for estimating saponins in quinoa that enabled both the total saponin content and composition to be assayed in quinoa plant tissue. The saponin composition was determined according to the three main groups of saponins found in quinoa, which contained oleanolic acid, hederagenin and phytolaccagenic acid as the aglycone in each group. The method was used to measure the saponin content of 15 ecotypes being used in a breeding programme in the UK and to compare the saponins present in sweet and bitter varieties.

The total saponin content and sapogenol composition was estimated in white and yellow ecotypes of quinoa using fast atom bombardment-mass spectrometry (FAB-MS) of the saponin extracts and gas chromatography (GC) analysis (Cuadrado *et al.*, 1995). Saponin composition in quinoa seeds comprised oleanolic acid and three other sapogenols identified as hederagenin, phytolaccagenic acid and deoxyphytolaccagenic acid. Oleanolic acid saponins were found to be the main class of saponin in quinoa seeds sampled for the study. The yellow ecotype of quinoa presented a significantly higher content of saponins and of oleanolic acid compared with the white ecotypes.

Mastebroek *et al.* (2000) analysed the content of sapogenins in the leaves of sweet and bitter quinoa genotypes at successive stages of plant development and in the seeds. Detectable amounts of sapogenins were found as early as 82 days after sowing in the leaves of both sweet and bitter quinoa genotypes. The total sapogenin content in leaves of sweet and bitter genotypes increased during plant development but remained lower than the content found in the seeds. The sapogenin content in seeds of sweet genotypes varied from 0.2 to 0.4 g/kg dry matter and in seeds of bitter genotypes from 4.7 to 11.3 g/kg dry matter. The difference in sapogenin content between leaves and seeds was much higher in bitter genotypes than in sweet genotypes. Hederagenin was the major sapogenin found in the leaves, and oleanolic acid in the seeds. In the field experiment it was found that the content of sapogenins in the leaves of F2 plants of crosses between both quinoa types did not differ between sweet and bitter genotypes. The results demonstrated that sweet genotypes cannot be selected before anthesis on the basis of the sapogenin content in the leaves.

Dini *et al.* (2001a) isolated six triterpenoid saponins from the edible grain quinoa and characterized their structures on the basis of hydrolysis and spectral evidence, including 1D- and 2D-NMR (Heteronuclear Multiple-Quantum Correlation [HMQC] and Heteronuclear Multiple-Bond Correlation [HMBC]) and electrospray ionization multi-stage (ESI-MS) analyses. The following structures were worked out: phytolaccagenic acid 3-O-[α-L-arabinopyranosyl-(1″→3′)-β-D-glucuronopyranosyl]-28-O-β-D-glucopyranoside (1); phytolaccagenic acid 3-O-[β-D-glucopyranosyl-(1″→3′)-α-L-arabinopyranosyl]-28-O-β-D-gluco-pyranoside (2); phytolaccagenic acid 3-O-[β-D-glucopyranosyl-(1‴→3″)-β-D-xylo-pyranosyl-(1″→2′)-β-D-glucopyranosyl]-28-O-β-D-glucopyranoside(3); phytolaccagenic acid 3-O-[β-D-glucopyranosyl-(1‴→2″)-β-D-glucopyranosyl-(1″→3′)- α-L-arabinopyranosyl]-28-O-β-D-glucopyranoside (4); oleanolic acid 3-O-[α-L-arabinopyranosyl-(1″→3′)-β-D-glucuronopyranosyl]-28-O-β-D-glucopyranoside (5); and oleanolic acid 3-O-[β-D-glucopyranosyl-(1″→3′)-α-L-arabinopyranosyl]- 28-O-β-D-glucopyranoside (6). The oleanane-type saponins (5, 6) were isolated for the first time in quinoa, two of the phytolaccagenane (1, 3) were new compounds and two (2, 4) were previously found in the crop.

Dini *et al.* (2001b) isolated six triterpenoid saponins from the seeds of *C. quinoa*. Their structures were as follows: phytolaccagenic acid 3-O-[α-L-arabinopyranosyl-(1″→3′)-β-D-glucuronopyranosyl]-28-O-β-D-glucopyranoside (1); spergulagenic acid 3-O-[β-D-glucopyranosyl-(1→2)-β-D-glucopyranosyl-(1→3)-α-L-arabinopyranosyl-28-O-β-D-glucopyranoside (2); hederagenin 3-O-[β-D-glucopyranosyl-(1→3)-α-L-arabinopyranosyl]-28-O-β-D-glucopyranoside (3); phytolaccagenic acid 3-O-[β-D-glucopyranosyl-(1→4)-β-D-glucopyranosyl-(1→4)-β-D-glucopyranosyl]-28-O-β-D-glucopyranoside (4); hederagenin 3-O-[β-D-glucopyranosyl-(1→4)-β-D-glucopyranosyl-(1→4)-β-D-glucopyranosyl]-28-O-β-D-glucopyranoside (5); and spergulagenic acid 3-O-[α-L-arabino-pyranosyl-(1″→3′)-β-D-glucuronopyranosyl]-28-O-β-D-glucopyranoside (6). Saponins 5 and 6 were new and reported for the first time.

At least 16 saponins were isolated and characterized in the seeds of quinoa using mainly NMR spectroscopy, mass spectrometry and chemical methods by Woldemichael and Wink (2001). The five previously isolated major saponins,

3-*O*-β-D-glucuronopyranosyl oleanolic acid 28-*O*-β-D-glucopyranosyl ester, 3-*O*-α-L-arabinopyranosyl hederagenin 28-*O*-β-D-glucopyranosyl ester, 3-*O*-β-D-glucopyranosyl-(1→3)-α-L-arabinopyranosyl hederagenin 28-*O*-β-D-glucopyranosyl ester, 3-*O*-α-L-arabinopyranosyl phytolaccagenic acid 28-*O*-β-D-glucopyranosyl ester, 3-*O*-β-D-glucopyranosyl-(1→3)-α-L-arabinopyranosyl phytolaccagenic acid 28-*O*-β-D-glucopyranosyl ester and the new saponin 3-*O*-β-D-glucopyranosyl-(1→3)-α-L-arabinopyranosyl phytolaccagenic acid were isolated and characterized. The antifungal activity of these compounds and derived monodesmosides were evaluated against *Candida albicans* and haemolytic activity on erythrocytes. Both bidesmosides and derived monodes-mosides showed little or no antifungal activity, whereas a comparatively higher degree of haemolytic activity could be determined for monodesmosides.

Twelve triterpene saponins were isolated from the debittered quinoa seeds by Zhu *et al.* (2002) and their structures characterized on the basis of hydrolysis and spectral data, especially NMR evidence. Among them, three compounds, viz. 3-*O*-β-D-glucuropyranosyl oleanolic acid (**1**), 3-*O*-β-D-glucopyranosyl-(1→3)-α-L-arabinopyranosyl hederagenin (**2**) and the new compound 3-*O*-β-D-glucopyranosyl-(1→3)-α-L-arabinopyranosyl-30-*O*-methyl spergulagenate 28-*O*-β-D-glucopyranosyl ester (**3**), were identified for the first time from quinoa seeds. The other isolated saponins were previously reported in the crop.

Flores *et al.* (2005) reported the separation and identification of two oleanane-type triterpenes: oleanolic acid (**1**) and deoxyphytolaccagenic acid (**2**), as well as the synthesis of two derivatives of the obtained oleanolic acid: 28-Omethyl and 3-*O*-acetyloleanolic acid, as one of the potential chemical uses of oleanolic acid. Oleanolic acid (**1**) is an interesting and valuable secondary metabolite with several pharmacological activities, nowadays commercially produced in China as a drug for liver disorders (Flores *et al.*, 2004, 2005).

A nano-HPLC electrospray ionization multi-stage tandem mass spectrom-etry (nLC-ESI-MS/MS) approach was applied to a complex crude triterpene saponin extract of quinoa seed coats, and 87 triterpene saponins comprising 19 reported and 68 novel components were identified (Madl *et al.*, 2006). In addition to the four reported, five novel triterpene aglycones were detected and characterized according to their fragmentation reactions in ESI-MS/MS and electron ionization-mass spectrometry (EI-MS). Since the relative distribution and composition of saponins varied between different cultivars and soils, a suit-able strategy was presented that allowed a rapid and complete analysis of qui-noa saponin distribution and composition, and could be of utility for quality control and screening of extracts designated for pharmaceutical, agricultural and industrial applications.

In a comprehensive study, Kuljanabhagavad *et al.* (2008) isolated 20 trit-erpene saponins (1–20) from different parts of quinoa plant (flowers, fruits, seed coats and seeds) and elucidated their structures by analysis of chemical and spectroscopic data including 1D- and 2D-NMR. Four compounds (1–4) were identified: 3β-[(*O*-β-D-glucopyranosyl-(1→3)-α-L-arabinopyranosyl)oxy]-23-oxo-olean-12-en-28-oic acid β-D-glucopyranoside acid (**1**), 3β-[(*O*-β-D-glucopyranosyl-(1→3)-α-L-arabinopyranosyl)oxy]-27-oxo-olean-12-en-28-oic β-D-glucopyranoside (**2**), 3-*O*-α-L-arabinopyranosyl serjanic acid 28-*O*-β-D-glucopyranosyl ester (**3**) and 3-*O*-β-D-glucuronopyranosyl serjanic acid 28-*O*-β-D-glucopyranosyl ester (**4**). Two bidesmosides of serjanic acid (**5**, **6**), four bidesmosides of oleanolic acid (**7–10**), five bidesmosides of phytolaccagenic acid (**11–15**), four bidesmosides of hederagenin (**16–19**) and one bidesmoside of 3β,23,30-trihydroxy olean-12-en-28-oic acid (**20**) were reported as saponin constituents from the flowers and the fruits of quinoa for the first time. The cytotoxicity of these saponins and their aglycones was tested in HeLa (cervix adenocarcinoma) cell line using MTT assay. Induction of apoptosis in Caco-2 cells by bidesmosidic saponins 1–4 and their aglycones I–III was determined by flow cytometric DNA analysis. The saponins with an aldehyde group were found to be most active.

Verza *et al.* (2012) studied the aggregates formed by self-association in aqueous solutions by two quinoa saponin fractions (FQ70 and FQ90), as well as several distinctive nanostructures obtained after their complexation with different ratios of cholesterol (CHOL) and phosphatidylcholine (PC). The FQ70 and FQ90 fractions were obtained by reversed-phase preparative chromatography. The structural features of their resulting aggregates were determined by dynamic light scattering and transmission electron microscopy (TEM). Novel nanosized spherical vesicles formed by self-association with mean diameter about 100–200 nm were observed in FQ70 aqueous solutions, whereas worm-like micelles of about 20 nm wide were detected in FQ90 aqueous solutions. Under experimental conditions similar to those reported for the preparation of *Q. saponaria*, ISCOM matrices, tubular and ring-like micelles arose from FQ70:CHOL:PC and FQ90:CHOL:PC formulations, respectively (Verza *et al.*, 2012). However, under these conditions no cage-like ISCOM matrices were observed. Phytolaccagenic acid, predominant in FQ70 and FQ90 fractions, was accountable for the formation of the nanosized vesicles and tubular structures observed by TEM in the aqueous solutions of both samples. Conversely, ring-like micelles observed in FQ90:CHOL:PC complexes were attributed to the presence of less polar saponins present in FQ90, in particular those derived from oleanolic acid.

13.7 Removal of Saponins from Quinoa Seeds

The removal of saponins from quinoa seeds can be carried out by three methods: moist, dry or combined.

13.7.1 Moist method

This method has been traditionally used by farmers and is quite efficient for saponin removal. It entails a successive washing of the grain using friction by hand or a stone to eliminate the episperm, which is the outer layer where the saponins are located. Rinsing quinoa thoroughly with water easily washes the saponin from the seeds. In Australia, a method has been recommended to remove the saponin wherein the grain is soaked in water for a few hours, followed by changing the water and resoaking or rinsing in ample running water either in a fine strainer or in a cheesecloth. Equipment for washing of quinoa seed is now available for industrial/commercial purposes. However, washing quinoa is currently not a viable option in developed countries because of water pollution (Johnson and Ward, 1993). Another risk is that the seed may begin germinating during the washing and drying process because of quinoa's high germinating ability (Repo-Carrasco *et al.*, 2003). A novel method developed by the National Agrarian University of La Molina (UNALM), Lima (Peru), involving a soak period of 30 minutes, a stirring period of 20 minutes and a water temperature of 70°C, is considered ideal for saponin removal (Torres and Minaya, 1980). This process can significantly reduce saponins down to 0.04–0.25%.

13.7.2 Dry method

The dry methods, viz. scarification and abrasive dehulling, require machinery to polish the grains in order to remove the saponins. In South America a 'dry' system involving a flotation cell is used. In this method, the seed is wetted and dirt, saponin and foreign matter are removed and thereafter the seed is subjected to forced air drying. Foam breakdown and drying costs are serious problems with this method (Johnson and Ward, 1993). In some communities of the Salares in Bolivia, a perforated stone of about 50 cm in diameter is used into which quinoa preheated in a thick sand layer called 'pokera' is placed. Quinoa grain and sand are then rubbed with the feet (Tapia *et al.*, 1979). High-saponin cultivars require more abrasion than low-saponin types. In general, dry methods are more economical, simple and do not lead to contamination, but they are relatively inefficient, only eliminating 80% of the saponin (Mujica, 1993). If the efficiency is increased, and the grain is burnished more intensely, some nutrients are lost, because the proteins are mainly present in the exterior layer of the grain (Repo-Carrasco *et al.*, 2003). Saponin removal by the dry method reduces the vitamin and mineral content to some extent, the loss being significant in the case of potassium, iron and manganese (Ruales and Nair, 1992). However, mechanical abrasion has been found to significantly increase α-amylase activity in quinoa seed (Lorenz and Nyanzi, 1989). This is due to the removal of the pericarp (relatively low in α-amylase) during abrasion milling.

13.7.3 Combined method

Apart from these methods, a combined method has been developed that is considered more suitable for removal of saponins. In this method, the seed is first quickly burnished and briefly washed thereafter. With the brief washing, the costs of drying are lower, and the previous burnishing lowers the concentration of saponin in the wastewater (Repo-Carrasco *et al.*, 2003).

13.8 Applications of Quinoa Saponins

Saponins have immense industrial importance. Quinoa saponins, due to their foaming capabilities, may have application in soaps, detergents, shampoos, cosmetics, beer production and fire extinguishers (Johnson and Ward, 1993).

Quinoa saponins have been reported to have substantial antifungal activity. The pure individual saponins showed little or no antifungal activity, while the crude saponin mixture inhibited the growth of *Candida albicans* at 50 µg/ml (Woldemichael and Wink, 2001) suggesting a synergistic effect between different saponins. Bidesmosidic saponins like 3-*O*-β-D-glucopyranosyl-(1→3)-α-L-arabinopyranosyl hederagenin 28-*O*-β-D-glucopyranosyl ester (**12**) and 3-*O*-β-D-glucopyranosyl-(1→3)-α-L-arabinopyranosyl phytolaccagenic acid 28-*O*-β-D-glucopyranosyl ester (**19**) exhibit antimicrobial activities (Woldemichael and Wink, 2001), while phytolaccagenic acid F with different carbohydrate chains did not show any antifungal activity (Escalante *et al.*, 2002). The monodesmosidic saponin named 3-*O*-β-D-glucopyranosyl-(1→3)-α-L-arabinopyranosyl phytolaccagenic acid (**6**) showed antifungal activity with MIC ≥ 100 l g/ml (Woldemichael and Wink, 2001), underlining the fact that the antifungal activity is more pronounced in monosaccharides than with the bidesmosides like 3-*O*-β-D-glucopyranosyl-(1→3)-α-L-arabinopyranosyl phytolaccagenic acid 28-*O*-β-D-glucopyranosyl ester (**19**) with MIC ≥ 500 µg/ml (Woldemichael and Wink, 2001; Kuljanabhagavad and Wink, 2009). Recently a study was carried out to test if quinoa saponins had antifungal properties against *Botrytis cinerea* and if this activity could be enhanced after alkaline treatment (Macarena and Ricardo, 2008). Six products, viz. non-purified quinoa extract, purified quinoa extract, alkali-treated non-purified quinoa extract, alkali-treated purified quinoa extract, non-purified quinoa extract treated with alkali but without thermal incubation, and purified quinoa extract treated with alkali but without thermal incubation were tested against the fungus. Untreated quinoa extracts showed minimum activity against mycelial growth of *B. cinerea*. Also, no effects were observed against conidial germination, even at 7 mg saponins/ml. However, when the saponin extracts were treated with alkali, mycelial growth and conidial germination were significantly inhibited. At doses of 5 mg saponins/ml, 100% of conidial germination inhibition was observed, even after 96 hours of incubation. Fungal membrane integrity experiments based on the uptake of the fluorogenic dye SYTOX green showed that alkali-treated saponins generate membrane disruption, while non-treated saponins had no effects. The higher antifungal activity of alkaline-treated saponins was probably due to the formation of more hydrophobic saponin derivatives that may have a higher affinity with the sterols present in cell membranes (Macarena and Ricardo, 2008). The antifungal property of quinoa saponins can be utilized for control and/or prevention of plant diseases, especially fungal diseases (Dutcheshes and Danyluk, 2002).

Quinoa saponins also exhibit moluscicidal activity, which is significantly enhanced on alkali treatment (Fleming and Galwey, 1995; San Martin *et al.*, 2008). Quinoa husks treated with alkali convert the bidesmosidic saponins to more active monodesmosides that cause 100% mortality in *Pomacea canaliculata*

(Golden apple snail). This activity has been linked to a monodesmosidic saponin as the 3-O-β-D-glucopyranosyl-(1→3)-α-L-arabinopyranosyl phytolaccagenic acid (6) (Zhu *et al.*, 2002; Woldemichael and Wink, 2001). The binding of the monodesmosidic saponins to the gill membranes, resulting in an increase of their permeability and a subsequent loss of important physiological electrolytes, might be the most feasible explanation for this activity (Treyvaud *et al.*, 2000). Surprisingly, no toxicity to fish, such as goldfish or tilapia, was observed up to the highest concentration tested. This is an advantage in relation to commercial molluscicides like niclosamide that kills fish at product doses lower than those that kill *P. canaliculata*.

Saponins have been studied because of their beneficial properties to health and are being explored as raw materials for the pharmaceutical industry (Fleming and Galwey, 1995). Saponins in general possess a broad variety of biological effects: analgesic, anti-inflammatory, antimicrobial, antioxidant, antiviral and cytotoxic activity. They affect the absorption of minerals and vitamins and animal growth, have haemolytic and immuno-stimulatory effects, increase the permeability of the intestinal mucosa, have a neuroprotective action and reduce fat absorption (Güçlü-Üstündag and Mazza, 2007). Saponins have the ability to induce changes in intestinal permeability (Johnson *et al.*, 1986; Gee *et al.*, 1989), which aids in the absorption of particular drugs (Basu and Rastogi, 1967). Saponins are also known to lower blood cholesterol levels (Oakenfull and Sidhu, 1990). Research has proved that quinoa saponins may have the potential to serve as adjuvants for mucosally administered vaccines (Estrada *et al.*, 1998). The saponins extracted from the seed of quinoa were studied for their ability to act as mucosal adjuvants on their intragastric or intranasal administration together with model antigens in mice. Quinoa saponins, co-administered intragastrically or intranasally with cholera toxin or ovalbumin, potentiated specific IgG and IgA antibody responses to the antigens in serum, intestinal and lung secretions (Estrada *et al.*, 1998).

Apoptosis or programmed cell death is responsible for pathological mechanisms related to human diseases. The bidesmosidic saponins named 3-O-α-L-arabinopyranosyl serjanic acid 28-O-β-D-glucopyranosyl ester (28), 3-O-β-D-glucuronopyranosyl serjanic acid 28-O-β-D-glucopyranosyl ester (29), 3β-[(O-β-D-glucopyranosyl-(1→3)-α-L-arabinopyranosyl)oxy]-23-oxo-olean-12-en-28-oic acid β-D-glucopyranoside (31), and 3β-[(O-β-D-glucopyranosyl-(1→3)-α-L-arabinopyranosyl)oxy]-27-oxo-olean-12-en-28-oic acid β-D-glucopyranoside (32) have an IC50 at ~100 µg/ml or 100 µM and showed low apoptosis inducing activity (~13–26%), while the aglycones named serjanic acid (IV), 3β-hydroxy-23-oxo-olean-12-en-28-oic acid (V), and 3β-hydroxy-27-oxo-olean-12-en-28-oic acid (VI) showed low apoptosis-inducing activity with IC50 at ~25–50 µg/ml (Kuljanabhagavad *et al.*, 2008).

Inflammation is a pathophysiological process characterized by redness, oedema, pain, fever and loss of function. It has been suggested that quinoa saponins possess anti-inflammatory activity (Mujica, 1994). A monodesmosidic saponin named 3-O-β-D-glucopyranosyl oleanolic acid (2) isolated from the seeds of *Randia dumetornm* Lam. showed significant anti-inflammatory activity in the exudative and proliferative phases of inflammation. This activity can be linked to a similar monodesmoside found in quinoa (Ma *et al.*, 1989).

Similarly the monodesmosidic saponin named 3-O-β-D-glucopyranosyl hedera-genin isolated from the fruits of *Hedera colchica* showed antioxidant activity (Gulcin *et al.*, 2006). A similar activity could be related to monodesmoside named 3-O-β-D-glucopyranosyl hederagenin in *C. quinoa* (Kuljanabhagavad and Wink, 2009).

13.9 Concluding Remarks

The wide chemical diversity of saponins, especially triterpene saponins, has resulted in renewed interest and investigations of these compounds in recent years, particularly as potential phytotherapeutic and chemotherapeutic agents. The low acute toxicity of saponins and hints from structure–activity relationship studies, which suggest that defined substitutents on the lipophilic 5-ring back-bone can increase selectivity, indicate that triterpenes might provide a rich natural resource of lead compounds for drug development (Kuljanabhagavad and Wink, 2009). Information about the phytochemistry and various biological properties of quinoa saponins might provide the incentive for further use of the plant in medicine and in agriculture. Seeing the pharmaceutical potential of saponins, efforts should be made towards the utilization of quinoa saponins for diverse purposes (Bhargava *et al.*, 2006).

References

Aguilar, R.H., Guevara, L. and Alvarez, J.O. (1979) A new procedure for the quantitative deter-mination of saponins and its application to several types of Peruvian quinoa (in Spanish). *Acta Cientifica Venezolana* 30, 167–171.

Basu, N. and Rastogi, R.P. (1967) Triterpenoid saponins and sapogenins. *Phytochemistry* 6, 1249–1270.

Bhargava, A., Shukla, S. and Ohri, D. (2006) *Chenopodium quinoa* – an Indian perspective. *Industrial Crops and Products* 23, 73–87.

Bilalis, D., Kakabouki, I., Karkanis, A., Travlos, I., Triantafyllidis, V. and Hela, D. (2012) Seed and saponin production of organic quinoa (*Chenopodium quinoa* Willd.) for different tillage and fertilization. *Notulae Botanicae Horti Agrobotanici Cluj-Napoca* 40, 42–46.

Burnouf-Radosevich, M., Delfel, N.E. and England, R. (1985) Gas chromatography-mass spectrometry of oleanane- and ursane-type triterpenes – application to *Chenopodium quinoa* triterpenes. *Phytochemistry* 24, 2063–2066.

Chappell, J. (2002) The genetics and molecular genetics of terpene and sterol origami. *Current Opinions in Plant Biology* 5, 151–157.

Chauhan, G.S., Eskin, N.A.M. and Tkachuk, R. (1992) Nutrients and antinutrients in quinoa seed. *Cereal Chemistry* 69, 85–88.

Connolly, J.D. and Hill, R.A. (2005) Triterpenoids. *Natural Product Reports* 22, 487–503.

Cuadrado, C., Ayet, G., Burbano, C., Muzquiz, M., Camacho, L., Cavieres, E., Lovon, M., Osagie, A. and Price, K.R. (1995) Occurrence of saponins and sapogenols in Andean crops. *Journal of the Sciences of Food and Agriculture* 67, 169–172.

Dini, I., Schettino, O., Simioli, T. and Dini, A. (2001a) Studies on the constituents of *Chenopodium quinoa* seeds: isolation and characterization of new triterpene saponins. *Journal of Agricultural and Food Chemistry* 49, 741–746.

Dini, I., Tenore, G.C., Schettino, O. and Dini, A. (2001b) New oleanane saponins in *Chenopodium quinoa*. *Journal of Agricultural and Food Chemistry* 49, 3976–3981.

Dutcheshes, J.M. and Danyluk, T.A. (2002) *Method and Composition for Protecting Plants from Disease*. U.S. Pat. No. 6,482,770.

Escalante, A.M., Santecchia, C.B. and López, S.N. (2002) Isolation of antifungal saponins from *Phytolacca tetramera*, an Argentinian species in critic risk. *Journal of Ethnopharmacology* 82, 29–34.

Estrada, A., Li, B. and Laarveld, B. (1998) Adjuvant action of *Chenopodium quinoa* saponins on the introduction of antibody responses to intragastric and intranasal administered antigens in mice. *Comparative Immunology Microbiology and Infectious Diseases* 21, 225–236.

Fleming, J.E. and Galwey, N.W. (1995) Quinoa (*Chenopodium quinoa*). In: Williams, J.T. (ed.) *Cereals and Pseudocereals*. Chapman and Hall, London, pp. 2–83.

Flores, Y., Sterner, O. and Almanza, G.R. (2004) Oleanoic acid, presence and importance in highland Bolivian plants. *Revista Boliviana de Química* 21, 31–35.

Flores, Y., Díaz, C., Garay, F., Colque, O., Sterner, O. and Almanza, G.R. (2005) Oleanane type triterpenes and derivatives from seed coat of Bolivian *Chenopodium quinoa* genotype 'Salar'. *Revista Boliviana de Química* 22, 71–77.

Francis, G., Kerem, Z., Makkar, H.P. and Becker, K. (2002) The biological action of saponins in animal systems: a review. *British Journal of Nutrition* 88, 587–605.

Galwey, N.W., Leakey, C.L.A., Price, K.R. and Fenwick, G.R. (1990) Chemical composition and nutritional characteristics of quinoa (*Chenopodium quinoa* Willd). *Food Science and Nutrition* 42, 245–261.

Gandarillas, H. (1979) Genetica y origen. In: *Quinua y Kaniwa*. Instituto Interamericano de Ciencias Agricolas, Bogota, Colombia, pp. 45–64.

Gee, J.M., Price K.R., Ridout, C.L., Johnson, I.T. and Fenwick, G.R. (1989) Effect of some purified saponins on the transmural potential difference in the mammalian small intestine. *Toxicology In Vitro* 3, 85–90.

Gómez-Caravaca, A.M., Segura-Carretero, A., Fernández-Gutiérrez, A. and Caboni, M.F. (2011) Simultaneous determination of phenolic compounds and saponins in quinoa (*Chenopodium quinoa* Willd) by a liquid chromatography-diode array detection-electrospray ionization-time-of-flight mass spectrometry methodology. *Journal of Agricultural and Food Chemistry* 59, 10,815–10,825.

Gómez-Caravaca, A.M., Lafelice, G., Lavini, A., Pulvento, C., Caboni, M.F. and Marconi, E. (2012) Phenolic compounds and saponins in quinoa samples (*Chenopodium quinoa* Willd.) grown under different saline and nonsaline irrigation regimens. *Journal of Agricultural and Food Chemistry* 60, 4620–4627.

Güçlü-Üstündag, Ö. and Mazza, G. (2007) Saponins: properties, applications and processing. *Critical Reviews in Food Science and Nutrition* 47, 231–258.

Gulcin, I., Mshvildadze, V., Gepdiremen, A. and Elias, R. (2006) Antioxidant activity of a triterpenoid glycoside isolated from the berries of *Hedera colchica*: 3-O-(b-D-glucopyranosyl)-hederagenin. *Phytotherapy Research* 20, 130–134.

Hanus, L.O., Rezanka, T. and Dembitsky, V.M. (2003) A trinorsesterterpene glycoside from the North American fern *Woodwardia virginica* (L.) Smith. *Phytochemistry* 63, 869–875.

Holstein, S.A. and Hohl, R.J. (2004) Isoprenoids: remarkable diversity of form and function. *Lipids* 39, 293–309.

Jacobsen, S.-E., Hill, J. and Stolen, O. (1996) Stability of quantitative traits in quinoa (*Chenopodium quinoa*). *Theoretical and Applied Genetics* 93, 110–116.

Johnson, D.I. and Ward, S.M. (1993) Quinoa. In: Janick, J. and Simon, J.E. (eds) *New Crops*. Wiley, New York, pp. 219–222.

Johnson, I.T., Gee, J.M., Price, K.R., Curl, C.L. and Fenwick, G.R. (1986) Influence of saponins on gut permeability and active nutrient permeability in vitro. *Journal of Nutrition* 116, 2270–2277.

Kalinin, V.I., Silchenko, A.S., Avilov, S.A., Stonik, V.A. and Smirnov, A.V. (2005) *Phytochemistry Reviews* 4, 221.

Kim, J.H. and Han, Y.N. (2011) Dammarane-type saponins from *Gynostemma pentaphyllum*. *Phytochemistry* 72, 1453–1459.

Koziol, M.J. (1991) Afrosimetric estimation of threshold saponin concentration for bitterness in quinoa (*Chenopodium quinoa* Willd.). *Journal of the Sciences of Food and Agriculture* 54, 211–219.

Koziol, M.J. (1992) Chemical composition and nutritional value of quinoa (*Chenopodium quinoa* Willd.). *Journal of Food Composition and Analysis* 5, 35–68.

Kuljanabhagavad, T. and Wink, M. (2009) Biological activities and chemistry of saponins from *Chenopodium quinoa* Wild. *Phytochemistry Reviews* 8, 473–490.

Kuljanabhagavad, T., Thongphasuk, P., Chamulitrat, W. and Wink, M. (2008) Triterpene saponins from *Chenopodium quinoa* Willd. *Phytochemistry* 69, 1919–1926.

Lorenz, K. and Nyanzi, F. (1989) Enzyme activities in quinoa (*Chenopodium quinoa*). *International Journal of Food Science and Technology* 24, 543–551.

Ma, W.W., Heinstein, P.F. and McLaughlin, J.L. (1989) Additional toxic bitter saponins from the seeds of *Chenopodium quinoa*. *Journal of Natural Products* 52, 1132–1135.

Macarena, S. and Ricardo, S.M. (2008) Antifungal properties of quinoa (*Chenopodium quinoa* Willd) alkali treated saponins against *Botrytis cinerea*. *Industrial Crops and Products* 27, 296–302.

Madl, T., Sterk, H., Mittelbach, M. and Rechberger, G.N. (2006) Tandem mass spectrometric analysis of a complex triterpene saponin mixture of *Chenopodium quinoa*. *Journal of the American Society of Mass Spectrometry* 17, 795–806.

Marston, A., Wolfender, J.L. and Hostettmann, K. (2000) Analysis and isolation of saponins from plant material. In: Oleszek, W. and Marston, A. (eds) *Saponins in Food, Feedstuffs and Medicinal Plants.* Annual Proceedings of the Phytochemical Society, Clarendon Press, Oxford, pp. 1–12.

Martinez, E.A., Veas, E., Jorquera, C., San Martín, R. and Jara, P. (2009) Re-Introduction of quinoa into arid Chile: cultivation of two lowland races under extremely low irrigation. *Journal of Agronomy and Crop Science* 195, 1–10.

Mastebroek, H.D., Limburg, H., Gilles, T. and Marvin, H.J.P. (2000) Occurrence of sapogenins in leaves and seeds of quinoa (*Chenopodium quinoa* Willd.). *Journal of the Sciences of Food and Agriculture* 80, 152–156.

Meyer, B.N., Heinstein, P.F., Burnouf-Radosevich, M., Delfel, N.E. and McLaughlin, J.L. (1990) Bioactivity-directed isolation and characterization of quinoside A: one of the toxic/bitter principles of quinoa seeds (*Chenopodium quinoa* Willd.). *Journal of Natural Products* 38, 205–208.

Mizui, F., Kasai, R., Ohtani, K. and Tanaka, O. (1988) Saponins from the bran of quinoa, *Chenopodium quinoa* Willd. I. *Chemical and Pharmaceutical Bulletin (Tokyo)* 36, 1415–1418.

Mizui, F., Kasai, R., Ohtani, K. and Tanaka, O. (1990) Saponins from the bran of quinoa, *Chenopodium quinoa* Willd. II. *Chemical and Pharmaceutical Bulletin (Tokyo)* 38, 375–377.

Muir, A.D., Ballantyne, K.D. and Hall, T.W. (2000) LC-MS and LC-MS/MS analysis of saponins and sapogenins: comparison of ionization techniques and their usefulness in compound identification. In: Oleszek, W. and Marston, A. (eds) *Saponins in Food, Feedstuffs and Medicinal Plants.* Annual Proceedings of the Phytochemical Society, Clarendon Press, Oxford, pp. 35–42.

Mujica, A. (1993) *Cultivo de Quinua*. Instituto Nacional de Investigacion Agraria (INIA), Direccion General de Investigacion Agraria, Lima, Peru.

Mujica, A. (1994) Andean grains and legumes. In: Hernando Bermujo, J.E. and Leon, J. (eds) *Neglected Crops: 1492 From a Different Perspective*. FAO, Rome, Italy, pp. 131–148.

Ng, K.G., Price, K.R. and Fenwick, G.R. (1994) A TLC method for the analysis of quinoa (*Chenopodium quinoa*) saponins. *Food Chemistry* 49, 311–315.

Oakenfull, D. (1981) Saponins in foods: a review. *Food Chemistry* 6, 19–40.

Oakenfull, D. and Sidhu, G.S. (1990) Could saponins be a useful treatment for hypercholester-olemia. *European Journal of Clinical Nutrition* 44, 79–88.

Oleszek, W.A. (2002) Chromatographic determination of plant saponins. *Journal of Chromatography A* 967, 147–162.

Osbourn, A., Goss, R.J.M and Field, R.A. (2011) The saponins: polar isoprenoids with important and diverse biological activities. *Natural Product Reports* 28, 1261–1268.

Phillips, D.R., Rasbery, J.M., Bartel, B. and Matsuda, S.P. (2006) Biosynthetic diversity in plant triterpene cyclization. *Current Opinions in Plant Biology* 9, 305–314.

Price, K.R., Johnson, I.T. and Fenwick, G.R. (1987) The chemistry and biological significance of saponins in foods and feedstuffs. *Critical Reviews in Food Science and Nutrition* 26, 27–135.

Reichert, R.D., Tatarynovich, J.T. and Tyler, R.T. (1986) Abrasive dehulling of quinoa (*Chenopodium quinoa*): effect on saponin content was determined by an adapted hemolytic assay. *Cereal Chemistry* 63, 471–475.

Repo-Carrasco, R., Espinoza, C. and Jacobsen, S.-E. (2003) Nutritional value and use of the Andean crops quinoa (*Chenopodium quinoa*) and kañiwa (*Chenopodium pallidicaule*). *Food Reviews International* 19, 179–189.

Reynolds, D.J. (2009) Genetic dissection of triterpenoid saponin production in *Chenopodium quinoa*. MS thesis. Brigham Young University, Provo, Utah.

Ricks, M.D. (2005) Genetic mapping of the bitter saponin production locus (BSP locus) in *Chenopodium quinoa* Willd. MS thesis, Brigham Young University, Provo, Utah.

Riguera, R. (1997) Isolating bioactive compounds from marine organisms. *Journal of Marine Biotechnology* 5, 187–193.

Ruales, J. and Nair, B.M. (1992) Nutritional quality of the protein in quinoa (*Chenopodium quinoa* Willd.) seeds. *Plant Foods for Human Nutrition* 42, 1–11.

San Martin, R., Ndjoko, K. and Hostettmann, K. (2008) Novel molluscicide against *Pomacea canaliculata* based on quinoa (*Chenopodium quinoa*) saponins. *Crop Protection* 27, 310–319.

Schopke, T. (2000) Non-NMR methods for structure elucidation of saponins. In: Oleszek, W. and Marston, A. (eds) *Saponins in Food, Feedstuffs and Medicinal Plants*. Annual Proceedings of the Phytochemical Society, Clarendon Press, Oxford, pp. 106–113.

Simmonds, N.W. (1965) The grain chenopods of the tropical American highlands. *Economic Botany* 19, 223–235.

Soliz-Guerrero, J.B., Jasso de Rodriguez, D., Rodriguez-Garcia, R., Angulo-Sanchez, J.L. and Mendez-Padilla, G. (2002) Quinoa saponins: concentration and composition analysis. In: Janick, J. and Whipkey, A. (eds) *Trends in New Crops and New Uses*. ASHS Press, Alexandria, Virginia, pp. 110–114.

Szakiel, A., Paczkowski, C. and Henry, M. (2011) Influence of environmental abiotic factors on the content of saponins in plants. *Phytochemistry Reviews* 10, 471–491.

Tang, K., Nie, R., Jing, L. and Chen, Q. (2009) Anti-athletic fatigue activity of saponins (Ginsenosides) from American ginseng (*Panax quinquefolium* L.). *African Journal of Pharmacy and Pharmacology* 3, 301–306.

Tapia, M., Gandarillas, H., Alandia, S., Cardozo, A., Mujica, A., Ortiz, R., Otazu, V., Rea, J., Salas, B. and Zanabria, E. (1979) *La Quinua y la Kaniwa*. Centro Internacional de Investigaciones para el Desarrollo, Instituto Internacional de Ciencias Agricolas, Bogota, Colombia.

Torres, H. and Minaya, I. (1980) *Escarificadora de Quinua. Diseño y Construcción.* Publicaciones Miseláneas No. 243, Instituto Interamericano de Ciencias Agricolas, Lima, Peru.

Treyvaud, V., Marston, A., Dyatmiko, W. and Hostettmann, K. (2000) Molluscicidal saponins from *Phytolacca icosandra. Phytochemistry* 55, 603–609.

van Raamsdonk, L.W.D., Pinckaers, V., Ossenkoppele, J., Houben, R., Lotgering, M. and Groot, M.J. (2010) Quality assessments of untreated and washed quinoa (*Chenopodium quinoa*) seeds based on histological and foaming capacity investigations. In: Méndez-Vilas, A. and Díaz, J. (eds) *Microscopy: Science, Technology, Applications and Education.* Formatax Research Centre, Badajoz, Spain.

Varriano-Marston, E. and De Francisco, A. (1984) Ultrastructure of quinoa fruit (*Chenopodium quinoa* Willd.). *Food Microstructure* 3, 165.

Verza, S.G., de Resende, P.E., Kaiser, S., Quirici, L., Teixeira, H.F., Gosmann, G., Ferreira, F. and Ortega, G.G. (2012) Micellar aggregates of saponins from *Chenopodium quinoa*: characterization by dynamic light scattering and transmission electron microscopy. *Pharmazie* 67, 288–292.

Vincken, J.P., Heng, L., de Groot, A. and Gruppen, H. (2007) Saponins, classification and occurrence in the plant kingdom. *Phytochemistry* 68, 275–297.

Ward, S.M. (2001) A recessive allele inhibiting saponin synthesis in two lines of Bolivian quinoa (*Chenopodium quinoa* Willd.). *Journal of Heredity* 92, 83–86.

Woldemichael, G.M. (2000) Phytochemical investigation of four triterpene, saponin, and alkaloid containing plants. PhD thesis, University of Heidelberg, Heidelberg, Germany.

Woldemichael, G.M. and Wink, M. (2001) Identification and biological activities of triterpenoid saponins from *Chenopodium quinoa. Journal of Agricultural and Food Chemistry* 49, 2327–2332.

Yendo, A.C.A., de Costa, F., Gosmann, G. and Fett-Neto, A.G. (2010) Production of plant bioactive triterpenoid saponins: elicitation strategies and target genes to improve yields. *Molecular Biotechnology* 46, 94–104.

Zhu, N.Q., Sheng, S.Q., Sang, S.M., Jhoo, J.W., Bai, N.S., Karwe, M.V., Rosen, R.T. and Ho, C.T. (2002) Triterpene saponins from debittered quinoa (*Chenopodium quinoa*) seeds. *Journal of Agricultural and Food Chemistry* 50, 865–867.

14 Transparency from Production to Consumption: New Challenges for the Quinoa Market Chain

Enrique A. Martínez and Pablo Olguín

14.1 Introduction

The international market for quinoa has been growing in close relationship with a growing global market for organic foods. For instance, the value of quinoa exports in 2011 in Bolivia was US$64 million, 36% more than in 2010, while in Peru over the same period the value almost doubled, to US$23 million. Concomitantly, production has risen from 7000 tons a year in the 1980s to 42,500 tons in 2011 in Peru (*The Economist*, 2012). What is also happening is that the transformation of quinoa into staple food in developed countries has diversified so much that every year new products are being proposed to the general public in formats that greatly facilitate widespread consumption (Table 14.1).

In this chapter we describe why quinoa is in such demand in the health food market and the particular situation of Chile compared with other countries in the international market quinoa cluster. We propose a new, transparent type of market to increase the sustainability and resilience of local and global market networks, including for quinoa.

14.2 Quinoa: Knowledge-Based Food Consumption in a Modern Society

People desire to maintain their health as they get older, and often consume better quality foods in order to achieve this. The stressful conditions of our modern society can lead to cardiovascular problems and cancer (Markus, 2008), so the nutritional properties of quinoa, along with other health-related habits, can be considered useful. Among the properties of this Andean staple food, we highlight not only its high protein content (12–20%), higher than most known cereals, but also that it contains 20 amino acids, making it of

Table 14.1. Quinoa products found in one organic food store in Montpellier, France. (From authors' survey 2011.)

Quinoa form	Packet quantity
Dehusked and washed-dried grains (single and mixed varieties)	1 kg or bulked
Mixed with other grains (ex. with lentils)	1 kg
Flour (single and mixed with other type of gluten-free flours)	500 g
Flavoured juice	1 litre
Grains to be germinated	200 g
Cooked and chilled salad (ready to eat)	160 g
Tortilla (ready to prepare)	350 g
Noodles (several combinations)	250–500 g
Precooked (for breakfast or soups)	250 g
Popped with honey	250 g
Muesli	250 g
Pre-cooked *flakes* (for breakfast)	250 g
Tomato soup thickener (ready to eat)	60 g

superior quality (Vega-Galvez *et al.*, 2010). It contains tryptophan, an important amino acid that improves the brain's resilience to stress, with implications for helping with mood disturbances and their consequences (Markus, 2008). Quinoa also contains several vitamins (vitamin E, B complex), plus antioxidants such as isoflavones and other flavonoid compounds that are related to cell membrane protection. There is a low prevalence of cancer in populations consuming isoflavone-rich foods, with the additional potential of better milk production in women with new-born children (Zhengkang *et al.*, 2006). Good quality oils and minerals (Ca, Fe, K, Mg, Mn, P, Zn) are also present in valuable quantities in quinoa grains, serving as a good source for human physiological needs. Some synergic interactions among the nutritive content of quinoa could be also proposed. For instance, it has been suggested that isoflavones could improve calcium absorption, although this might vary according to their source (Spence *et al.*, 2005). Thus quinoa isoflavones and its high calcium content could improve the mineral's accessibility for humans, as has been demonstrated for iron bioavailability from quinoa (Valencia *et al.*, 1999). The absence of gluten in quinoa seed makes it more accessible for people with coeliac disease, who have an allergy to this protein. The saponins of the outer grain cell layer (epicarp) could also have beneficial effects because of its potential for improving cholesterol metabolism, hyperlipidaemia and hyperglycaemia (Oakenfull and Sidhu, 1990; Bhavsar *et al.*, 2009), and ameliorating obesity, hepatic steatosis and glucose intolerance (Hu *et al.*, 2012).

14.3 Contrasts Between the Chilean Case and Other Latin American Models for the Quinoa Market

There is a big difference between Chile and other countries that export quinoa to international markets, not just because of the cultivated areas involved (45,000 ha in Bolivia compared with less than 1500 ha in Chile), but also the export volumes and the efforts to cultivate quinoa under certified organic agriculture. The Chilean law on organic production and commerce was approved in 2008, but it has not been followed by any official organization aiming to increase capacity or any official outreach programme for small-scale farmers. There has been no official financial support for farmers or for organic product processors. This lack of support is probably the reason why Chile is not competitive in the international market for quinoa.

In the high Andes the natural agroecological conditions help organic farming at 4000 m above sea level: extreme cold temperatures and acid volcanic soils with poor organic matter content mean the absence of weeds or competing plants and the absolute lack of genetically modified crops around the quinoa fields. Under these natural conditions of the Andean highlands, a large number of small-scale farmers cultivate wide areas. However, this is not the case in Chile, where international migration trends mean that more than 50% of the population live in towns and small-scale farmers are decreasing in number because strong economic forces have compelled farmers to dedicate more land to a growing cellulose-driven forest industry and international agro-industrial companies have provoked great inequalities in land distribution (Garin and Ortega, 2009). Farmers are becoming agro-industrial workers in the export of fruit for world counter-season markets (Martínez *et al.*, 2010), and earn less than half of the US$14,000/year that is the mean salary in Chile (Valdes and Foster, 2005).

Quinoa is still cultivated in Chile, but by small-scale farmers for their own consumption or for informal markets. These farmers belong to ancient indigenous communities that also keep these agricultural practices for traditional reasons, such as the Aymaras and Quechuas people in the north (Altiplano between 18°S to 20°S in the high Andes over 2000 m above sea level) and the Mapuches people in the south (below 30°S). They are isolated farmers, like those in central Chile (around 34°S) who cultivate quinoa on a small scale (<5 ha) on coastal soils, often clayish or salty-sandy ones, where no other crops can be cultivated. These three quinoa cultivation zones surviving in Chile have been described by Bazile and Negrete (2009) and articles therein.

Under this scenario the recovery of quinoa production in Chile is far from being resolved and needs much more network integration. This has to involve not just farmers, geographically very isolated, but also scientists, agronomists, nutritionists and world politicians, most of whom have never seen quinoa even though it has been cultivated for 3000 years in Chile (Tagle and Planella, 2002). However, one of the most important challenges is market sustainability, which needs to be considered for the whole production chain, from the very basic production management in the fields to the interactions between the final seller and consumer, as pointed out recently by Altieri *et al.* (2012).

14.4 Transparent Markets: A New Model for Quinoa and Other Trade Chains

Transparent market is a type of fair trade market model. Fair Trade, as defined by the World Fair Trade Organization (WFTO, 2012), is a tangible contribution to the fight against poverty, climate change and the economic crisis that has hit the world's most vulnerable populations hardest: the one-third of the world's population who live on less than US$2 a day across underdeveloped and developing countries. Fair Trade has started in Chile, but it has not been officially promoted by the Chilean government. However, there have been some local experiences since 2005. One of these involved an export chain for two different final products (honey and olive oil) and three different agricultural resources (bees, olive trees, and smoked and ground hot chilli peppers that were mixed with the olive oil for gourmet markets). Bees and hot chilli peppers were produced by small-scale farmers in the south of the country (below 39°S) and olive oil was produced by small-scale farmers in the north of the country (a region at 32°S). They were not certified farmers and were not formally organized in a cooperative. The buyer of their products was Paulina Peñaloza (Agr. Eng.). She simply agreed with them a price and requested a particular product quality. Farmers were informed about the product transformation quality, and more interestingly about the prices of the transformed products once they reached the port (FOB prices), and the prices given to the products in the foreign country along all the intermediaries until reaching the supermarket shelves where they were sold to the final consumers (in this case, in the USA). The model worked very well because Paulina Peñaloza gained the trust of the product providers and they refused to sell at higher prices to people not offering transparency to product transformations and price changes along the commercial chain. The model is still valid for Paulina Peñaloza (personal communication, 2012), but she has had problems with the USA dealers who were not engaged with this transparent market model, and their bad payment terms made the model fail. This is a model we would like to propose for the quinoa market chain in Chile and ideally for any market of any product in the world that starts with production by small-scale farmers.

In 2012 in Chile organic quinoa grain (without processing other than dehusking) imported from Bolivia sold in supermarkets for US$10–18/kg, while farmers in Chile received less than US$1/kg. Such ten-fold difference in prices points towards lack of confidence in the market chains. Production costs in Chile should be higher than in the Altiplano of Bolivia because the increasing drought means irrigation is likely to be needed, something not required for commercial quinoa production elsewhere.

Many farmers might not currently have the skills to transform the products, but with time this could be a possibility, or it could be for their children if they acquire a foreign language in the next generation. If this is made possible, it would give confidence and sustainability to the whole economic system. The exporter will not lose his job because not all farmers would want to get involved in product transformations or in selling too far from their communities. Becoming richer is not everyone's aim. Sustainability of local markets, and not only export ones, would be the benefit from such a transparent trade model. For instance, in Chile the number of schools lunches given to those under high-school age is 2 million per day. Not enough quinoa can be produced (and bought) in Chile in 2012 under the formal private–government agreement system. This is due to the above described trends leading to the transformation of our agricultural system into an export-biased model (including cellulose industry-driven forestry) and many other historical reasons that led to lack of trust in cooperative systems. One of the reasons is lack of transparency, without mentioning corruption. The rules of free market alone, even though favouring excellent international trading prices, might lead to conflict among farmers (*The Economist*, 2012) or even between scientists who have worked with them (Winkel, 2012; WFTO, 2012).

Finally, the proposed transparency would help small-scale farmers to gain self-confidence and other skills that could help them to diversify their incomes. For instance, new horizons opened through modern tourism activities associated with ancient cultures and traditions.

14.5 Concluding Remarks

Quinoa is in great demand in the health food market because of its nutraceutical benefits. The plant is cultivated in Chile by small-scale farmers for their own consumption as well as for informal markets. One of the most important challenges with respect to quinoa is market sustainability, which needs to be considered for the whole production chain, from the very basic production management in the fields to the interactions between the final seller and consumer. Transparent market, a type of fair trade market model, is ideally suited for the quinoa market chain in Chile and for any market of any product in the world that starts with production by small-scale farmers. The proposed transparency would help small-scale farmers to gain self-confidence and other skills that could help them to diversify their incomes.

Acknowledgements

ANR-France financial aid (2008–2012) to discuss these ideas under the IMAS project aimed to study seed biodiversity and survey on dynamic systems for farmer seed exchanges and markets as compared in Chile (South America) and Mali (East Africa).

References

Altieri, M.A., Funes-Monzote, F.R. and Petersen, P. (2012) Agroecologically efficient agricultural systems for smallholder farmers: contributions to food sovereignty. *Agronomy and Sustainable Development* 32, 1–13.

Bazile, D. and Negrete, J. (2009) Quínoa y biodiversidad: ¿cuáles son los desafíos regionales? *Revista Geográfica de Valparaíso* 42, 1–141.

Bhavsar, S.K., Singh, S., Giri, S., Jain, M.R. and Santani, D.D. (2009) Effect of saponins from *Helicteres isora* on lipid and glucose metabolism regulating genes expression. *Journal of Ethnopharmacology* 124, 426–433.

Garin, A. and Ortega, E. (2009) Transformaciones socioterritoriales a partir de la implantación del modelo neoliberal, en el espacio rural de la Región de la Araucanía entre los años 1975–2007. *Anales de la Sociedad Chilena de Ciencias Geográficas* 2008, 171–177.

Hu, X., Li, Z., Xue, Y., Xu, J., Xue, C., Wang, J. and Wang, Y. (2012) Dietary saponins of sea cucumber ameliorate obesity, hepatic steatosis, and glucose intolerance in high-fat diet-fed mice. *Journal of Medicinal Food* 15, 909–916.

Markus, C.R. (2008) Dietary amino acids and brain serotonin function; implications for stress-related affective changes. *Euromolecular Medicine* 10, 47–58.

Martínez, E.A., Bazile, D., Thomet, M., Delatorre, J., Salazar, E., Leon-Lobos, P., Von Baer, I. and Nuñez, L. (2010) Neo-liberalism in Chile and its impacts on agriculture and biodiversity conservation: the experience with the re-start of quinoa crop cultivation. Innovation and Sustainable Development, ISDA 2010, 28 June–1 July 2010, Montpellier, France. Available at: http://hal.archives-ouvertes.fr/ISDA2010 (accessed 1 October 2012).

Oakenfull, D. and Sidhu, G. (1990) Could saponins be a useful treatment for hypercholesterolaemia? *European Journal of Clinical Nutrition* 44, 79–88.

Spence, L.A., Lipscomb, E.R., Cadogan, J., Martin, B., Wastney, M.E., Peacock, M. and Weaver, C.M. (2005) The effect of soy protein and soy isoflavones on calcium metabolism in postmenopausal women: a randomized crossover study. *American Journal of Clinical Nutrition* 81, 916–922.

Tagle, M.B. and Planella, M.T. (2002) *La Quinoa en la Zona Central de Chile: Supervivencia de una Tradición pre-Hispana.* Editorial IKU, Santiago, Chile.

The Economist (2012) Quinoa selection. The Andes' new cash crop. Foreign interest grows in an old highland staple. *The Economist* May 2012.

Valdes, A. and Foster, W. (eds) (2005) *Externalidades de la Agricultura Chilena.* Universidad Católica de Chile and FAO, Santiago, Chile.

Valencia, S., Svanberg, U., Sandberg, A.S. and Ruales, J. (1999) Processing of quinoa (*Chenopodium quinoa* Willd.): effects on *in vitro* iron availability and phytate hydrolysis. *International Journal of Food Sciences and Nutrition* 50, 203–211.

Vega-Gálvez, A., Miranda, M., Vergara, J., Uribe, E., Puente, L. and Martínez, E.A. (2010) Nutrition facts and functional potential of quinoa (*Chenopodium quinoa* Willd.), an ancient Andean grain: a review. *Journal of the Sciences of Food and Agriculture* 90, 2541–2547.

WFTO (World Fair Trade Organization) (2012) *About Fair Trade.* Available at: http://www.wfto.com/ (accessed 1 October 2012).

Winkel, T., Bertero, H.D., Bommel, P., Bourliaud, J., Chevarría-Lazo, M., Cortes, G., Gasselin, P., Geerts, S., Joffre, R., Léger, F., Martinez-Visa, B., Rambal, S., Riviere, G., Tichit, M., Tourrand, J.F., Vassas-Toral, A., Vacher, J.J. and Vieira-Pak, M. (2012) The sustainability of quinoa production in Southern Bolivia: from misrepresentations to questionable solutions. Comments on Jacobsen (2011, *J. Agron. Crop Sci.* 197, 390–399). *Journal of Agronomy and Crop Science* 198, 314–319.

Zhengkang, H., Wang, G., Yao, W. and Zhu, W.-Y. (2006) Isoflavonic phytoestrogens: new prebiotics for farm animals: a review on research in China. *Current Issues in Intestinal Microbiology* 7, 53–60.

Index